Bergman's Linear Integral Operator Method in the Theory of Compressible Fluid Flow

By

M. Z. v. Krzywoblocki

Sc. D. (Lille), Ph. D. (Brooklyn), M. A. (Math., Stanford), M. S. (Appl. Math., Brown)
M. Aer. En. (Brooklyn), Dipl. Ing. (Lemberg), Professor, University of Illinois

With an Appendix

by

Dr. Ph. Davis and Dr. Ph. Rabinowitz
U. S. Department of Commerce, National Bureau of Standards

With 3 Figures

Springer-Verlag Wien GmbH 1960

This book is based on the Author's papers
"Bergman's Linear Integral Operator Method in the Theory of Compressible Fluid Flow"
having been published in "Österreichisches Ingenieur - Archiv" Vol. VI (1952), No. 5,
Vol. VII (1953), No. 4, Vol. VIII (1954), No. 4, and Vol. X (1956), No. 1

ISBN 978-3-7091-3995-0
DOI 10.1007/978-3-7091-3994-3

ISBN 978-3-7091-3994-3 (eBook)

To my beloved Wife

Preface

The reader who is somewhat acquainted with the field of compressible fluid flow hears much about Stefan Bergman's method of integral operators. It took many years for him to develop this method which is based primarily on the theory of analytic functions and particularly on the theory of functions of two complex variables. The method, as a whole, is scattered throughout many papers in mathematical journals, and as a matter of fact, in its present state, is accessible only to those who are fully acquainted with mathematical literature. In one of their papers, Professors R. von Mises and M. Schiffer greatly simplified the method in the subsonic case. The purpose of the present work is to represent the method in all its variations in such a way that a theoretical engineer or an applied aerodynamicist can use it in practical applications. A professional mathematician will find the discussion too elementary for him. The parts of Bergman's presentation which are most interesting mathematically—the proofs—are mostly omitted in the present work. The emphasis was put upon the simplified representation of the final results and formulas, rather than upon the derivation of those formulas.

In the preliminary remarks the author discusses various types of singularities in a very elementary way. The first two parts of the work deal with the subsonic case. In these sections the author followed mostly the paper of von Mises and Schiffer. Part III contains the supersonic flow, Part IV the transonic flow and Part V the axially symmetric flow. These sections are based on Bergman's original papers and form a condensation of several of his papers on the subject. Part VI describes singularities which play so important a role in all hodograph transformations. Part VII contains a review of other methods available in the theory of compressible fluid flow. The subsequent parts contain: Part VIII, a review of tables in Bergman's method; Part IX, general remarks; Part X, a list of tables; Part XI, examples; and Part XII, errata in previous papers. Part XIII contains generalization to diabatic flow, Part XIV three-dimensional flow and Part XV by P. Davis and P. Rabinowitz from U. S. National Bureau contains some computations of subsonic fluid flows by Bergman's method of integral operators carried out on the U. S. National Bureau of Standards Eastern Automatic Computer (SEAC), Washington, D. C.

The present work is the reproduction of the research report[1] prepared by the author in the years 1949–1951 for Harvard University and the late Professor von Mises, done under Navy Contract Nord 10-449, Task 3, Bureau of Ordnance, Operator Methods in the Theory of Compressible Fluids (Report 18). The work was done under the supervision of the late Professor R. v. Mises and Professor S. Bergman. The author is greatly indebted to the late Professor v. Mises for encouraging him to undertake this task and for the permission of using freely all the works of

---

[1] M. Z. v. Krzywoblocki: Bergman's Linear Integral Operator Method in the Theory of Compressible Fluid Flow. Bureau of Ordnance, Operator Method in the Theory of Compressible Fluids (Report 18), Contract Nord 10-449, Task 3, Harvard University, Cambridge, Massachusetts, March, 1951.

Professor v. Mises as the source of information. He expresses his deep thanks to Professor S. Bergman for numerous discussions on the subject of the work and for permission to use all the papers and works pertinent to the integral method published by Professor S. Bergman in numerous journals. His special thanks are due to the Headquarters, Wright Air Development Center, for permission to publish the results of the calculation applying Bergman's method, performed by Dr. P. Davis and Dr. P. Rabinowitz, National Bureau of Standards, sponsored by the Wright Air Development Center; to the National Bureau of Standards, U. S. Department of Commerce for the permission to incorporate the calculations in the work; to Dr. P. Davis and Dr. P. Rabinowitz for their kind agreement to publish their calculations in the present work; to Professor Dr. Hilda von Mises for her kind permission, as the successor in the rights of Professor von Mises, to use in Part I and II the research report by R. von Mises and M. M. Schiffer, done under Navy Contract Nord 8555-Task at Harvard University, published subsequently in "Advances in Applied Mechanics" in 1948; to Professor Dr. M. M. Schiffer for his kind permission to make any use of the above mentioned report and paper.

The author is particularly grateful to the Academic Press Inc., Publishers, New York, N. Y., for their kind permission for quoting verbatim some sentences and paragraphs from the paper by R. v. Mises and M. Schiffer "On Bergman's Integration Method in Two-Dimensional Compressible Fluid Flow", published in "Advances in Applied Mechanics", Volume I, 1948, pp. 249–285, Academic Press Inc., New York.

His thanks are also due to Dr. G. S. S. Ludford, previously of Harvard University, at present with the University of Maryland for reading the manuscript and making valuable comments concerning the material.

Fluid Dynamics Panel University of Illinois,
    Urbana, Illinois, U. S. A.
      April, 1960

                  M. Z. v. Krzywoblocki

Table of Contents

Page

List of Symbols

F, compressibility function for stream eq. [see (15.2.16)].

$F_1 = Re\,F$.

$F_2 = Im\,F$.

F^*, real approximation to F.

$$h = \left(\frac{k-1}{k+1}\right)^{1/2}.$$

$$H = (1 - M^2)^{-1/4}\left(1 + \frac{k-1}{2}\,M^2\right)^{-\frac{1}{2(k-1)}}.$$

k, gas constant.

$l = (1 - M^2)/\varrho^2$.

M, local Mach number.

p, pressure.

P, compressibility function for potential eq. [see (15.2.17)].

$P(z^N)$, Integral operator [see (15.3.26)].

q, speed.

S^*, fundamental singularity of eq. (15.2.14).

$T = (1 - M^2)^{1/2}$.

$$w = \frac{1}{2}(z - \lambda_0).$$

$$w^* = \frac{1}{2}(z^* - \lambda_0).$$

W_1, W_2, singular points in the logarithmic plane.

$z = \lambda + i\,\theta$.

$z^* = \lambda - i\,\theta$.

θ, angle of flow.

$\vartheta = -\,Im\,z^*$.

λ, pseudologarithmic variable [see (15.2.5)].

$\Lambda = Re\,z^*$.

ϱ, density.

$\sigma = \arg T$.

$\tau = |T|$.

Φ, potential function.

$\Phi^* = \Phi H$, modified potential function.

χ, singular stream function.

ψ, stream function.

$\psi^* = \psi/H$, modified stream function.

ψ_N^*, particular solutions of (15.2.14).

$\psi^{(L,\,2)}$, stream function of source.

Part 0

Preliminary Remarks. Singularities

1. General Properties and Remarks on Analytic Functions

In this section, we shall remind the reader of some basic definitions and properties of analytic functions, singular points, etc. First of all, let us repeat Cauchy's definition of an analytic function of a complex variable.

Let a two-dimensional region in the z-plane be given, where $z = x + iy$, $\bar{z} = x - iy$, and let u be a function of z defined uniquely at all points of the region. Let z, $z + \delta z$ be values of the variable z at two points, and u, $u + \delta u$, the corresponding values of u. Then, if at any point z within the area, $\delta u/\delta z$ tends to a limit when $\delta x \to 0$, $\delta y \to 0$, independently (where $\delta z = \delta x + i\,\delta y$), i. e., if the limit is independent of the path by which the point $z + \delta z$ tends toward coincidence with z, u is said to be a function of z which is monogenic or analytic at the point (the words regular and holomorphic are sometimes used). If the function is analytic and one-valued at all points of the region, we say that the function is analytic throughout the region. For almost every one of the elementary functions, however, there will be certain points z at which this property will cease to hold good. Thus, it does not hold for the function $f(z) = 1/(z - a)$ at the point $z = a$, since in this case

$$\lim_{h \to 0} 1/h \,\{f(z + h) - f(z)\} = \lim_{h \to 0} 1/h \,\{(z - a + h)^{-1} - (z - a)^{-1}\}, \qquad (0.1.1)$$

does not exist when $z = a$. Similarly, it does not hold for the function $\log z$ and $z^{1/2}$ at the point $z = 0$. These exceptional points are called singular points, or singularities, of the function $f(z)$ under consideration.

Let x and y be two real functions of a real variable t which are continuous for every value of t such that $a \leq t \leq b$. We denote the dependence of x and y on t by writing

$$x = x(t), \quad y = y(t); \quad (a \leq t \leq b). \qquad (0.1.2)$$

The functions $x(t)$, $y(t)$ are supposed to be such that they do not assume the same pair of values for any two different values of t in the range $a < t < b$. Then the set of points with coordinates (x, y) corresponding to these values of t is called a simple curve. If

$$x(a) = x(b), \quad y(a) = y(b), \qquad (0.1.3)$$

the simple curve is said to be closed.

A two-dimensional continuum is a set of points in a plane possessing, among others, the property that any two points of the set can be joined by a simple curve consisting entirely of points of the set. A simple closed curve divides the plane into two continua (the "interior" and the "exterior"). A continuum formed by the interior of a simple curve is sometimes called an open two-dimensional region, and the curve is called its boundary; such a continuum with its boundary is then called a closed two-dimensional region, or, briefly, a closed region or domain. A simple closed curve C in the plane of the variable z is sometimes called a closed one-dimensional region, or a contour; a simple curve with its end-points omitted is then called an open one-dimensional region.

Cauchy's theorem on the integral of a function around a contour may be stated as follows: If $f(z)$ is a function of z, analytic at all points on and inside a contour C, then

$$\int_C f(z)\, dz = 0. \tag{0.1.4}$$

Another important theorem on the value of a function is the following one: The value of a function $f(z)$ (which is analytic on and inside C) at any point "a" within a contour C, may be expressed in terms of an integral which depends only on the value of $f(z)$ at points on the contour itself:

$$f(a) = (2\pi i)^{-1} \int_C (z-a)^{-1} f(z)\, dz. \tag{0.1.5}$$

Laurent's theorem can be expressed in the following form: If $f(z)$ is analytic on the concentric circles C and C' of center a, and throughout the annulus between them, then at any point z of the annulus, $f(z)$ can be expanded in the form

$$f(z) = a_0 + a_1(z-a) + a_2(z-a)^2 + \ldots + b_1(z-a)^{-1} + b_2(z-a)^{-2} + \ldots, \tag{0.1.6}$$

where

$$a_n = (2\pi i)^{-1} \int_C (t-a)^{-(n+1)} f(t)\, dt, \tag{0.1.6a}$$

$$b_n = (2\pi i)^{-1} \int_{C'} (t-a)^{n-1} f(t)\, dt. \tag{0.1.6b}$$

An important case of Laurent's theorem arises when there is only one singularity within the inner circle C', namely at the center a. In this case, the circle C' can be taken as small as we please, and thus Laurent's expansion is valid for all points in the interior of the circle C, except the center a.

Consider a function $f(z)$ which is analytic throughout a closed region S, except at a single point a inside the region. Let it be possible to define a function $\Phi(z)$ such that

I. $\Phi(z)$ is analytic throughout S,

II. when $z \neq a$, $f(z) = \Phi(z) + B_1(z-a)^{-1} + B_2(z-a)^{-2} + \ldots B_n(z-a)^{-n}$.

Then $f(z)$ is said to have a pole of order n at a; and the terms $\sum_i^n B_i(z-a)^{-}$ are called the principal part of $f(z)$ near a. By the definition of a singularity, a pole is a singularity. If $n = 1$, the singularity is called a simple pole.

Any singularity of a one-valued function other than a pole is called an essential singularity. If the essential singularity, a, is isolated (i. e., if the region, of which a is an interior point, can be found containing no singularities other than a), then a Laurent expansion can be found, in ascending and descending powers of $(z-a)$. Hence, the "principal part" of a function near an isolated essential singularity consists of an infinite series.

It should be noted that a pole is, by definition, an isolated singularity, so that all singularities which are not isolated are essential singularities.

The behavior of a function $f(z)$ as $|z| \to \infty$, can be treated in a similar way to its behavior as z tends to a finite limit. If we write $z = 1/z'$, so that the large values of z are represented by small values of z' in the z'-plane, there is a one-to-one correspondence between z and z', provided that neither is zero; and to make the correspondence complete, it is sometimes convenient to say that when z' is the origin, z is the point at infinity. But the reader must be careful to observe that this is not a definite point, and any proposition about it is really a proposition concerning the point $z' = 0$.

Let $f(z) = \Phi(z')$. Then $\Phi(z')$ ist not defined at $z' = 0$, but its behavior near $z' = 0$ is determined by its Taylor (or Laurent) expansion in powers of z'; and we define $\Phi(0)$ as $\lim_{z' \to 0} \Phi(z')$ if that limit exists. For instance, the function $\Phi(z')$ may have a zero of order m at the point $z' = 0$; in this case, the Taylor expansion of $\Phi(z')$ will be of the form

$$\Phi(z') = A z'^m + B z'^{m+1} + C z'^{m+2} + \ldots, \tag{0.1.7}$$

and so the expansion of $f(z)$ valid for sufficiently large values of $|z|$ will be of the form

$$f(z) = A z^{-m} + B z^{-m-1} + C z^{-m-2} + \ldots. \tag{0.1.8}$$

In this case, $f(z)$ is said to have a zero of order m at infinity.

Again, the function $\Phi(z')$ may have a pole of order m at the point $z' = 0$; in this case,

$$\Phi(z') = A/z'^m + B/z'^{m-1} + C/z'^{m-2} + \ldots + L/z' + M + N z' + P z'^2 + \ldots; \tag{0.1.9}$$

and so, for sufficiently large values of $|z|$, $f(z)$ can be expanded in the form

$$f(z) = A z^m + B z^{m-1} + C z^{m-2} + \ldots + L z + M + N/z + P/z^2 + \ldots. \tag{0.1.10}$$

In this case, $f(z)$ is said to have a pole of order m at infinity.

Similarly, $f(z)$ is said to have an essential singularity at infinity, if $\Phi(z')$ has an essential singularity at the point $z' = 0$. Thus, the function $\exp z$ has an essential singularity at infinity, since the function $\exp(1/z')$ or

$$1 + 1/z' + 1/(2! \, z'^2) + 1/(3! \, z'^3) + \ldots \tag{0.1.11}$$

has an essential singularity at $z' = 0$.

In all the previous considerations, the functions have had a unique value (or limit) corresponding to each value (other than singularities) of z. But functions may be defined which have more than one value for each value of z; thus $z^{1/2}$ has two values, and the function arc tan x (x real) has an unlimited number of values, viz. $n\pi + \text{arc tan } x$, where $-\frac{1}{2}\pi < \text{arc tan } x < \frac{1}{2}\pi$ and n is any integer; further examples of many-valued functions are $\log z$, $z^{-5/3}$, $\sin(z^{1/2})$. This is easy to understand if we recall that $z = r \exp(i\theta) = r(\cos\theta + i\sin\theta) = r[\cos(\theta + 2n\pi) + i\sin(\theta + 2n\pi)] = r \exp[i(\theta + 2n\pi)]$. Thus, $\log z = \log r + i(\theta + 2n\pi)$, which is a many-valued function, depending upon the value of the integer n.

Either of the functions which $z^{1/2}$ represents is, however, analytic except at $z = 0$, and we can apply to them the theorems of this section; the two functions are called "branches" of the many-valued function $z^{1/2}$. There will be certain points in general at which two or more branches coincide or at which one branch has an infinite limit; these points are called "branch-points". Thus, $z^{1/2}$ has a branch-point at 0; and, if we consider the change in $z^{1/2}$ as $z = r e^{i\theta}$ describes a circle counterclockwise around 0, we see that θ increases by 2π, r remains unchanged, and either branch of the function passes over into the other branch. This will be found to be a general characteristic of branch-points. Thus, we cannot apply Cauchy's theorem to such a function as $z^{3/2}$ when the path of integration is a circle surrounding the origin; but it is permissible to apply it to one of the branches of $z^{3/2}$ under special conditions (for more details, the reader may refer to[1]).

Geometrically, a many-valued function represents the so-called "Riemann surface". Each branch of a many-valued function represents geometrically a branch or a sheet of the Riemann surface. Seemingly, all the sheets of a Riemann surface lie in one and the same plane. Actually, they are separate geometric elements, connected together only by means of branch points or other singular points.

An extension of the theorem of (0.1.5) is represented by the theorem of residues. If the function $f(z)$ has a pole of order m at $z = a$, then, by the definition of a pole, an equation of the form

$$f(z) = \frac{a_{-m}}{(z-a)^m} + \frac{a_{-m+1}}{(z-a)^{-m+1}} + \ldots + \frac{a_{-1}}{z-a} + \Phi(z), \qquad (0.1.12)$$

where $\Phi(z)$ is analytic near and at a, is true near a. The coefficient a_{-1} in this expansion is called the residue of the function $f(z)$ relative to the pole a. Now let C be any contour, containing in the region interior to it a number of poles $a, b, c, \ldots$, of a function $f(z)$, with residues $a_{-1}, b_{-1}, c_{-1}, \ldots$, respectively; and suppose that the function $f(z)$ is analytic throughout C and its interior, except at these poles. Then the theorem of residues states that: If $f(z)$ be analytic throughout a contour C and its interior except at a number of poles inside the contour, then

$$\int_C f(z)\,dz = 2\pi i \sum R, \qquad (0.1.13)$$

where $\sum R$ denotes the sum of the residues of the function $f(z)$ at those of its poles which are situated within the contour C.

Let us note that if a is a simple pole of $f(z)$, the residue of $f(z)$ at that pole is $\lim_{z \to a} \{(z-a)f(z)\}$.

We shall briefly mention the singularities of linear differential equations. The standard form of a linear ordinary differential equation of the second order will be taken to be

$$\frac{d^2 u}{dz^2} + p(z)\frac{du}{dz} + q(z)\,u = 0, \qquad (0.1.14)$$

and it will be assumed that there is a domain S in which both $p(z)$ and $q(z)$ are analytic except at a finite number of poles. Any point of S at which $p(z)$, $q(z)$ are both analytic will be called an ordinary point of the equation; other points of S will be called singular points. The property that the solution of a linear differential equation is analytic except at singularities of the coefficients of the equation is common to linear equations of all orders.

When two particular solutions of an equation of the second order are not constant multiples of each other, they are said to form a fundamental system.

Suppose that a point c of S is such that, although $p(z)$ or $q(z)$ or both have poles at c, the poles are of such order that $(z-c)\,p(z)$, $(z-c)^2\,q(z)$ are analytic at c. Such a point is called a regular point for the differential equation. Any poles of $p(z)$ or of $q(z)$ which are not of this nature are called irregular points.

---

[1] E. Whittaker and G. N. Watson: A course of modern analysis. Cambridge: University Press, 1902, 1915, 1920, 1927.

One may distinguish a few basic singular points: branch points (double or multiple), cusps of the first order, if both branches have the same tangents at the vertex, or of the second order, if each branch has a different tangent, isolated points, end points, corner points (discontinuous first derivatives), etc.

A domain is called simply-connected if it is bounded by one curve. If a domain is bounded by more than one curve we call it a multiply-connected domain.

To illustrate better the nature of some singularities, let us discuss a few particular singular points. Let us note that the general character of the solutions in the hodograph plane can be easily examined by investigating the behavior of the transition function for an incompressible fluid. To begin with the simplest case first, consider for illustrative purposes a steady, irrotational flow in an infinite, simply-connected domain D bounded by a curve C in the z-plane ($z = x + i\,y$), with a parallel flow at infinity. At every point z of D there is one and only one velocity vector $\vec{q}$. Assume that the z-plane is mapped onto a hodograph w-plane, where $w = u - i\,v$. If the curve C is mapped into $\underline{C}$ and the flow at infinity corresponds to a point $\underline{P}$ on the real axis of the w-plane within $\underline{C}$, then the domain D is mapped onto $\underline{D}$ by a mapping function

$$w = w\,(z), \qquad (0.1.15)$$

where $w\,(z)$ is an analytic function of z. The inverse function

$$z = z_0\,(w), \qquad (0.1.16)$$

will set up a continuous one-to-one correspondence between the w- and z-planes, provided the mapping is conformal. As one may easily verify by means of the fundamentals of analysis, this requires that $w\,(z)$ is analytic and simple within D and $w'\,(z) \neq 0$. However, for most problems these conditions cannot be satisfied throughout the field of flow. In the first place, the function $w\,(z)$ is generally non-simple. For example, in the case of a uniform flow, $w\,(z)$ is constant, i. e., $w\,(z) = $ constant, thus $w'\,(z) = 0$, and the whole z-plane corresponds to a point in the w-plane. Furthermore, the complex velocity for a two-dimensional problem generally can be put in the following form

$$w = w_\infty + w^*(z), \qquad (0.1.17)$$

where w_∞ is a constant. Since w_∞ represents the flow at infinity, the boundary conditions at infinity require that $w^*(z) = 0$, as z tends to infinity. As a consequence $w^{*\prime}(z) = 0$, as z tends to infinity. Therefore, in all cases, the point $\underline{P}$ in the w-plane, is a singular point. It is a branch-point at w_∞ if $z_0\,(w)$ is many-valued; or a pole, if otherwise.

In practice, there are two kinds of singularities that play a dominant rôle in the problem of a two-dimensional flow. These singularities will presently be discussed briefly.

The function $z_0\,(w)$ is said to have a branch-point of order k at $w = w_\infty$, if its inverse $w\,(z)$ contains the part w^* which has a zero of order $(k + 1)$ at $z = \infty$. One of the most important singular points in two-dimensional flow problems is the branch-point of order 1. As is known, when a closed body is present in a uniform flow, there always exist two stagnation points S_1 and S_2, both of which correspond to the origin of the w-plane, since both the velocity components, u and v, at those points are equal to zero. If a streamline $P\,S$ is followed, for instance, from $+ \infty$ to S_1, then along the surface of the body, to S_2, and then from S_2 to $- \infty$, a curve $\underline{P\,S}$ in the w-plane would be described twice. This indicates that the function $z_0\,(w)$

possesses two branches of Riemann surface joining together about the branch point $\underline{P}$. In order to make the domain $\underline{D}$ single-valued, a cut is put along the real axis from the branch point to $+\infty$. Then one portion of the z-plane is mapped onto a definite branch of the Riemann surface in the w-plane, and this will be defined as the domain D. If the body is symmetrical with respect to the coordinate axes with parallel flow at infinity, then the domain D, containing the points z which possess the negative imaginary part, i. e., Im $\{z\} < 0$, will be mapped conformally into $\underline{D}$ on one branch of the Riemann surface and the points z with Im $\{z\} > 0$ on the other (Im denotes the imaginary part).

Another singularity of importance is the logarithmic singularity. Let us consider the flow over a wavy surface, placed parallel to a uniform stream. Such a flow has a periodic nature. For such flows there are infinitely many points in the physical plane that have the same velocity. Hence, there is an infinite number of branches in the w-plane, each of which corresponds to a definite portion of the z-plane. The function $z_0(w)$ must have a term of the form $\ln\left(1 - \frac{w}{U}\right)$ and the point $\underline{P}$ now is a logarithmic singularity. If, however, a proper cut is introduced from the singular point to $+\infty$, then the domain $\underline{D}$ is again made single-valued.

Generally, a function in the hodograph plane may have more than one singularity. There may exist other singularities than the one at $w = w_\infty$. However, such singularities usually lie outside the region of interest and thus need not be investigated. Often those other singularities correspond to points inside the body. Of course, only the flow outside the body is of interest.

The author believes that the reader is acquainted with such common types of flow as source (or sink), doublet and vortex. In a source (or sink) the streamlines coincide with the radii, and the flow conditions on all the streamlines are identical. The mass of fluid per unit time, flowing out from (or into) the center of the source (or sink), is called the strength of the source (or sink). A doublet is constructed from a source and sink of equal strength, located on an axis (called axis of the doublet), after the distance between the centers of those two elements shrinks to zero, so that both centers coincide. The streamlines are circles, each passing through the center of the doublet and being tangent to the axis of the doublet. The centers of all the streamlines are located on the line perpendicular to the axis of the doublet. The streamlines of a vortex are concentric circles. It is quite obvious that a representation of all those elements (sink, doublet, vortex) in the hodograph plane is connected with singularities of various kinds, and that the passage from the hodograph into the physical plane cannot be achieved by a one-to-one correspondence. This is caused by the fact that in a source, say, the velocity distribution along each streamline (radius) is the same. Thus in the hodograph plane all the streamlines (radii) are represented by one and the same curve (hodograph).

The expressions "in the small" and "in the large", used in the present work, refer, as is known, to infinitesimal and finite regions, respectively.

Part I

General Theory of Subsonic Flow

1. Physical and Hodograph Planes, System of Equations, Chaplygin's Equation

The law of conservation of matter and the condition of irrotationality lead to two well-known equations

$$(\varrho\,u)_x + (\varrho\,v)_y = 0, \qquad u_y - v_x = 0, \tag{1.1.1}$$

where the known notations are used and subscripts denote partial differentiation. These equations are satisfied if the notions of a potential Φ and a stream-function Ψ are introduced such that

$$u = \Phi_x = \varrho^{-1}\,\Psi_y, \quad v = \Phi_y = -\,\varrho^{-1}\,\Psi_x. \tag{1.1.2}$$

For illustrative purposes we take an isentropic flow for which the relation $p = c\,\varrho^k$ holds. By combining that relation with the Bernoulli equation $q^2/2 + \int \varrho^{-1}\,dp = $ $= \text{const.}$, we obtain the result [with $\int \varrho^{-1}\,dp = k\,(k-1)^{-1}\,p\,\varrho^{-1}$]:

$$\varrho = [\varrho_0^{(k-1)} - (k-1)\,(kc)^{-1}\,q^2/2]^{1/(k-1)}, \tag{1.1.3}$$

where ϱ_0 denotes the density at a stagnation point (at rest). Thus the density is a function of $q^2 = \Phi_x^2 + \Phi_y^2$, and the equations (1.1.2) and (1.1.3) represent a system of three non-linear partial differential equations for Φ, Ψ and ϱ. An important simplification in the study of the motion of such a fluid has been achieved by Molenbroeck[2] and Chaplygin[3]. That is, by means of transformation from the (x, y)-plane (physical plane) onto the (q, θ)-plane (hodograph plane) we obtain a system of linear equations. The symbol θ denotes the angle between the direction of the resultant velocity q and the horizontal axis. If Φ and Ψ are considered as functions of q and θ instead of x and y, they satisfy the system

$$\Phi_\theta = q\,\varrho^{-1}\,\Psi_q, \quad \Phi_q = -\,(1 - M^2)\,(\varrho\,q)^{-1}\,\Psi_\theta, \tag{1.1.4}$$

where M is the Mach number q/a, and $a = (dp/d\varrho)^{1/2}$ is the local sound velocity. Consider both M and ϱ to be given functions of q. Then equations (1.1.4) are linear. Differentiating the first equation (1.1.4) with respect to q and the second one with respect to θ and comparing, results in the well-known equation of Chaplygin, which is a linear one:

$$\Psi_{\theta\theta} + \varrho\,q\,(1 - M^2)^{-1}\,(q\,\varrho^{-1}\,\Psi_q)_q = 0. \tag{1.1.5}$$

2. Transformation of Chaplygin's Equation, Pseudo-logarithmic Plane

For subsonic flow let us replace the variable q by a new, dimensionless variable λ, which is defined by[4]

$$\frac{d\lambda}{dq} = (1 - M^2)^{1/2}\,q^{-1} = (q^{-2} - a^{-2})^{1/2}, \tag{1.2.1}$$

with $\lambda = 0$ for $q = a$, i. e., for $M = 1$, or

$$\lambda = -\int_q^a [1 - M^2\,(x)]^{1/2}\,x^{-1}\,dx. \tag{1.2.1'}$$

While q goes from zero to a, the variable λ varies from $-\infty$ to 0. From (1.2.1) we may express the derivative Ψ_q in terms of Ψ_λ, that is, $\Psi_q = (1 - M^2)^{1/2}\,q^{-1}\,\Psi_\lambda$, and by introducing the quantity $l^{1/2} = (1 - M)^{1/2}\,\varrho^{-1}$, we easily obtain the equality $q\,\varrho^{-1}\,\Psi_q = l^{1/2}\,\Psi_\lambda$, and the relation

$$(q\,\varrho^{-1}\,\Psi_q)_q = (1 - M^2)^{1/2}\,q^{-1}\,(l^{1/2}\,\Psi_\lambda)_\lambda. \tag{1.2.2}$$

[2] P. Molenbroek: Über einige Bewegungen eines Gases mit Annahme eines Geschwindigkeitspotentials. Arch. Math. Physik (2) 9, 157—195 (1890).

[3] S. A. Chaplygin: On gas jets. Scientific Memoirs, Moscow University, Phys.-Math. Section 21, 1—127 (1904), or N. A. C. A., T. M. No. 1063 (1944).

[4] R. von Mises and M. Schiffer: On Bergman's integration method in two-dimensional compressible fluid flow. Advances in Applied Mechanics, Vol. I. New York: Academic Press Inc. 1948.

Introducing (1.2.2) into (1.1.5), Chaplygin's equation becomes:

$$s^{1/2}\, \Psi_{\theta\theta} + s^{-1/2} (s\, \Psi_\lambda)_\lambda = 0, \quad s = l^{1/2}. \tag{1.2.3}$$

In (1.2.3) we replace Ψ by $\Psi^* = s^{1/2}\,\Psi$ and calculate $(s\,\Psi_\lambda)_\lambda = s^{1/2}\,\Psi^*_{\lambda\lambda} - \Psi^*\,\dfrac{d^2}{d\lambda^2}\,(s^{1/2})$. Equation (1.2.3) takes the form:

$$\Delta\,\Psi^* + f(\lambda)\,\Psi^* = 0, \tag{1.2.4}$$

where the symbol Δ denotes the Laplacian, i. e., $\Delta = \partial^2/\partial\lambda^2 + \partial^2/\partial\theta^2$, and the dimensionless function $f(\lambda)$,

$$f(\lambda) = -\,s^{-1/2}\,\frac{d^2}{d\lambda^2}\,(s^{1/2}) = -\,[s'\,(2\,s)^{-1}]^2 - \frac{d}{d\lambda}\,[s'\,(2\,s)^{-1}], \tag{1.2.5}$$

depends on λ only. In (1.2.5) the symbol s' denotes the derivative of s with respect to λ. Each function $\Psi^* = \Psi^*\,(\lambda,\theta)$ that satisfies the partial differential equation (1.2.4) obviously defines a stream-function $\Psi = s^{-1/2}\,\Psi^*$. The plane (λ,θ) is called the pseudo-logarithmic plane (see section I.4).

3. Isentropic Flow

The transformation of Chaplygin's equation, as presented so far, applies to any fluid for which both M and ϱ are certain functions of q only and for which a definite relation between pressure and density holds. Let us specify this relation now, by assuming the isentropic conditions. As it is known, for such a flow the relation $a^2 = k\,p\,\varrho^{-1} = k\,c\,\varrho^{k-1}$ holds. Using that relation we obtain from (1.1.3) the formula

$$\frac{1}{2}\,(k-1)\,q^2 + a^2 = a_0^2, \quad \text{or} \quad q^2\Big[1 + \frac{1}{2}\,(k-1)\,m\Big] = m\,a_0^2, \tag{1.3.1}$$

where $m = M^2 = q^2\,a^{-2}$. Differentiating the second equation (1.3.1) with respect to m we get the relation

$$\frac{d}{dm}\,(q^2) = q^4\,(a_0\,m)^{-2}. \tag{1.3.2}$$

From the equation for the sound velocity we have the relation

$$(k-1)\,\log\varrho = \log\,(a^2) + \text{const.}, \tag{1.3.3}$$

which, differentiated with respect to m, leads to:

$$\frac{d}{dm}\,(\log\varrho) = -\,\frac{1}{2}\,q^2\,m^{-1}\,a_0^{-2}. \tag{1.3.4}$$

In the derivation of equation (1.3.4) the first formula (1.3.1) and formula (1.3.2) were used. In (1.2.5) we apply the substitution $s'\,(2\,s)^{-1} = \dfrac{1}{2}\dfrac{d}{d\lambda}\,(\log s) = n$, and obtain

$$f = -\,n^2 - \frac{dn}{d\lambda}. \tag{1.3.5}$$

By taking the logarithm of s (section I.2) we obtain the relation

$$\log s = \frac{1}{2}\,\log\,(1-m) - \log\varrho, \tag{1.3.6}$$

from which we get, after some elementary transformations by means of the second equation (1.3.1) and (1.3.4),

$$\frac{d}{dm}\,(\log s) = -\,\frac{1}{4}\,(k+1)\,(1-m)^{-1}\,q^2\,a_0^{-2}. \tag{1.3.7}$$

From the definition of λ given in (1.2.1), it follows that

$$\frac{d\lambda}{d(q^2)} = (1-m)^{1/2}\,(2\,q^2)^{-1}, \tag{1.3.8}$$

and thus, using (1.3.2), (1.3.7) and (1.3.8) we obtain after some elementary calculations:

$$n = \frac{d}{2\,d\lambda}\,(\log s) = -\frac{1}{4}\,(k+1)\,m^2\,(1-m)^{-3/2}. \tag{1.3.9}$$

By differentiating, using the second equation (1.3.1) and equations (1.3.2) and (1.3.8), we obtain

$$\frac{dn}{d\lambda} = -\frac{1}{4}\,(k+1)\,m^2\,(1-m)^{-3}\left[4 - (3 - 2k)\,m - \frac{1}{2}\,(k-1)\,m^2\right]. \tag{1.3.10}$$

Then, by substituting (1.3.10) and the square of the right-hand side of (1.3.9) into equation (1.3.5), we obtain the function f in terms of m:

$$f = \frac{1}{16}\,(k+1)\,m^2\,(1-m)^{-3}\,[16 - 4\,(3 - 2k)\,m - (3k-1)\,m^2]. \tag{1.3.11}$$

To obtain the relation between λ and m, we multiply the right-hand sides of (1.3.2) and (1.3.8), and by considering the second equation (1.3.1),

$$\frac{d\lambda}{dm} = \frac{1}{2}\,(1-m)^{1/2}\left[m^{-1} + \frac{1}{2}\,(k-1)\right]^{-1} m^{-2}. \tag{1.3.12}$$

Therefore, by integrating the expresion (1.3.12),

$$\lambda = \frac{1}{2}\,\{\log\,(1-T) - \log\,(1+T) + h^{-1}\,[\log\,(1+h\,T) - \log\,(1-h\,T)]\}, \tag{1.3.13}$$

$$T^2 = 1 - m, \qquad h = (k-1)^{1/2}\,(k+1)^{-1/2}. \tag{1.3.13a}$$

Using the substitutions $(1+T) = m\,(1-T)^{-1}$, $(1-T) = 1 - (1-m)^{1/2}$, $(1+h\,T) = (1 - h\,T)^{-1}\,[2\,(k+1)^{-1} + h^2\,m]$, etc., we obtain from (1.3.13) formula (11′) in [4], p. 254.

Combining (1.3.11) and (1.3.13a) we easily obtain another formula for $f(\lambda)$:

$$f = \frac{1}{16}\,(k+1)\,[5\,(1+k)\,T^{-6} - 12\,k\,T^{-4} + 2\,(3k-7)\,T^{-2} + 4\,(2+k) - (3k-1)\,T^2],$$
$$\tag{1.3.14}$$

or

$$f = \frac{1}{16}\,(k+1)\,(1 - T^2)^2\,T^{-6}\,[5\,(k+1) + 2\,(5-k)\,T^2 - (3k-1)\,T^4]. \tag{1.3.14a}$$

The relations (1.3.11) and (1.3.13) between f, λ and m obviously do not involve any integration constants, since f, λ and m are dimensionless.

4. The Duality between the Flows of an Incompressible and a Compressible Fluid. Logarithmic Plane

In considering the two-dimensional flows of a perfect fluid, it is convenient to introduce a certain correspondence between subsonic flows of a compressible fluid and flows of an incompressible fluid. Naturally, such a correspondence can be defined in various ways. On the other hand, one can formulate certain requirements which will simplify the forms of this correspondence. First, it is convenient to consider the flows in a plane where Φ and Ψ satisfy a linear homogeneous equation, since in this case the principle of superposition holds. Second, it is natural to require that the equations for Φ and Ψ have the simplest form possible, which for linear equations is the so-called "canonical form". In the case of a compressible fluid, these conditions lead to considering the potential and the stream-function in the pseudo-logarithmic plane, where both requirements are fulfilled. In the case of an incompressible fluid, Φ and Ψ satisfy in various planes (that is, in the physical, hodograph, etc.) the same equation, namely, the Laplace equation. Since, however, we consider the compressible fluid motion in the pseudo-logarithmic plane, it is natural to introduce the correspondence with incompressible fluid flows defined in the logarithmic plane. In the latter case the equations for Φ and Ψ have the "canonical form" (as may be easily verified), and the transition from the pseudo-logarithmic plane to the logarithmic plane means only one-dimensional stretching: Thus one of the coordinates, θ, is the same, and for the second coordinate we take in the case of the incompressible fluid $\tilde{\lambda} = \log q$, and in the case of the compressible fluid the quantity λ, defined by (1.3.13), which reduces to $\log q$ as the compressibility effect goes to zero. The plane $(\tilde{\lambda}, \theta)$ is the so-called logarithmic plane. This duality between the flow of an incompressible and a compressible fluid may also be extended to the considerations covering the methods of integration of the partial differential equations of flows of those two types of fluid. Let us consider partial differential equation (1.2.4). The corresponding equation in the case of an incompressible fluid evidently must have the form $\varDelta \Psi^* = 0$, i. e. Laplace's form. Chaplygin had already determined an infinity of solutions of equation (1.2.4) by the simple well-known procedure of separating the variables. Thus, every solution Ψ^* of (1.2.4) which is regular in a given domain (see section 0.1) of the (λ, θ)-plane may be approximated uniformly in each closed subdomain (see section 0.1) by a series (finite or infinite)

$$\Psi_N{}^*(\lambda, \theta) = \sum_{n=1}^{N} (a_n C_n + b_n S_n), \qquad (1.4.1)$$

to an arbitrary degree of precision. The symbols a_n and b_n denote the constant coefficients and C_n and S_n denote a set of solutions

$$C_n(\lambda, \theta) = \alpha_n(\lambda) \cos n\,\theta, \quad S_n(\lambda, \theta) = \alpha_n \sin n\,\theta, \quad n = 0, 1, 2, \ldots, \quad (1.4.2)$$

where the α_n are expressed in a simple way by means of hypergeometric functions. The particular solutions (1.4.2) evidently correspond to the set

$$c_n(\lambda, \theta) = \lambda^n \cos n\,\theta, \quad s_n(\lambda, \theta) = \lambda^n \sin n\,\theta, \quad n = 0, 1, 2, \ldots \quad (1.4.2')$$

of solutions for Laplace's equation $\varDelta \Psi^* = 0$. Every solution (1.4.1) may be associated with the harmonic function (i. e., a function which satisfies Laplace's equation)

$$P_N(\lambda, \theta) = \sum_{n=1}^{N} \lambda^n (a_n \cos n\,\theta + b_n \sin n\,\theta), \qquad (1.4.1')$$

and will, in fact, have many properties similar to those of the latter. Hydrodynamically, this means that $\Psi_N = s^{-1/2} \Psi_N{}^*$ and P_N will be stream-functions of compressible and incompressible flows, respectively, bearing more or less similarity to each other. Thus, the great experience in the theory of ideal incompressible flows may be used in order to create similar compressible flow patterns by the association $(1.4.1) \leftrightarrow (1.4.1')$. This transition may be conceived as an "operator" transforming a solution of the one differential equation into a corresponding solution of the other. Let us mention that determining the incompressible flow around the given obstacle and obtaining the complex potential as a function of $(\log q, \theta)$ is always a solvable problem by the use of Theodorsen's method[5].

But often[4] there appear in the theory of incompressible fluid flows functions that are neither regular nor single-valued when expressed in the hodograph plane. Consider for example a flow originating from a doublet at infinity and passing around a circle; one easily verifies that in the hodograph plane the stream-function is a two-valued function possessing a point with branch character corresponding to the point at infinity in the physical plane. In general, even in the simplest examples of flows around closed curves we may expect stream-functions that are multivalued and possess singularities of various kinds. Moreover, the infinite series representing a solution by Chaplygin's method will converge in general only within the part of the

<hr>

[5] Th. Theodorsen: Theory of wing sections of arbitrary shape. N. A. C. A., Rep. No. 411 (1931).

domain in which the flow is defined. Therefore several expressions are needed in order to obtain the whole flow. Bergman[6] derived another method which yields certain types of particular solutions Ψ_ν, the stream-function of a compressible fluid flow. Bers and Gelbart[7] obtained the same solutions independently of Bergman. They denote the functions $\Phi_\nu + i\,\Psi_\nu$ as Σ-monogenic functions. Here Φ_ν is the potential function. But the flow patterns generated by the functions Ψ_ν mentioned above (Bergman, Bers) or a linear combination of them $\Sigma\,\alpha_\nu\,\Psi_\nu$, are of rather special character. In particular, the flow patterns with stream-functions $\Sigma\,\alpha_\nu\,\Psi_\nu$, cannot (in general) represent an entire flow around a closed body (a similar situation to Chaplygin's method). Frequently, in the theory of analytic functions of a complex variable in a similar situation (i. e., when one expression of a certain kind—e. g., power series—does not represent the function, say, f, in the entire domain B in which the function has to be considered), the procedure employed is to decompose B into smaller regions, say, B_K, $K = 1, 2, \ldots, n$, $\sum_{K=1}^{n} B_K = B$, such that it is possible to find in every region B_K, another analytic expression which represents f in that region. Generalizing this method of representation of a function of a complex variable, Bergman described in [8] a method for representing the stream-function by means of decomposing the domain B into parts B_K, and representing Ψ in every B_K by another analytical expression. This method will be given in the part on transonic flow.

But in many instances it is important to have a single expression represent the whole flow. Thus the problem arises to develop another procedure in order to make harmonic functions correspond to solutions of Chaplygin's equation. This procedure has to be defined in the hodograph plane and must be applied in domains situated on Riemann surfaces (see 0.1) over this plane. Only such methods will permit the translation of incompressible fluid flow patterns into flow patterns of a compressible fluid in the large (see 0.1). Such a method has been elaborated by Bergman[6, 9, 10] through [15]. One of the fundamental ideas of this method is the application of "operators" transforming the solutions of flow around closed curves in the domain of an incompressible fluid into the solutions of flow around closed curves in the domain of a compressible fluid. This will be discussed more thoroughly in the next section.

5. Method of Generating Stream-functions

According to Bergman's method, we integrate the transformed Chaplygin's equation (1.2.4) by setting[4] the function Ψ^* equal to an infinite series:

$$\Psi^* = \sum_{n=0}^{\infty} g_n G_n, \tag{1.5.1}$$

where each G_n depends on λ only and each g_n is a harmonic function of λ and θ. This series will be subjected to certain formal transformations. The justification of these

[6] Bergman: A formula for the stream function of certain flows. Proc. nat. Acad. Sci. USA **29**, 276—281 (1943).

[7] L. Bers and A. Gelbart: On a class of differential equations in mechanics of continua. Quart. Appl. Math. **1**, 168—188 (1943). — On a class of functions defined by partial differential equations. Trans. Amer. Math. Soc. **56**, No. 1, 67—93 (1944).

[8] S. Bergman: On two-dimensional flows of compressible fluids. N. A. C. A., T. N. No. 972 (1945).

[9] S. Bergman: Zur Theorie der Funktionen, die eine lineare partielle Differentialgleichung befriedigen, I. Recueil Mathématique, Nouvelle Série **2**, 1169—1198 (1937).

[10] S. Bergman: Review of the paper: Two-dimensional irrotational transonic flows of a compressible fluid. By Y. H. Kuo: N. A. C. A., T. N. No. 1445 (1948); in Mathem. Reviews **11**, No. 3, 223 (1950).

transformations will be given in the next section by studying the convergence of the series considered. We find from the identity which is well-known:

$$\Delta (A B) = A \Delta B + B \Delta A + 2 (A_\lambda B_\lambda + A_\theta B_\theta),$$

and

$$(G_n)_\theta = 0, \quad \Delta g_n = 0,$$

that

$$\Delta (g_n G_n) = g_n G_n'' + 2 G_n' (g_n)_\lambda.$$

Introducing the value of $\Delta (g_n G_n)$ in (1.5.1), we obtain obviously:

$$\Delta \Psi^* = \sum_0^\infty [g_n G_n'' + 2 G_n' (g_n)_\lambda],$$

so that

$$\Delta \Psi^* + f \Psi^* = \sum_0^\infty [g_n (G_n'' + f G_n) + 2 G_n' (g_n)_\lambda]. \tag{1.5.2}$$

In order to have the right-hand side vanish as required in (1.2.4), we take:

$$G_0 = 1, \quad G'_{n+1} = G_n'' + f G_n, \quad n = 0, 1, 2, \ldots, \tag{1.5.3a}$$

$$2 (g_n)_\lambda = - g_{n-1}, \tag{1.5.3b}$$

where g_0 is harmonic, but otherwise arbitrary. Then (1.5.2) becomes clearly:

$$\Delta \Psi^* + f \Psi^* = \sum_0^\infty [g_n (G'_{n+1} - f G_n + f G_n) - g_{n-1} G_n'] = \sum_0^\infty [g_n G'_{n+1} - g_{n-1} G_n'] = 0,$$

as is easily seen by rearrangement of the series and in view of the fact that $G_0' = 0$. Thus, (1.5.1) with the conditions (1.5.3a) and (1.5.3b) is a solution of (1.2.4). We now solve the recursion formula (1.5.3b) for the g_n functions in the following form: The functions g_n must be harmonic functions of λ and θ. Therefore we take analytic functions $\Phi_n (Z), Z = \lambda + i \theta$, such that $\Phi_0 (Z)$ is arbitrary analytic, and

$$\Phi_n (Z) = -\frac{1}{2} \int_0^Z \Phi_{n-1} (t) \, dt, \tag{1.5.4}$$

and let g_0 be the real part of $\Phi_0 (Z)$ and g_n the real part of Φ_n, i. e., $g_n = \mathrm{Re} \{\Phi_n (Z)\}$. Then we have, in view of (1.5.4), as one can easily verify:

$$2 (g_n)_\lambda = \mathrm{Re} \{2 \Phi_n'\} = - \mathrm{Re} \{\Phi_{n-1}\} = - g_{n-1},$$

which shows that the g_n satisfy in fact the recursion formula (1.5.3b). On the other hand, we may easily derive from (1.5.4), as one can again verify:

$$\Phi_n (Z) = \frac{(-1)^n}{(n-1)! \, 2^n} \int_0^Z \Phi_0 (t) (Z - t)^{n-1} dt, \quad n > 0, \tag{1.5.4'}$$

whence

$$g_n (\lambda, \theta) = \frac{(-1)^n}{(n-1)! \, 2^n} \mathrm{Re} \left\{ \int_0^Z \Phi_0 (t) (Z - t)^{n-1} dt \right\}. \tag{1.5.5}$$

Consequently, the development (1.5.1) takes the form:

$$\Psi^* (\lambda, \theta) = g_0 (\lambda, \theta) + \sum_1^\infty G_n (\lambda) \mathrm{Re} \left\{ \int_0^Z \Phi_0 (t) \frac{(-1)^n (Z - t)^{n-1}}{2^n (n-1)!} \, dt \right\}. \tag{1.5.6}$$

6. The Domain of Convergence of the Development for Ψ^*

Assume in (1.5.6) a uniform convergence for t in a certain domain[4]. This assumption allows us to interchange the process of summation and integration, so that

$$\Psi^*(\lambda, \theta) = g_0(\lambda, \theta) + \mathrm{Re}\left\{\int_0^Z \Phi_0(t)\, U(t; \lambda, \theta)\, dt\right\}, \tag{1.6.1}$$

with

$$U(t; \lambda, \theta) = \sum_{n=1}^{\infty} G_n(\lambda)\, \frac{(-1)^n\, (Z-t)^{n-1}}{2^n\, (n-1)!}. \tag{1.6.1a}$$

The transformation of the series of $\Psi^*(\lambda, \theta)$ is valid in exactly the same domain of the complex t-plane, where (1.6.1a) converges uniformly. To find a proper domain of uniform convergence of $U(t; \lambda, \theta)$, we shall apply the following procedure: we shall find a series which dominates the series $U(t; \lambda, \theta)$ and shall study its convergence. Because the development of U contains the functions G_n, we have first to find functions $Q_n(\lambda)$ such that, for each n, $Q_n(\lambda)$ dominates $G_n(\lambda)$. This means that, for $-\infty < \lambda < 0$, all Q_n and their derivatives are larger than or equal to the moduli (absolute values) of the G_n and of their corresponding derivatives; by means of symbols, the condition that $G_n \ll Q_n$, will imply:

$$|G_n| \leqq Q_n, \quad \text{and} \quad \left|\frac{d^k G_n}{d\lambda^k}\right| \leqq \frac{d^k Q_n}{d\lambda^k}. \tag{1.6.2}$$

It is easy to prove that: (A) For each partial differential equation of the type of equation (1.2.4), where $f(\lambda)$ can be dominated by an expression of the form $C(\varepsilon - \lambda)^{-2}$, which mathematically may be presented in the form

$$f \ll F = C(\varepsilon - \lambda)^{-2}, \quad C > 0, \quad \varepsilon < 0, \tag{1.6.3}$$

the corresponding series $U(t; \lambda, \theta)$ is uniformly convergent in the domain[4]

$$\left|\frac{Z-t}{2(\varepsilon - \lambda)}\right| \leqq a < 1. \tag{1.6.3'}$$

(B) A suitable constant C in (1.6.3) can be calculated in the special case we are dealing with, namely, the function f derived from the isentropic conditions[4].

(A) We shall present only an outline of the proof. Namely, under the assumption (1.6.3) we easily find that Q_n satisfying the condition (1.6.2) is of the form

$$Q_n = n!\, \mu_n (\varepsilon - \lambda)^{-n}, \tag{1.6.4}$$

with the following recursion formula for the constants μ_n:

$$\mu_0 = 1,\ \mu_{n+1} = \mu_n (n + \alpha)(n + \beta)(n + 1)^{-2}, \tag{1.6.5}$$

where $\alpha = \dfrac{1}{2} - \left(\dfrac{1}{4} - C\right)^{1/2}$, $\beta = \dfrac{1}{2} + \left(\dfrac{1}{4} - C\right)^{1/2}$.

As one may easily notice, the coefficients μ_n obey the same recursion formula as the coefficients of the hypergeometric series

$$H(\alpha, \beta, \gamma; x) = 1 + \frac{\alpha\beta}{1!}\, x + \frac{\alpha(\alpha+1)\beta(\beta+1)}{2!\,\gamma(\gamma+1)}\, x^2 + \ldots = \sum_0^{\infty} \mu_\nu x^\nu, \tag{1.6.5a}$$

with $\gamma = 1$. Because none of the parameters α, β, γ, is zero or a negative integer, according to the theory of hypergeometric series, the series H, together with all its derivatives, is uniformly convergent for $|x| \leqq a < 1$. Substituting (1.6.4) into (1.6.1a) in place of G_n we obtain after some transformations

$$U(t; \lambda, \theta) \ll \frac{1}{2}(\varepsilon - \lambda)^{-1}\frac{d}{dx}H_1, \tag{1.6.6}$$

where H_1 denotes the hypergeometric series of the form

$$H_1 = H_1(\alpha, \beta, 1; x), \qquad x = \frac{1}{2}\frac{|Z - t|}{|\varepsilon - \lambda|}. \tag{1.6.6a}$$

As mentioned above, the hypergeometric series is a uniformly convergent series for $|x| \leqq a < 1$. From the fundamental concepts of analysis it is known that a series whose dominant is a convergent series of positive terms (this last condition in the present case is not essential, as one may easily show), converges uniformly. Hence the uniform convergence of U and, therefore, of the development for Ψ^*, is assured for each region $\frac{1}{2}\frac{|Z - t|}{|\varepsilon - \lambda|} \leqq a < 1$. In this domain, the series for $U(t; \lambda, \theta)$ as well as those for all derivatives of this function converge uniformly and absolutely. This obviously justifies all the transformations performed in section I.5, and proves that Ψ^* satisfies (1.2.4), i. e., is a solution.

(B) We shall briefly outline the proof that in equation (1.6.3) a suitable constant C can be determined in the case of the function f that was derived in section I.3 on the basis of the isentropic relation. Introducing the variable T into the expression for f (1.3.11) we obtain

$$f(T) = \alpha_1 T^{-6} + \beta_1 T^{-5} + \ldots, \tag{1.6.7}$$

where $\alpha_1, \beta_1, \ldots$ are constant coefficients. As is obvious from (1.6.7), f is an analytic function of T and the development (1.6.7) is regular for all complex values of T finite and different from zero. The formula (1.3.13) shows that for complex values of T the variable λ assumes complex values also. From equation (1.3.13) we see that λ is a regular analytic function of T if $T \neq \pm 1, \pm h^{-1}$, and that for these exceptional values of T, λ becomes infinite. For $T = \infty$ every determination of λ is clearly purely imaginary. Of course, vice versa, T is a certain function of λ and $f(T) = f(T(\lambda))$. By a suitable transformation[4] it is possible to show that

$$\left|\frac{d^n f(T(\lambda))}{d\lambda^n}\right| \leqq n! K(\varepsilon - \lambda)^{-n}, \tag{1.6.8}$$

where K is a constant. On the other hand, we have from the definition of F in (1.6.3)

$$\frac{d^n F}{d\lambda^n} = C(n + 1)!(\varepsilon - \lambda)^{-(n+2)}.$$

Hence, for real $\lambda < \varepsilon$

$$\frac{d^n f(T(\lambda))}{d\lambda^n} \leqq K C^{-1}(n + 1)^{-1}(\varepsilon - \lambda)^2 \frac{d^n F}{d\lambda^n},$$

and taking

$$C = K A^2(n + 1)^{-1}, \qquad A \geqq -\lambda, \tag{1.6.9}$$

we easily obtain

$$\left|\frac{d^n f}{d\lambda^n}\right| \leqq \frac{d^n F}{d\lambda^n}, \tag{1.6.10}$$

as required, since $(\varepsilon - \lambda)^2 A^{-2}$ is greater than or equal to unity. The discussion[4] of the inequalities (1.6.3), (1.6.3'), and of the condition (1.6.9) as well as that of $\varepsilon \geqq \lambda \geqq -A$, leads to the determination of the domain of the convergence of the series (1.5.6). This domain is contained on the left-hand side of the (λ, θ) plane between two vertical lines; $-\varepsilon = \text{const.}$ and $-A = \text{const.}$, and two straight lines

$\theta = \pm\, 3^{1/2}\,\lambda$ through the origin forming an angle of 120° with each other. In the interior of this domain all the inequalities (1.6.3), (1.6.3′), the condition (1.6.9) and $\varepsilon \geqq \lambda \geqq -A$ are insured and hence we may assume the absolute and uniform convergence of the series (1.5.6) in each closed subdomain of the angular domain considered, as long as $\Phi_0(Z)$ itself is bounded in this domain.

7. Discussion of the Domain of Convergence of the Series Representing Ψ^*

We have outlined the proof that the series (1.5.6) converges absolutely and uniformly in each closed subdomain of the angular domain $|\theta| < 3^{1/2}\,|\lambda|$. The problem arises whether this series might not converge outside this domain, too. If we insert into the recursion formula (1.5.3a), instead of f, the comparison function $\bar{f}(\lambda) = C/\lambda^2$, we obtain instead of the G_n-factors the terms (1.6.4) with $\varepsilon = 0$ and the general term of our series is, obviously:

$$a_n = 2^{-n}\, n\, \mu_n\, \lambda^{-n}\, \mathrm{Re}\left\{\int_0^Z \Phi_0(t)\,(Z-t)^{n-1}\,dt\right\}, \tag{1.7.1}$$

instead of (1.5.6). In particular, if $\Phi_0(Z) = 1$, we obtain correspondingly:

$$a_n = \mu_n \cdot \mathrm{Re}\left\{Z^n\,(2\,\lambda)^{-n}\right\}. \tag{1.7.2}$$

Let us write simply:

$$Z = |Z|\,e^{i\alpha}, \qquad \lambda = |Z|\cos\alpha. \tag{1.7.3}$$

Then we obtain

$$|a_n| = \mu_n\,|2\cos\alpha|^{-n} \cdot |\cos n\,\alpha|. \tag{1.7.4}$$

We may now select a point Z outside our domain of proved convergence, but arbitrarily close to it by taking $\alpha = \dfrac{2}{3}\pi - \dfrac{\pi}{N}$, where N is a sufficiently large integer. Thus for all integers n' which are multiples of N we obtain

$$|a_{n'}| \geqq \frac{1}{2}\,\mu_{n'}\,|2\cos\alpha|^{-n'}. \tag{1.7.5}$$

Since $|2\cos\alpha| < 1$, and $\lim\limits_{n' = \infty}\left|\dfrac{\mu_{n'+1}}{\mu_{n'}}\right| = 1$, which means that the series $\mu_{n'}$ is a divergent one, one easily sees that the sequence $|a_{n'}|$ will increase beyond every bound. Thus in this case, the series for Ψ^* does not converge outside the angular domain $|\theta| < 3^{1/2}\,|\lambda|$. Since $\bar{f}(\lambda)$ behaves asymtotically like $f(\lambda)$ at $\lambda = 0$, there is obviously little chance that the conditions of convergence of (1.5.1) can be improved. In general, we shall need solutions of the differential equation (1.2.4) that are defined not only in the aforementioned domain of convergence. As it is well-known, there are two ways to improve our results given above.

First of all, we remark that in every domain $\mathrm{Re}\{Z\} < b < 0$ the function $f(\lambda)$ may be approximated to any degree of precision by a series of the form

$$f_m(\lambda) = \sum_{\nu=1}^m c_\nu \exp(\lambda\,\nu), \tag{1.7.6}$$

which is, as an analytic function of the real variable λ, regular in the whole finite λ-plane. Hence, as in part B of section I.6, we may prove that the development for a solution Ψ_m^* of the differential equation

$$\Delta\,\Psi_m^* + f_m(\lambda)\,\Psi_m^* = 0, \quad \text{or} \quad \frac{1}{4}\Delta\,\Psi_m^* + F_m(\lambda)\,\Psi_m^* = 0, \tag{1.7.7}$$

where $f_m = 4\,F_m$, converges in the entire half-plane $\mathrm{Re}\{Z\} < 0$. Therefore, we replace the function $f(\lambda)$ by an approximating $f_m(\lambda)$, to obtain a good agreement between the actual behavior of the gas described by the function $f_m(\lambda)$ and that of the compressible gas described by $f(\lambda)$ up to a certain Mach number, as well as a series development for the solution $\Psi_m{}^*$ that converges absolutely in the whole domain $\mathrm{Re}\,\{Z\} < 0$. On the other side, one should not exaggerate the importance of this result. The computation of the $G_n(\lambda)$ for the real $f(\lambda)$ is still quite a difficult task. If, for every degree m of approximation one had to carry out new computations for the corresponding $G_n{}^{(m)}(\lambda)$, the amount of work will be excessive. Moreover, we know that the convergence in the half-plane $\mathrm{Re}\,\{Z\} < 0$, will become poor for increasing approximation outside the sector $|\theta| < 3^{1/2}\,|\lambda|$. Let us mention that—as was shown by Bergman[8]—the actual solution of (1.2.4) may be written in the form

$$\Psi^* = \lim_{m \to \infty} \Psi_m{}^*. \tag{1.7.7a}$$

We may, therefore, to point out another way of continuing the development of Ψ^* outside the original domain of convergence. To this end let us describe briefly the process of analytic continuation. Consider a power series in $Z = \lambda + i\,\theta$. This series converges and represents a function for certain values of Z only, $Z < 1$, say. This means that this series converges at all points located in the interior of the circle of the radius $Z = 1$. The question arises, whether it is possible to define the function at points outside the circle in such a way that the function possesses a derivative and is single-valued at each point (i. e., is analytic) in a larger domain than the interior of the circle. For this purpose choose any point P_1 within the circle. We know the value of the function and all its derivatives at P_1, from the series, and so we can form the Taylor series for the same function with P_1 as origin, which will define a function analytic throughout some circle of the center P_1. This new circle will usually lie partly outside the old circle of convergence, and for points in the region which are included in the new circle but not in the old circle, the new series may be used to define the values of the function, although the old series failed to do so. Similarly, we can take any other point P_2, in the region for which the values of the function are now known, and form the Taylor series with P_2 as origin. This will, in general, enable us to define the function at other points at which its values were not previously known and so on. This process is called analytic continuation.

Consequently, if the function exists outside the domain of convergence $|\theta| < 3^{1/2}\,|\lambda|$, then the values of Ψ^* can be obtained by use of various methods of analytic continuation. For example, the Borel summation method[1] or that of Lindelöf can be used. Borel's result represents an analytic function in a more extended region than the interior of the circle of convergence. Using the Borel method, we put down[4]

$$s_n(\lambda,\,\theta) = \sum_{\nu=0}^{n} G_\nu(\lambda)\,g_\nu(\lambda,\,\theta), \tag{1.7.8}$$

and obtain as a consequence

$$\Psi^*(\lambda,\,\theta) = \lim_{\varkappa=\infty} \left\{ \exp(-\varkappa) \sum_{0}^{\infty}{}' \frac{\varkappa^{n+1}}{(n+1)!}\, s_n(\lambda,\,\theta) \right\}. \tag{1.7.9}$$

This expression coincides with the original value of Ψ^* in the whole domain of convergence. It might converge in a much larger domain. If we again replace $f(\lambda)$ by the function of comparison $\bar{f}(\lambda) = C/\lambda^2$, we see easily for instance that this expansion (1.7.9) is defined in the entire half-plane $\mathrm{Re}\,\{Z\} < 0$. It is, obviously, a useful transformation for the original function $f(\lambda)$ too, and leads, in fact, in many cases, to the representation of $\Psi^*(\lambda,\,\theta)$ throughout the domain in question, as the condition may require.

8. Transformation to the Physical Plane

After a stream-function $\Psi = s^{-1/2}\,\Psi^*$ has been determined in the λ, θ-plane by means of the method discussed above, the task still remains of determining the flow in the physical plane—i. e., to express Ψ as a function of x and y. The transition is almost immediate, for the mapping of the λ, θ-plane upon the x, y-plane is given by the following formulas:

$$x = \int \varrho^{-1}\{[-A\,\Psi_\theta \cos\theta - q^{-1}\Psi_\lambda \sin\theta]\,d\lambda + [A\,\Psi_\lambda \cos\theta - q^{-1}\Psi_\theta \sin\theta]\,d\theta\}, \qquad (1.8.1\,\text{a})$$

$$y = \int \varrho^{-1}\{[-A\,\Psi_\theta \sin\theta + q^{-1}\Psi_\lambda \cos\theta]\,d\lambda + [A\,\Psi_\lambda \sin\theta + q^{-1}\Psi_\theta \cos\theta]\,d\theta\}, \qquad (1.8.1\,\text{b})$$

$$A = q^{-1}(1 - M^2)^{1/2}. \qquad (1.8.1\,\text{c})$$

9. Alternative Formulas, Additional Remarks

Let us mention that various representations of Bergman's procedure and his final results are possible. Below, we shall try to represent all possible phases and variations of his method. This will enable the reader to follow not only the original papers of Bergman, if he wishes to do so, but also the numerous tables of various functions which are so important in applications of this method to practical problems. In previous sections we explained the meaning of a few functions such as Ψ, Ψ^*, etc. We discussed the behavior of the compressibility equation in the pseudo-logarithmic plane (λ, θ) or in the Gaussian plane $Z = \lambda + i\,\theta$. We note, without showing it in a more precise way, that one may use various systems of coordinates such as: $\lambda + i\,\theta$, $\lambda - i\,\theta$, $\theta + i\,\lambda$, or $\theta - i\,\lambda$. Theoretically, it makes no difference which system we use and in the subsequent sections we shall refer to all these systems; similarly, the functions Ψ, Ψ^*, etc., will be used, depending upon the situation. To avoid complicating the present representation of Bergman's method, we shall restrict ourselves in many instances to a simple citation of the final results and formulas, referring the reader who is mathematically more advanced to his original papers, where all the proofs are given.

The differential equation for Ψ in the pseudo-logarithmic plane, corresponding to (1.2.4) is given by the formula

$$\Psi_{\lambda\lambda} + \Psi_{\theta\theta} + 4N\,\Psi_\lambda = 0, \qquad N = -\frac{1}{8}(k+1)\frac{M^4}{(1-M^2)^{3/2}}. \qquad (1.9.1)$$

An alternative solution of the second equation (1.7.7) is given by the formula

$$\Psi_m{}^* = \operatorname{Im}\left\{\left[g(\bar{Z}) + \sum_{n=1}^{\infty} \frac{(2n)!}{2^{2n}\,n!}\,g^{(n)}Q_m{}^{(n)}\right]\right\}, \qquad (1.9.2)$$

where $\bar{Z} = \lambda - i\,\theta$, and

$$g^{(0)}(\bar{Z}) = g(\bar{Z}), \qquad g^{(n)}(\bar{Z}) = \int_0^{\bar{Z}} g^{(n-1)}(\bar{Z})\,d\bar{Z}, \qquad n = 1, 2, \ldots \qquad (1.9.2\,\text{a})$$

The solution $\Psi_m{}^*$ of (1.7.7) is the imaginary part of the series (1.9.2) which converges in the interior of the wedge-shaped region $|\theta| \leq 3^{1/2}|\lambda|$ of the half-plane $\lambda < 0$. The function $g(\bar{Z})$ is any arbitrary analytic function of $\bar{Z}$, which contains the origin and is regular (i. e., does not possess singularities) in the entire domain in which we wish to obtain a solution. As the lower limit of the integral in (1.9.2a) any fixed value of $\bar{Z}$, α, say, $\operatorname{Re}\alpha \leq 0$, may be taken. If $\alpha = 0$, we obtain the above mentioned region $|\theta| \leq 3^{1/2}|\lambda|$. The functions $Q^{(n)}(2\lambda)$ are defined by the recurrence formula:

$$(2n+1)\,Q_\lambda^{(n+1)} + Q_{\lambda\lambda}^{(n)} + 4\,F_m(\lambda)\,Q^{(n)} = 0, \qquad (1.9.3)$$

or

$$(2n+1)\,Q^{(n+1)} = -\,Q_\lambda^{(n)} - 4\int_{-\infty}^{\lambda} F_m(\lambda)\,Q^{(n)}(\lambda_1)\,d\lambda_1, \quad \text{with } n = 1, 2, \ldots \qquad (1.9.4)$$

$$Q^{(1)} = -\int_{-\infty}^{\lambda} 4\,F_m(\lambda)\,d\lambda,$$

$$Q^{(0)} = 1,$$

and

$$Q^{(n)}(-\infty) = 0.$$

From (1.9.3) we may easily obtain:

$$Q^{(2)} = -\,\frac{4}{3}\,F_m(\lambda) + \frac{1}{6}\,Q^{(1)2}, \qquad (1.9.5)$$

$$Q^{(3)} = -\,\frac{4}{15}\,(F_m)_\lambda + \frac{4}{15}\,F_m Q^{(1)} - \frac{16}{15}\int_{-\infty}^{\lambda} F_m{}^2\,d\lambda + \frac{1}{90}\,Q^{(1)3}, \text{ etc.} \qquad (1.9.6)$$

Equation (1.9.3) determines each $Q^{(n)}$ to within a constant; this constant can be specified by the requirement that for some fixed value, say, $q = q_0$ (and, therefore, for some λ, say, $\lambda = \lambda_0$), $Q^{(n)}$ vanishes. Bergman[11] proposed as the most natural choice that all $Q^{(n)}$ should vanish for $\lambda = -\infty$, i. e., for $q = 0$. This assumption gives the lower limit of the integral in the formula (1.9.4).

According to definition $\Psi^* = s^{1/2}\,\Psi$ (see I.2); hence $\Psi = H\,\Psi^*$, where $H = s^{-1/2}$ or

$$H = (1 - M^2)^{-1/4}\left[1 + \frac{1}{2}\,(k-1)\,M^2\right]^{-\frac{1}{2(k-1)}}, \qquad (1.9.7)$$

as one can easily verify, using the identities $\varrho = \left[1 - \frac{1}{2}\,(k-1)\,q^2\right]^{1/(k-1)}$ and $M^2 = q^2\,[1 - (k-1)\,q^2/2]$, where the units of mass and velocity are so chosen that at a stagnation point $\varrho = 1$, and $dp/d\varrho = 1$. Hence a solution of (1.9.1) is given by $H\,\Psi^*$, where Ψ^* is defined by (1.9.2). It is often more convenient to use another operator which can be obtained from (1.9.2) by setting

$$g(\overline{Z}) = \int_{-1}^{+1} f\left(\frac{1}{2}\,\overline{Z}\,(1-t^2)\right)(1-t^2)^{-1/2}\,dt, \qquad (1.9.8)$$

namely

$$\Psi = \operatorname{Im}\left\{ H\left[\int_{-1}^{+1} f\left(\frac{1}{2}\,\overline{Z}\,(1-t^2)\right)(1-t^2)^{-1/2}\,dt + \int_{-1}^{+1} \overline{Z}\,Q^{(1)}\,t^2\,f\left(\frac{1}{2}\,\overline{Z}\,(1-t^2)\right)(1-t^2)^{-1/2}\,dt + \right.\right.$$

$$\left.\left. +\int_{-1}^{+1} (\overline{Z})^2\,Q^{(2)}\,t^4\,f\left(\frac{1}{2}\,\overline{Z}\,(1-t^2)\right)(1-t^2)^{-1/2}\,dt + \ldots\right]\right\}. \qquad (1.9.9)$$

Both (1.9.2) and (1.9.9) represent two forms of the so-called "integral operator of the second kind". The disadvantage of this operator is that it converges only in a certain portion of the subsonic region (i. e., a portion of the half-plane $\lambda < 0$). In addition to this operator there exists, among others, the so-called integral operator of the first kind[6, 12], which possesses the advantage of converging in the whole

---

[11] S. Bergman: On supersonic and partially supersonic flows. N. A. C. A., T. N. No. 1096 (1946).

[12] S. Bergman: Two-dimensional subsonic flow of a compressible fluid and their singularities. Trans. Amer. Math. Soc. 61, 452—498 (1947).

subsonic region, but which on the other hand does not appear to be suitable for actual numerical computation. Both these operators will be discussed more thoroughly in the section on transonic flow.

In equations (1.5.1), (1.9.2) and others, the functions G_n, H and $Q^{(n)}$ are functions of λ only and are the same for all stream-functions. They may be computed once and for all and the results used in all future computations. If any analytic function $g^{(0)}$ of a complex variable Z (regular at $Z = 0$) is substituted into one of the formulas given above, the result is a stream-function of a subsonic compressible fluid flow pattern. On the other hand, it is well-known that an analytic function g ($\log q - i\,\theta$) may be regarded as defining a flow pattern of an incompressible fluid. Consequently, those formulas may be interpreted as operators which distort an incompressible flow pattern into a compressible one. The operator formulas are of such a form that a knowledge of the incompressible flow pattern used leads directly to both qualitative and quantitative knowledge of the behavior of the stream-function of the corresponding compressible flow.

As mentioned above, from the purely mathematical point of view, an operator is a rule which is applied to a function of one class to obtain a function belonging to another class. A simple example of an operator is the process of obtaining a harmonic function (i. e., a function satisfying Laplace's equation) of two real variables by taking the real (or imaginary) part of an analytic function of a complex variable. This operator preserves many properties of analytic functions and is often used as a tool for the investigation of harmonic functions.

An obvious objection is that, whereas any incompressible flow will lead to some compressible flow, there seems to be no indication of how to find precisely the g to be substituted into the operator so that we may have a specified obstacle in the compressible flow. A procedure to solve this problem approximately can be developed[13]. In practical application the procedure described below may be applied. However, it is easy to see why the same analytic function g which solves the corresponding incompressible problem can be expected to yield a first approximation to the compressible problem in the case of low speeds; for in such a case the solution (1.9.2) is easily shown to yield a flow very similar to the incompressible flow, which agrees with the physical state of affairs.

The fundamental problem to be faced in the theory of airfoils is to determine the flow with a certain velocity at infinity around an obstacle given in the physical plane. This leads to a very complicated non-linear problem in the hodograph plane since the domain where the flow is defined is determined by the flow itself. But, as mentioned above, we may overcome this difficulty by choosing a function g which solves the corresponding incompressible problem. With this g let $\overline{\Psi}(\lambda, \theta)$ be the solution of (1.9.1) obtained from (1.9.2). A solution $\Psi(\lambda, \theta)$ of (1.9.1) is further determined such that $\overline{\Psi} + \Psi$ assumes a constant value on the boundary h (of the hodograph of the incompressible flow). This particular streamline $\overline{\overline{\Psi}} = \overline{\Psi} + \Psi$ defines the shape of the body, the image of which in the physical plane will give a distorted (in a certain way) shape of the actual body; if the given profile is distorted in opposite directions and if the procedure described is repeated to the distorted profile, then in many instances a better approximation is obtained. This method may be repeated until the desired degree of accuracy is attained.

The procedure explained below (proposed by Bergman), may be more suitable in practical applications. To explain the method more clearly, let us consider the

---

[13] S. Bergman: The hodograph method in the theory of compressible fluids. Supplement to "Fluid Dynamics" by R. v. Mises and K. O. Friedrichs. Brown University. 1941/42.

flow of an incompressible fluid, as a starting point. Then a rather large class of symmetric flows around closed obstacles can be obtained[14] by putting:

$$\Psi^+ = \sum A_\nu \Psi_\nu^+(\lambda^+, \theta), \qquad (1.9.10)$$

where A_ν are arbitrary constants and the symbol Ψ^+ denotes a stream-function in the domain of incompressible fluids. For the set $\{\Psi^+\}$ it is advisable to choose the real and imaginary parts of the functions:

$$\Psi_{2\varkappa}^+(\lambda^+, \theta) + i\,\Psi_{2\varkappa+1}^+(\lambda^+, \theta) = (\exp \overline{Z}^+ - \exp \alpha)^{\varkappa/2}, \qquad (1.9.11)$$

$$\varkappa = -1, 0, 1, 2, \ldots \text{etc.},$$

$$\overline{Z}^+ = \lambda^+ - i\,\theta, \qquad \lambda^+ = \log q,$$

where α is a constant, whose value may be calculated from the conditions at infinity. Namely, $\exp \alpha$ is equal to the velocity vector at infinity. In analogy to the above set we may choose in the compressible case the class of functions $\Psi_\nu(\lambda, \theta)$ given by

$$\Psi_{2\varkappa}(\lambda, \theta) + i\,\Psi_{2\varkappa+1}(\lambda, \theta) = P\left[(\exp \overline{Z} - \exp \alpha)^{\varkappa/2}\right], \qquad (1.9.12)$$

$$\varkappa = -1, 0, 1, 2, \ldots, \qquad \overline{Z} = \lambda - i\,\theta,$$

where the symbol P denotes the "integral operator of the second kind", which, as explained above, generates solutions of the compressibility equations having many properties similar to those of the analytic functions to which the operator is applied. For each fixed value of the parameter α, it is possible, by applying the operator P, to generate a set $(1.9.12)$ of solutions of the compressibility equation. From this set of solutions it is then possible to determine, by the use of $(1.8.1)$, the corresponding functions $x(\lambda, \theta)$ and $y(\lambda, \theta)$, so that by the use of $(1.9.10)$ it would be possible to construct a broad class of compressible flow patterns. A slight modification of this procedure is the following:

Suppose that we are interested in determining flows past certain types of profiles, say Joukowski profiles or those of elliptical shape. By solving the corresponding incompressible flow problems, determining the complex potentials in the logarithmic plane $(\overline{Z}^+ = \lambda^+ - i\,\theta)$, and applying the P-operator to these functions (with λ^+ replaced by λ), there are obtained complex solutions of the compressibility equation whose imaginary parts give the stream-functions (in the $\overline{Z} = \lambda - i\,\theta$-plane) of a compressible fluid flow past an obstacle bearing a certain resemblance to that appearing in the corresponding incompressible flow. By adding to this stream-function, which we may designate as a "basic solution", the stream-function $\Psi_{2\varkappa}$, $\Psi_{2\varkappa+1}$, corresponding to $\varkappa = 0, 1, 2, \ldots$ in $(1.9.12)$ one can, by taking various choices of the coefficients A_ν in $(1.9.10)$, construct a class of compressible fluid flows which may be looked upon as modifications of that given by the "basic solution". It may be noted[14] that for even values of $\varkappa$ the functions to which the P-operator is applied become polynomials in $\exp \overline{Z}$, so that it suffices to determine the functions $P(\exp n\overline{Z})$, $n = 0, 1, 2, \ldots$. For odd values of $\varkappa$ this simplification is not possible. Thus we obtain three sets of solutions: A. Basic solutions; B. Functions obtained by applying the P-operator to integral powers of $(\exp \overline{Z} - \exp \alpha)^{1/2}$, i. e. $(\exp \overline{Z} - \exp \alpha)^{\varkappa/2}$ where $\varkappa = 1, 3, 5, 7, \ldots$, and for α assuming a sufficiently dense set of values (a two-parameter family of functions); C. Functions obtained from $\exp(n\overline{Z})$ in the same manner as described in B (a one-parameter family of functions). We

---

[14] S. Bergman and B. Epstein: Determination of a compressible fluid flow past an oval-shaped obstacle. J. Math. Physics 26, 195—222 (1948).

note that in order to avoid the necessity of adding too many terms of solutions listed under B and C, it is useful to prepare not one basic flow pattern, but for each type of obstacle a number of "basic solutions" corresponding to different values of any parameter, e. g., for an ellipse for a number of values of the eccentricity.

The operator method, described above, involves the following elements:

1. It is necessary to have considerable knowledge about incompressible flow patterns, in particular about their image in the hodograph and logarithmic planes.

2. The operator formula has to be derived, the functions H and $Q^{(n)}$ have to be determined, and rules must be developed for interpreting properties of incompressible fluid motions as properties of the corresponding compressible flows given by the operator.

If the origin is moved and λ is replaced by $\lambda^* - \alpha$, say, that is, we set[8]

$$\lambda = \lambda^* - \alpha, \qquad \theta = \theta^*, \qquad \alpha \text{ real}, \tag{1.9.13}$$

then the second equation (1.7.7) assumes the form

$$\frac{1}{4} \Delta^{(\alpha)} \Psi_m{}^* + F_m{}^{(\alpha)}(\lambda^*) \Psi_m{}^* = 0, \tag{1.9.14}$$

where

$$\Delta^{(\alpha)} = \frac{\partial^2}{\partial \lambda^{*2}} + \frac{\partial^2}{\partial \theta^2}, \qquad F_m{}^{(\alpha)} = F_m(2\lambda^* - 2\alpha). \tag{1.9.15}$$

The solution of (1.9.14) is given by the following formula (1.9.16), where the substitution $\Psi = H \Psi^*$ is used

$$\Psi = H \Psi^* = H(2\lambda^* - 2\alpha) \operatorname{Im}\left[g(\overline{Z}^*) + \sum_{n=1}^{\infty} \frac{(2n)!}{2^{2n} n!} Q^{(n)}(2\lambda^* - 2\alpha) g^{(n)}(\overline{Z}^*)\right], \tag{1.9.16}$$

$$\overline{Z}^* = \overline{Z} + \alpha = \lambda^* - i\,\theta, \tag{1.9.17a}$$

$$g^{(n+1)}(\overline{Z}^*) = \int_0^{\overline{Z}^*} g^{(n)}(\overline{Z}_1{}^*)\, d\overline{Z}_1{}^*. \tag{1.9.17b}$$

The transformation from the pseudo-logarithmic plane to the physical plane may be achieved in two ways: either directly by means of (1.8.1), or indirectly. In the latter case, we pass at first from the pseudo-logarithmic plane (λ, θ) to the hodograph plane (q, θ), using the relation (1.3.13), and next from the hodograph plane to the physical plane.

The transformation from the hodograph plane (q, θ) to the physical plane may be achieved by means of formulas, given below (analogous to 1.8.1):

$$x = -\int (\varrho q^2)^{-1} \cos\theta\, (\Psi_\Theta)^{-1} \left[(1 - M^2)\, \Psi_\Theta{}^2 + q^2\, \Psi_q{}^2\right] dq, \tag{1.9.18a}$$

$$y = -\int (\varrho q^2)^{-1} \sin\theta\, (\Psi_\Theta)^{-1} \left[(1 - M^2)\, \Psi_\Theta{}^2 + q^2\, \Psi_q{}^2\right] dq, \tag{1.9.18b}$$

where the integration is carried out along a streamline, $\Psi(q, \theta) = \text{const.}$ The local Mach number is defined by the formula

$$M = \frac{q}{a} = \frac{q}{\left[a_0{}^2 - \dfrac{1}{2}(k - 1)\, q^2\right]^{1/2}}, \tag{1.9.19}$$

where a_0 is the speed of sound at rest, i. e., $a_0{}^2 = k\, R\, T_0$.

The application of Riemann's function to the solution of equation (1.7.7) is briefly discussed in[15]. The latter result does not differ essentially from the function introduced by Bergman in equation (7.4) of reference[13].

[15] S. Bergman: Methods for determination and computation of flow patterns of a compressible fluid. N. A. C. A., T. N. No. 1018 (1946).

If there are instances in which the whole flow is subsonic and therefore it is only necessary to consider particular solutions in this range, (as in the present section), then it is useful sometimes to consider certain other expressions which will be derived in the following. The procedure is based on the method of separation of variables. As these variables, (λ, θ) are employed, so that it becomes necessary to consider equation (1.2.4) with the function $f(\lambda)$ represented in the form $f(\lambda) = \sum_{n=2}^{\infty} C_n \exp(2n\lambda)$. Let

$$\Psi^* = U_\nu \cos \nu \theta \quad (\text{or } U_\nu \sin \nu \theta), \tag{1.9.20}$$

where θ is bounded in a certain interval $(-L \leq \theta \leq L)$ and ν has the values

$$\nu = \nu^* L^{-1}, \quad \nu^* = 0, 1, 2, \ldots. \tag{1.9.21}$$

Below, we shall drop the subscript ν and let $U = U_\nu$; U will vary, of course, with the choice of ν, even though this is not indicated by the notation. Thus, we obtain for U the ordinary differential equation

$$U'' - \nu^2 U + 4F U = 0, \quad F = \frac{1}{4} f(\lambda). \tag{1.9.22}$$

The two independent solutions of (1.9.22) will be denoted by $U^{(i)}$, $i = 1, 2, \ldots$. Two cases must be distinguished, that is, whether ν is or is not an integer. In the latter case

$$U^{(1)} = \sum_{n=0}^{\infty} d_n^{(1)} \exp\left[(\nu + 2n)\lambda\right], \tag{1.9.23a}$$

$$U^{(2)} = \sum_{n=0}^{\infty} d_n^{(2)} \exp\left[(-\nu + 2n)\lambda\right], \tag{1.9.23b}$$

where $d_0^{(1)} = d_0^{(2)} = 1$, $d_1^{(1)} = -C_1(1+\nu)^{-1}$, $d_1^{(2)} = -C_1(1-\nu)^{-1}$, etc. The general formula for $d_n^{(i)}$, $i = 1, 2$, and the coefficients C_n are given in[11] (pp. 37 and 75). In case ν is an integer, we have:

$$U^{(1)} = \sum_{n=0}^{\infty} d_n^{(3)} \exp\left[(\nu + 2n)\lambda\right], \tag{1.9.24a}$$

$$U^{(2)} = \sum_{n=0}^{\infty} \left\{\lambda s_n \exp\left[(\nu + 2n)\lambda\right] + d_n^{(4)} \left[\exp(-\nu + 2n)\lambda\right]\right\}. \tag{1.9.24b}$$

The convergence of the series (1.9.23) and (1.9.24) may be demonstrated as follows: Let

$$\exp(2\lambda) = z. \tag{1.9.25}$$

Then the half-plane $\mathrm{Re}(\lambda) < 0$ corresponds to the circle $|z| < 1$. Since

$$\frac{dU}{d\lambda} = 2z\frac{dU}{dz}, \quad \frac{d^2U}{d\lambda^2} = 4z\left(z\frac{d^2U}{dz^2} + \frac{dU}{dz}\right), \tag{1.9.26}$$

the differential equation (1.9.22) may be written in the form

$$z^2 \frac{d^2U}{dz^2} + z\frac{dU}{dz} + \left(F - \frac{\nu^2}{4}\right)U = 0. \tag{1.9.27}$$

Since $F - \frac{\nu^2}{4}$ is a polynomial in z, the point $z = 0$ is a regular singular point (see Part 0.1) of the differential equation. The general solution of (1.9.27) may be expressed in the following form, provided ν is not an integer:

$$U = c_1 U_1 + c_2 U_2, \tag{1.9.28}$$

where c_1 and c_2 are arbitrary constants and

$$U_1 = z^{\nu/2}\left(1 + \sum_{n=1}^{\infty} \alpha_n\, z^n\right), \tag{1.9.29a}$$

$$U_2 = z^{\nu/2}\left(1 + \sum_{n=1}^{\infty} \beta_n\, z^n\right), \tag{1.9.29b}$$

with α_n, β_n properly chosen constants. Since the only (finite) singular point of equation (1.9.27) is at $z = 0$, the series converges for all values of z, as the theory shows[16]. However, if ν is an integer, as in the case under consideration, the above-mentioned method fails; the series (1.9.29a) may be retained, but the series (1.9.29b) must be replaced by

$$U_2^* = U_1\left\{\sum_{n=0}^{\nu-1} \gamma_n\, z^{n-\nu}\,(n-\nu)^{-1} + \gamma_\nu \log z + \sum_{n=\nu+1}^{\infty} \gamma_n\, z^{n-\nu}\,(n-\nu)^{-1}\right\}, \tag{1.9.30}$$

where the γ_n are properly chosen constants. Therefore, in the case under consideration, the general solution of (1.9.27), valid for all z, will be of the form

$$U = c_1\,U_1 + c_2\,U_2^*. \tag{1.9.31}$$

It is seen that by replacing z by $\exp{(2\,\lambda)}$, (1.9.29a) and (1.9.30) will assume the forms given in (1.9.24); the factor λ in the first term of the series (1.9.24b) arises from the logarithmic term in (1.9.30). Since the expansions (1.9.29a) and (1.9.30) are valid for all z, in particular for $|z| < 1$, the expansion (1.9.24) will be valid for $|\exp{(2\,\lambda)}| < 1$; that is, $\mathrm{Re}\,(\lambda) < 0$.

Part II

Simplified Pressure-Density Relation

1. Definition of a Series Representing a Stream-function Ψ^* Regular in the Whole Half-plane

The result of section I.6 was that the series for Ψ^* converges in an angular sector of the half-plane $\lambda < 0$. To extend the expansion of Ψ^* beyond that domain by analytic continuation is possible but involves complicated computations. What more, the procedure employed above has the disadvantage that the computation of the functions G_n is increasingly complicated as n increases, and that each G_n involves the previous G_m's up to $m = n - 1$. Above, we have started with the isentropic relation, based on the equation of state of a perfect gas, and developed a representation that can only be carried out in practice with a certain approximation. We may now reverse our method[4]. We may begin with an approximate magnitude concerning the physical function f, thus replacing it by another function $\overline{f}$; this leads to a simple mathematical representation for $\overline{\Psi}^*$, instead of the stream-function Ψ^*, which function $\overline{\Psi}^*$ converges absolutely and uniformly at least in the half-plane $\lambda < 0$, i. e., in the whole subsonic region. The computation may now be relatively easy, while we may have to justify our choice of $\overline{f}$ by showing that the hypothetical gas thus described is sufficiently similar to a real physical gas. We may begin with

$$\overline{f} = C\,\lambda^{-2}, \quad C > 0, \tag{2.1.1}$$

---

[16] E. T. Copson: Functions of a complex variable. Oxford: The Clarendon Press. 1935.

instead of the function f as defined and calculated in previous sections. Similarly, as in section I.6, we now obtain functions $\overline{G}_n$, such that one has:

$$\overline{G}'_{n+1} = \overline{G}_n'' + C \lambda^{-2} \overline{G}_n, \tag{2.1.2}$$

$$\overline{G}_0 = 1, \quad \overline{G}_n (-\infty) = 0, \quad n > 0,$$

whence the general formula:

$$\overline{G}_n = n! \; \mu_n (-\lambda)^{-n}. \tag{2.1.3}$$

Using the function U (1.6.1a) now takes the different form:

$$\overline{U} (t; \lambda, \theta) = (2 \lambda)^{-1} \sum_{n=1}^{\infty} \mu_n \, n \, (Z - t)^{n-1} (2 \lambda)^{-(n-1)}, \tag{2.1.4}$$

or as shown in section I.6 (which easily can be derived):

$$\overline{U} (t; \lambda, \theta) = (2 \lambda)^{-1} \frac{d}{dx} [H (\alpha, \beta, \gamma; x)], \tag{2.1.5}$$

where H is the hypergeometric series with α, β, γ from (1.6.5a) and $x = (2 \lambda)^{-1} (Z - t)$. Hence we obtain:

$$\overline{U} (t; \lambda, \theta) = - \frac{d}{dt} [H (\alpha, \beta, \gamma; x)], \tag{2.1.6}$$

and the series for $\overline{\Psi}^*$ takes the form which may be a final one:

$$\overline{\Psi}^* (\lambda, \theta) = g_0 (\lambda, \theta) - \mathrm{Re} \left\{ \int_0^Z \Phi_0 (t) \frac{d}{dt} [H (\alpha, \beta, \gamma; x)] \, dt \right\}. \tag{2.1.7}$$

Since $g_0 (\lambda, \theta) = \mathrm{Re} (\Phi_0 (Z))$, equation (2.1.7) can, after some elementary transformations, be written in the form:

$$\overline{\Psi}^* = \mathrm{Re} \left\{ \Phi_0 (Z) - \Phi_0 (Z) H (\alpha, \beta, \gamma; 0) + \Phi_0 (0) H [\alpha, \beta, \gamma; (2 \lambda)^{-1} Z] + \right.$$

$$\left. + \int_0^Z \Phi_0' (t) H [\alpha, \beta, \gamma; (2 \lambda)^{-1} (Z - t)] \, dt \right\}. \tag{2.1.8}$$

The considerations concerning the points of singularity of the hypergeometric series H show[4] that $\overline{U}$, and hence $\overline{\Psi}^*$, generated from the analytic function $\Phi_0 (Z)$, are regular in the whole subsonic region. A further simplification can be gained by the assumption $\Phi_0 (0) = 0$ which leads[4] to [with $H (\alpha, \beta, \gamma; 0) = 1$]

$$\overline{\Psi}^* = \mathrm{Re} \left\{ \int_0^Z \Phi_0' (t) H (\alpha, \beta, \gamma; x) \, dt \right\}. \tag{2.1.9}$$

If we start from a flow of an incompressible fluid, given by a stream-function $g_0 (\lambda, \theta) = \mathrm{Re} [\Phi_0 (Z)]$, the application of the operation (2.1.7) to $\Phi_0 (Z)$ furnishes a corresponding flow pattern of the compressible fluid considered that usually may preserve the type of the original pattern. But when a change of $\Phi_0 (Z)$ by an additive constant in the incompressible flow does not change the flow considered at all, the addition of every such constant [see equation (2.1.8)] leads to a different flow pattern in the compressible fluid flow. Among all the flows, corresponding to the same flow of an incompressible fluid, there exists one [characterized by the assumption $\Phi_0 (0) = 0$] for which the particularly simple representation (2.1.9) exists. That shows that the correspondence between incompressible and compressible flow patterns is not at all uniquely defined; the same generating procedure leads to an infinity of different compressible fluid stream-functions if one starts from the same incompressible fluid flow. As mentioned in section I.4, the procedure for having solutions of the transformed Chaplygin's equation may be interpreted in the way: One begins with a known flow pattern of an incompressible fluid and corrects it for the compressible case by applying the above operation to its stream-function $g_0\,(\lambda,\theta)$. The result is a certain distortion of this stream-function g_0. It is important to get a measure for the value of this distortion. We find some answer from (2.1.7). Since $g_0\,(\lambda,\theta) = \mathrm{Re}\,[\Phi_0\,(Z)]$, for a particular stream-function $g_0 = c$, equation (2.1.7) transforms into

$$\overline{\Psi}* = c - c\,\mathrm{Re}\,\{H\,(\alpha,\beta,\gamma;x)\}\big|_0^Z = c\,\mathrm{Re}\,\{H\,[\alpha,\beta,\gamma;(2\,\lambda)^{-1}Z]\}, \qquad (2.1.10)$$

$$\text{since } x\big|_0^Z = 0 - (2\,\lambda)^{-1}Z, \text{ and } H\,(0) = 1.$$

We see that for large values of $|\lambda|$, i. e., small velocities, $\overline{\Psi}*$ does not vary significantly. For small values of λ the deviation from constancy may become appreciable. This particular $\Psi*$ may serve as a criterion for the distortion in dependence on λ.

2. The Physical Relations Corresponding to the Assumption $\bar{f} = C\,\lambda^{-2}$

One has to study how far the hypothetical gas, described by the function $\bar{f}$, approximates the real one. A gas may be described by a curve in the space of four variables, q, p, ϱ and M. It is sufficient to choose three of the six possible relations between these variables and we choose $M\,(\varrho)$, $q\,(\varrho)$ and $p\,(\varrho)$.

(*a*) The relation $M\,(\varrho)$.

In equation (1.2.5), written in the form $(s^{1/2})'' + f\,(\lambda)\,s^{1/2} = 0$ we substitute $\bar{f}$ for f and therefore obtain for $s^{1/2}$ the following linear differential equation:

$$(s^{1/2})'' + C\,\lambda^{-2}\,s^{1/2} = 0. \qquad (2.2.1)$$

The general solution of (2.2.1) can easily be found in the form:

$$s^{1/2} = \gamma_1\,|\lambda|^{\delta_1} + \gamma_2\,|\lambda|^{\delta_2}, \qquad (2.2.2)$$

where the exponents δ_1, δ_2 are the solutions of the equation $\delta\,(\delta - 1) + C = 0$, i. e.

$$\delta_{1,\,2} = \frac{1}{2} \pm \left(\frac{1}{4} - C\right)^{1/2}. \qquad (2.2.3)$$

In the particular case $C = 0$, we obtain $\delta_1 = 1$, $\delta_2 = 0$, i. e.,

$$s^{1/2} = \gamma_1\,|\lambda| + \gamma_2. \qquad (2.2.4)$$

We have at our choice the two parameters γ_1 and γ_2. Chaplygin[3] had already realized that the equation (1.1.5) could be transformed into Laplace's equation if the $p - \varrho$ relation is assumed to be of the form $p = \alpha\,\varrho^{-1} + \beta$; in this case the whole theory of a compressible fluid flow can be reduced to the well developed theory of an incompressible fluid flow. This approach was later significantly developed by von Kármán[17] and Tsien[18] and the assumption of a linear $p - \varrho^{-1}$ relation is often known now as the Kármán-Tsien approximation. One easily finds in this case $s = \gamma$, i. e., a particular case of (2.2.4) with $\gamma_1 = 0$. This approximation is very useful over a wide range of λ; we may henceforth also assume $\gamma_1 = 0$ in our present approximation. This may simplify our computations considerably; but we should keep in our mind that, if necessary, we have at our disposal an additional parameter to

---

[17] Th. von Kármán: Compressibility effects in aerodynamics. J. aeronaut. Sci. 8, 337—356 (1941).

[18] H. S. Tsien: Two-dimensional subsonic flow of compressible fluids. J. aeronaut. Sci. 6, 399—407 (1939).

obtain a better agreement between the hypothetical and the physical gas. Hence, choosing $\gamma_1 = 0$ and dropping the indices, we obtain in (2.2.2):

and consequently:

$$s^{1/2} = \gamma \, |\lambda|^\delta, \quad \text{or} \; -\lambda = s^{1/(2\,\delta)} \, \gamma^{-1/\delta}, \tag{2.2.5}$$

$$\frac{d\lambda}{d\varrho} = -\,(2\,\delta)^{-1} \, \gamma^{-1/\delta} \, s^{1/(2\,\delta)\,-\,1} \, \frac{ds}{d\varrho}. \tag{2.2.6}$$

On the other hand, equation (1.2.1) and the quantity s (section I.2) lead to $\dfrac{d\lambda}{dq} = \varrho \, q^{-1} \, s$. From the Bernoulli equation we have $q \, \dfrac{dq}{d\varrho} + \varrho^{-1} \, \dfrac{dp}{d\varrho} = 0$, and thus we obtain $\dfrac{d\lambda}{d\varrho} = \dfrac{d\lambda}{dq} \, \dfrac{dq}{d\varrho} = -\, s \, q^{-2} \, \dfrac{dp}{d\varrho}$. Since $M = q \left(\dfrac{dp}{d\varrho} \right)^{-1/2} = (1 - s^2 \, \varrho^2)^{1/2}$, we have

$$\frac{d\lambda}{d\varrho} = -\, s \, M^{-2} = -\, s \, (1 - s^2 \, \varrho^2)^{-1}. \tag{2.2.7}$$

When one equates the right side of (2.2.6) and (2.2.7), the result gained is:

$$\frac{ds}{d\varrho} - 2 \, \delta \, \gamma^{1/\delta} \, s^{2\,-1/(2\,\delta)}(1 - s^2 \, \varrho^2)^{-1} = 0. \tag{2.2.8}$$

This differential equation yields s as a function of ϱ; because $M^2 = 1 - s^2 \, \varrho^2$, one can, from (2.2.8), calculate $M(\varrho)$, i. e., one of our desired relations between the variables of the gas. But the differential equation (2.2.8) cannot be solved in a closed form. It may be solved numerically. The extreme values of the interval $0 < \varrho \, s < 1$ correspond to the Mach numbers 1 and 0, respectively. The integral curves $s \, (\varrho)$, representing solutions of (2.2.8), increase monotonically in the subsonic region, because $1 - s^2 \, \varrho^2 > 0$, $\gamma^{1/\delta} > 0$, for positive s [equation (2.2.5)], and $\delta > 0$.

(b) The relation $q \, (\varrho)$.

In Bernoulli's equation $\dfrac{1}{2} \, q^2 + \displaystyle\int_{\varrho_0}^{\varrho} x^{-1} \, dp \, (x) = c_1$, we insert $q^2 = M^2 \, \dfrac{dp}{d\varrho} = M^2 \, p'$, and get the relation:

$$\frac{1}{2} \, \varrho \, M^2 \, \varrho^{-1} \, p' + \int_{\varrho_0}^{\varrho} x^{-1} \, p' \, (x) \, dx = c_1. \tag{2.2.9}$$

Taking $J \, (\varrho)$ for $\displaystyle\int_{\varrho_0}^{\varrho} x^{-1} \, p' \, (x) \, dx$, we get the differential equation for $J \, (\varrho)$, $\dfrac{1}{2} \, M^2 \, J' \, (\varrho) \, \varrho + J \, (\varrho) = c_1$; this furnishes:

$$J \, (\varrho) = c_1 - c_2 \exp \left[-\, 2 \int_{\varrho_0}^{\varrho} x^{-1} \, M^{-2}(x) \, dx \right]. \tag{2.2.10}$$

But $\dfrac{1}{2} \, q^2 = c_1 - J \, (\varrho)$; this results in the expression:

$$q^2 = 2 \, c_2 \exp \left[-\, 2 \int_{\varrho_0}^{\varrho} x^{-1} \, M^{-2}(x) \, dx \right], \tag{2.2.11}$$

which is the sought relation $q \, (\varrho)$.

(c) The relation $p \, (\varrho)$.

Differentiating the quantity $J \, (\varrho)$ with respect to ϱ (which is $\varrho^{-1} \, p'$) and comparing the result so obtained with the derivative J' obtained from equation (2.2.10) gives us the expression for p', which, when integrated, results in

$$p = 2 \, c_2 \int_{\varrho_0}^{\varrho} M^{-2}(x) \exp \left[-\, 2 \int_{\varrho_0}^{x} y^{-1} \, M^{-2}(y) \, dy \right] dx + c_3, \tag{2.2.12}$$

which is the sought relation $p \, (\varrho)$.

One may notice that the integration of the differential equation (2.2.8) defines $M(\varrho)$ only up to a constant of integration. One may calculate this constant by the requirement that the values $M = 0$ and $\varrho = 1$ shall correspond, which determines the units in which ϱ is measured. In formula (2.2.11), the value $\varrho_1 = 1$ of ϱ for which $M = 0$, is clearly of special interest. One may show that $\int_{\varrho_0}^{1} x^{-1}\,M^{-2}(x)\,dx$ is convergent, and this implies that q is different from zero at $M = 0$. Therefore, since $M = q\,(p')^{-1/2}$, the velocity of sound, $a = (p')^{1/2}$, is infinite for $M = 0$, i. e. at $M = 0$ the gas behaves like an incompressible fluid.

We first study the behavior of $s(\varrho)$ at $\varrho = 1$. The proof presented below is the author's modification of the proof derived in [4]. We use the substitution:

$$s = 1 + A_1\,(\varrho - 1)^{\varepsilon_1} + A_2\,(\varrho - 1)^{\varepsilon_1 + 1/2} + A_3\,(\varrho - 1)^{\varepsilon_1 + 1} + \ldots, \quad 0 < \varepsilon_1, \qquad (2.2.13)$$

with undetermined constants ε_1 and A_i, $i = 1, 2, \ldots$ Then, with the application of the binomial theorem, we have:

$$s^{2 - (2\delta)^{-1}} = 1 + [2 - (2\delta)^{-1}]\,A_1\,(\varrho - 1)^{\varepsilon_1} + [2 - (2\delta)^{-1}]\,A_2\,(\varrho - 1)^{\varepsilon_1 + 1/2} + \ldots, \qquad (2.2.14)$$

and

$$\frac{ds}{d\varrho} = \varepsilon_1\,A_1\,(\varrho - 1)^{\varepsilon_1 - 1} + \left(\varepsilon_1 + \frac{1}{2}\right) A_2\,(\varrho - 1)^{\varepsilon_1 - 1/2} + \ldots \qquad (2.2.15)$$

Let

$$(1 - s^2\,\varrho^2) = \{1 - s^2\,[1 + 2\,(\varrho - 1) + (\varrho - 1)^2]\}, \qquad (2.2.16)$$

so that applying the binomial theorem again gives the formula

$$(1 - s^2\,\varrho^2) = -\,2\,A_1\,(\varrho - 1)^{\varepsilon_1} - 2\,(\varrho - 1) - A_1^2\,(\varrho - 1)^{2\,\varepsilon_1} - $$
$$-\,2\,A_2\,(\varrho - 1)^{\varepsilon_1 + 1/2} - 2\,A_3\,(\varrho - 1)^{\varepsilon_1 + 1} + \ldots \qquad (2.2.17)$$

Substituting (2.2.14), (2.2.15) and (2.2.17) into (2.2.8) written in the form $(1 - s^2\,\varrho^2)\,\dfrac{ds}{d\varrho} = 2\,\delta\,\gamma^{1/\delta}\,s^{2 - (2\delta)^{-1}}$ gives the equation:

$$-\,2\,\varepsilon_1\,A_1^2\,(\varrho - 1)^{2\,\varepsilon_1 - 1} - 2\left(\varepsilon_1 + \frac{1}{2}\right) A_1\,A_2\,(\varrho - 1)^{2\,\varepsilon_1 - 1/2} - 2\,\varepsilon_1\,A_1\,(\varrho - 1)^{\varepsilon_1} + \ldots =$$
$$= 2\,\delta\,\gamma^{1/\delta}\,\{1 + [2 - (2\delta)^{-1}]\,A_1\,(\varrho - 1)^{\varepsilon_1} + [2 - (2\delta)^{-1}]\,A_2\,(\varrho - 1)^{\varepsilon_1 + 1/2} + \ldots\}. \qquad (2.2.18)$$

By comparing the smallest powers of $(\varrho - 1)$ on both sides, we obtain the condition

$$2\,\varepsilon_1 - 1 = 0, \quad \varepsilon_1 = \frac{1}{2}, \qquad (2.2.19)$$

$$A_1^2 = -\,2\,\delta\,\gamma^{1/\delta}. \qquad (2.2.19\,\mathrm{a})$$

By comparing the terms of the order $(\varrho - 1)^{1/2}$, we obtain

$$A_2 = -\,\delta\,\gamma^{1/\delta}\,[2 - (2\delta)^{-1}] - \frac{1}{2}, \text{ etc.} \qquad (2.2.19\,\mathrm{b})$$

The coefficient A_1 is imaginary; it must be so since in the range $0 \leq M \leq 1,\ 1 \geq \varrho > 0$, and consequently the expression $(\varrho - 1)^{1/2}$ is also imaginary. Hence we obtain:

$$s = 1 + A_1\,(\varrho - 1)^{1/2} + \text{higher powers of } (\varrho - 1). \qquad (2.2.20)$$

From the equality $M^2 = 1 - s^2\,\varrho^2$, we have M^2 given by (2.2.17), and from the expression for ϱ^2 in (2.2.16), we have

$$\varrho^{-1}\,M^{-2} = [1 - (\varrho - 1) + \ldots]\,M^{-2} = -\,(2\,A_1)^{-1}\,(\varrho - 1)^{-1/2} + \ldots, \qquad (2.2.21)$$

where the dots symbolize a series which is finite at the point $\varrho = 1$. Now we can integrate (2.2.21), term by term, thus obtaining:

$$\int_{\varrho_0}^{\varrho} x^{-1} M^{-2}(x)\, dx = -A_1^{-1}\left[(\varrho - 1)^{1/2} - (\varrho_0 - 1)^{1/2}\right] + \ldots, \qquad (2.2.22)$$

which confirms the convergence of the integral at the point $\varrho = 1$.

We arrive, therefore, at the result that the velocity q of our hypothetical gas is never zero; it has a minimum value $q_{\min}$ corresponding to $M = 0$. To understand the meaning of this fact, consider at first a physical gas coming from infinity with a finite speed upon an obstacle with a smooth boundary. The streamline $\Psi = 0$ which encloses this boundary splits up at a certain point of this boundary into two branches; they reunite at another point of the boundary from which the streamline continues to infinity. From continuity considerations, one concludes that the velocity at the two branch points must be zero. Applying the same considerations to the hypothetical fluid, which does not permit zero velocities, one arrives at a seeming contradiction. Its solution is in the fact that the fluid fans out a little before the body and closes in a little behind it; the branch points of the streamline are the vertices of two cusps formed by the streamline. At the cusp points the velocity can obviously remain finite. Hence, if one considers the flow of the real gas that is approximated by the hypothetical fluid, one may say that this approximation obviously breaks down in the region where the velocity becomes smaller than the minimum velocity. Here the hypothetical flow produces wakes that are only caused by the insufficient approximation. This is a serious deficiency of this kind of approximation. But it may be kept less significant if we choose the constant δ so small that $q_{\min}$ has a sufficiently small value. In this manner one may improve the Kármán-Tsien method for higher Mach numbers without any significant sacrifice in the approximation at the velocity zero.

3. Discussion of the Obtained Results

To study the approximation to the physical gas given by our theoretical assumption (2.1.1), we may discuss the $M(\varrho)$-, $q(\varrho)$-, and $p(\varrho)$-relations for a physical gas, for a gas based on the Kármán-Tsien theory, and for two gases based on (2.1.1). s was an important auxiliary variable in the theory; one may include the relation $s(\varrho)$. To simplify comparison, we choose the scale in the following so that we have at the point $M = 0$, $\varrho_0 = 1$ and $p_0 = 1$. This leads to a scale for q also, by the use of the Bernoulli equation, as one can verify.

For the physical gas, one obtains the following equations:

$$p = \varrho^k = \varrho^{1.4}, \qquad (2.3.1)$$

$$q^2 = 2\,k\,(k-1)^{-1}\,(1 - \varrho^{k-1}) = 7\,(1 - \varrho^{0.4}), \qquad (2.3.2)$$

$$M^2 = 2\,(k-1)^{-1}\,(\varrho^{1-k} - 1) = 5\,(\varrho^{-0.4} - 1), \qquad (2.3.3)$$

$$s^2 = (1 - M^2)\,\varrho^{-2} = 6\,\varrho^{-2} - 5\,\varrho^{-2.4}. \qquad (2.3.4)$$

The Kármán-Tsien approximation gives the following corresponding equations:

$$p = -A\,\varrho^{-1} + (1 + A), \quad A > 0, \qquad (2.3.1')$$

$$q^2 = A\,(\varrho^{-2} - 1), \qquad (2.3.2')$$

$$M^2 = 1 - \varrho^2, \qquad (2.3.3')$$

$$s = 1. \qquad (2.3.4')$$

Beginning with (2.1.1), we obtained (2.2.5) and (2.2.8). From the definition of s and the choice of scale, $M = 0$, $\varrho_0 = 1$ imply $s_0 = 1$; this fixes the constant of integration. But γ and δ are two parameters at our disposal. For the particular choice $\delta = 0$ and $\gamma = 1$ one gets the Kármán-Tsien case given above. It is clear that the higher degree of freedom in the general case leads to an improvement of the approximation. The integration of (2.2.8) was carried out in [4] for two different choices of the parameters γ and δ. In the first case, a small δ was chosen in order to stay in the neighborhood of the Kármán-Tsien approximation. In this way, we may study the improvement of the approximation under variation of δ. In the second case, the constant δ was chosen in such a way that the asymptotic behavior of the hypothetical gas at Mach number 1 is that of a real isentropic gas. The equations connecting p, q, M and ϱ have now the following forms:

$$M^2 = 1 - \varrho^2\, s^2, \tag{2.3.3''}$$

$$q^2 = q_0{}^2 \exp\left[- 2 \int_1^{\varrho} x^{-1}\, M^{-2}\,(x)\, dx\right], \tag{2.3.2''}$$

$$p = 1 + q_0{}^2 \int_1^{\varrho} M^{-2}\,(x)\, \exp\left[- 2 \int_1^{x} y^{-1} M^{-2}\,(y)\, dy\right] dx. \tag{2.3.1''}$$

Integrating (2.2.8), one obtains s as a function of ϱ; therefore, (2.3.1''), (2.3.2'') and (2.3.3'') give a representation of p, q, M as functions of ϱ. The new arbitrary constant of integration, q_0, determines the minimum velocity of the fluid.

A. First case: $\delta = \dfrac{1}{14}$. Equation (2.2.8) then has the form:

$$\frac{ds}{d\varrho} = D\, s^{-5}\, (1 - s^2\, \varrho^2)^{-1}, \quad D = 2\,\delta\, \gamma^{1/\delta} = \frac{1}{7}\, \gamma^{14}. \tag{2.3.5}$$

Using $L = s^2$, we have the result:

$$\frac{dL}{d\varrho} = 2\, D\, L^{-2}\, (1 - \varrho^2\, L)^{-1}. \tag{2.3.6}$$

This equation may be integrated numerically. In order to get a good agrement for smaller Mach numbers, the constant D may be chosen to be equal to 0,105. The constant $q_0{}^2$ in (2.3.2'') may be chosen more or less arbitrarily; to provide for a small value of the minimal velocity q_0 one may take $q_0 = 0,265$.

B. To obtain a model gas that behaves like a physical one at the point $M = 1$, we may replace $(1 - M^2)^{1/2}$ by T and obtain from (2.3.3) and (2.3.4) in the case of a physical gas the following relations:

$$\varrho = \left[\frac{1}{2}\,(k + 1) - \frac{1}{2}\,(k - 1)\, T^2\right]^{-\frac{1}{k-1}}, \tag{2.3.7}$$

$$s = T\, \varrho^{-1} = T\left[\frac{1}{2}\,(k + 1) - \frac{1}{2}\,(k - 1)\, T^2\right]^{\frac{1}{k-1}}. \tag{2.3.8}$$

But, as one can see, from (1.3.13) we easily get:

$$\lambda = \frac{1}{3}\,(h^2 - 1)\, T^3 + \ldots, \quad h^2 = \frac{1}{6}. \tag{2.3.9}$$

Thus, for small values of λ the development (2.3.10) seems to be valid:

$$T = \left[\frac{1}{3}\,(1 - h^2)\right]^{-1/2} |\lambda|^{1/3} + \ldots, \tag{2.3.10}$$

and using this in (2.3.8), we have:

$$s = \left[\frac{1}{3}\left(1 - h^2\right)\right]^{-1/2} |\lambda|^{1/3} \left[\frac{1}{2}\left(k + 1\right)^{\frac{1}{(k+1)}} + \dots\right] = K\,|\lambda|^{1/3} + \dots \qquad (2.3.11)$$

If one desires to have in (2.2.5) a choice of parameters that leads to the asymptotic behavior of the physical gas for $M = 1$, i. e., $\lambda = 0$, a comparison of (2.2.5) with (2.3.11) gives the value $\delta = \frac{1}{6}$. For this value of δ, (2.2.8) takes the different form:

$$\frac{ds}{d\varrho} = D\,s^{-1}\left(1 - s^2\,\varrho^2\right)^{-1}. \qquad (2.3.12)$$

Using again $L = s^2$, one gets:

$$\frac{dL}{d\varrho} = 2\,D\left(1 - L\,\varrho^2\right)^{-1}. \qquad (2.3.13)$$

Suppose we fix γ so that $D = \frac{1}{3}\,\gamma^6 = 0{,}606$; this choice leads to curves that agree with the corresponding curves for a physical gas quite well for higher Mach numbers. In this case the minimum velocity q_0 ought to be chosen greater ($= 0{,}471$), in order to get a good behaviour near $M = 1$. The reader can find in [4] a set of curves for the two cases discussed.

Part III

Supersonic Flow

1. Some Remarks on Different Types of Equations

Let us discuss briefly various simple cases of partial differential equations and the general solutions of these equations. The following three equations will be considered

$$\frac{1}{4}\left(\Psi_{\mu\mu} + \Psi_{\vartheta\vartheta}\right) = \Psi_{\mathfrak{z}\bar{\mathfrak{z}}} = 0, \qquad \mathfrak{z} = \mu + i\,\vartheta, \qquad \bar{\mathfrak{z}} = \mu - i\,\vartheta, \qquad (3.1.1)$$

$$\Psi_{xx} - \Psi_{yy} = 0, \quad \text{or} \quad 4\,\Psi_{\mu\vartheta} = 0, \qquad \mu = x + y, \qquad \vartheta = x - y, \qquad (3.1.2)$$

$$\Psi_{\vartheta\vartheta} + 4\left(1 - \mu\right)\Psi_{\mu\mu} - 2\,\Psi_{\mu} = 0. \qquad (3.1.3)$$

In equation (3.1.2) by the transformation $x + y = \mu$, $x - y = \vartheta$, $\Psi(x, y) = \Psi(\mu, \vartheta)$, equation $\Psi_{xx} - \Psi_{yy} = 0$, changes to $4\,\Psi_{\mu\vartheta} = 0$. The first equation is of elliptic type, the second of hyperbolic and the third of mixed type. In the first case, the general solution is given by

$$\Psi = f(\mathfrak{z}) + g(\bar{\mathfrak{z}}), \qquad (3.1.4)$$

and in the second one, by

$$\Psi = f(\mu) + g(\vartheta), \qquad (3.1.5)$$

where f and g are arbitrary (sufficiently many times differentiable) functions of one variable. As μ and ϑ are real variables, it is seen that in (3.1.4) there is an arbitrary function of one complex variable and in equation (3.1.5) two arbitrary functions of one real variable. The former remark follows, since in order to obtain real solutions, g must be chosen the conjugate to f; that is $g = \bar{f}(\mu - i\,\vartheta)$. By a transformation $\lambda_1 = (1 - \mu)^{1/2}$, equation (3.1.3) can be reduced to the form $\Psi_{\vartheta\vartheta} + \Psi_{\lambda_1\lambda_1} = 0$ and the general solution is

$$\Psi = f\left[i\left(1 - \mu\right)^{1/2} - \vartheta\right] + g\left[i\left(1 - \mu\right)^{1/2} + \vartheta\right]. \qquad (3.1.6)$$

2. Differential Equation in the Supersonic Flow

Introduce in (1.1.4) the variables $\log q$ and θ instead of q and θ, i. e., $\Psi_q = q^{-1} \Psi_{\log q}$ and $\Phi_q = q^{-1} \Phi_{\log q}$. Then we obtain:

$$\Phi_\theta = \varrho^{-1} \Psi_{\log q}, \quad \Phi_{\log q} = -(1 - M^2) \varrho^{-1} \Psi_\theta. \tag{3.2.1}$$

Differentiating the first equation (3.2.1) with respect to $\log q$ and the second one with respect to θ, and equating leads to

$$\varrho^{-1} (1 - M^2) \Psi_{\theta \theta} + (\varrho^{-1} \Psi_{\log q})_{\log q} = 0. \tag{3.2.2}$$

For convenience, an auxiliary variable $\bar{H} = \bar{H}(q)$ is introduced, given by

$$\frac{d\bar{H}}{dq} = \varrho\, q^{-1}, \quad \text{or} \quad \bar{H} = \int_0^q \varrho\, d(\log q). \tag{3.2.3}$$

Since $\partial/\partial(\log q) = \varrho\, \partial/\partial\bar{H}$, we obtain from (3.2.2) the expression

$$\varrho^{-2} (1 - M^2) \Psi_{\theta \theta} + \Psi_{\bar{H}\bar{H}} = l\,(\bar{H})\, \Psi_{\theta \theta} + \Psi_{\bar{H}\bar{H}} = 0. \tag{3.2.4}$$

We now introduce the quantity λ [compare with (1.2.1)], defined by

$$d\lambda/d\bar{H} = \varrho^{-1}(1 - M^2)^{1/2}, \quad \text{or} \quad d\lambda/dq = q^{-1}(1 - M^2)^{1/2}. \tag{3.2.5}$$

Then since M^2 is given by (1.9.19), we can find the value of λ by a purely formal computation. In the case of a subsonic flow, we obtain the equation (1.3.13), and in the case of a supersonic flow, the expression

$$\lambda\,(M) = +\,i\,\{\arctan\,[(M^2 - 1)^{1/2}] - h^{-1} \arctan\,[h\,(M^2 - 1)^{1/2}]\}. \tag{3.2.6}$$

If we introduce the new variable Λ defined by

$$\Lambda = i\,\lambda, \tag{3.2.7}$$

we get

$$\Lambda = h^{-1} \arctan\,[h\,(M^2 - 1)^{1/2}] - \arctan\,[(M^2 - 1)^{1/2}]. \tag{3.2.8}$$

In this case equation (3.2.4) becomes

$$\Psi_{\Lambda \Lambda} - \Psi_{\theta \theta} + 4\,N_1\,\Psi_\Lambda = 0, \tag{3.2.9}$$

$$N_1 = \frac{(k + 1)}{8}\, M^4\,(M^2 - 1)^{-3/2}. \tag{3.2.9a}$$

3. Solution of the Differential Equation

The solution of the differential equation (3.2.9) follows an analogous pattern to that one applied in the case of a subsonic flow. As has been indicated by Bergman[1], there exist various operators for transforming solutions of one equation into solutions of another equation. Often for different purposes it is convenient to use different operators or different forms of the same operator. In the present section, two operators are derived. They transform solutions of the relatively simple hyperbolic equation

$$\Psi_{\Lambda \Lambda} - \Psi_{\theta \theta} = 0, \tag{3.3.1}$$

---

[1] S. Bergman: On two-dimensional supersonic flows. N. A. C. A., T. N. No. 1875 (1949).

i. e. a pair of differentiable functions $g(\mathfrak{z})$ and $h(\eta)$, $\mathfrak{z} = \varLambda + \theta$, $\eta = \varLambda - \theta$, of one variable $\mathfrak{z}$ or η, into solutions of the compressibility equations. As soon as the function $\varPsi(\varLambda, \theta)$, satisfying equation (3.2.9) is obtained, the corresponding flow in the physical plane can be obtained exactly in the same manner as in the subsonic case. In contrast to the subsonic case, flows which in the logarithmic plane satisfy equation (3.3.1) have not been studied. Relatively little is known concerning how to choose g and h in order to obtain in the physical plane a flow around a prescribed obstacle, although there is reason to expect that this investigation should be simpler than in the subsonic case. On the other hand, it seems that this method of attack would not be of considerable interest in applications since most flows in which we are interested are mixed or supersonic with shocks. These combinations make necessary considerable modifications in the approach.

4. Operator Obtained by the Use of Riemann's Function

One method of generating the stream-function of a supersonic flow from expressions g and h [solutions of (3.3.1)] is by the use of Riemann's function. To this end, let us transform equation (3.2.9). If in (3.2.9) instead of $\varLambda$ and θ, the variables $\mathfrak{z}$ and η, defined in section III.3, are introduced, and, instead of $\varPsi$, the "reduced stream-function"

$$\varPsi^* = \varPsi \exp\left[-\int_a^{\mathfrak{z}+\eta} N_1(\tau)\,d\tau\right], \qquad (3.4.1)$$

(where a is an arbitrary constant) is considered, then a formal computation shows that $\varPsi^*$ satisfies the equation

$$\varPsi^*_{\mathfrak{z}\,\eta} + F_1\,(\mathfrak{z} + \eta)\,\varPsi^* = 0, \qquad (3.4.2)$$

$$F_1 = -\,N_1{}^2 - \frac{dN_1}{d(2\,\varLambda)} = \qquad (3.4.3)$$

$$= \frac{(k+1)}{64}\left[\frac{5\,(k+1)}{B^6} + \frac{12\,k}{B^4} + \frac{6\,k-14}{B^2} + (4\,k+8) - (3\,k-1)\,B^2\right], \quad B^2 = M^2 - 1.$$

Obviously it is allowable to take for the lower limit a of the integral in equation (3.4.1) an arbitrary point, since replacing a by a^*, $a^* \neq a$, means only that $\varPsi^*$ is multiplied by a constant, namely $\exp\left[\int_a^{a^*} N_1(\tau_1)\,d\tau_1\right]$. The so-called "Riemann Integration method"[2] essentially permits the derivation of an integration formula which expresses the solution of the hyperbolic type of equation in a clear way by means of the initial data. Consider the most general form of a linear, hyperbolic differential equation of the second order in two independent variables:

$$u_{x\,y} + a\,u_x + b\,u_y + c\,u = 0. \qquad (3.4.4)$$

Suppose that we know that a solution of (3.4.4) exists. The "characteristic curves" of the equation (3.4.4) are the lines $x = $ const., and $y = $ const. Assume two arbitrary characteristic lines $x = $ const. and two $y = $ const. The so-obtained parallelogram has four vertices: A and B on one diagonal and $P(\mathfrak{z}, \eta)$, $D(x_0, y_0)$ on another. Assume that we know the initial value of the function u along the "initial curve" ADB. By means of the theory of characteristics we can determine

---

[2] R. Courant and D. Hilbert: Methoden der Mathematischen Physik, Vol. I and II. Berlin: Julius Springer, Berlin (1931 and 1937).

also the value of u_x along AD and the value of u_y along BD. This is the so-called "characteristic initial value problem".

The function v, depending upon the arguments x, y and the parameters $\mathfrak{z}$, η, satisfying the following conditions

$$v_{xy} - (a\,v)_x - (b\,v)_y + c\,v = 0, \qquad (3.4.5\,\mathrm{a})$$

$$v_x\,(x, \eta;\, \mathfrak{z}, \eta) = b\,(x, \eta)\,v\,(x, \eta;\, \mathfrak{z}, \eta)\ \text{on}\ A\,P, \qquad (3.4.5\,\mathrm{b})$$

$$v_y\,(\mathfrak{z}, y;\, \mathfrak{z}, \eta) = a\,(\mathfrak{z}, y)\,v\,(\mathfrak{z}, y;\, \mathfrak{z}, \eta)\ \text{on}\ B\,P, \qquad (3.4.5\,\mathrm{c})$$

$$v\,(\mathfrak{z}, \eta;\, \mathfrak{z}, \eta) = 1\ \text{at}\ (P), \qquad (3.4.5\,\mathrm{d})$$

is called the "Riemann function" belonging to the differential equation (3.4.4). The function u satisfying (3.4.4) is then given by the expression

$$u\,(x, y) = u\,(D)\,v\,(D) + \int_D^A v\,(u_y + a\,u)\,dy + \int_D^B v\,(u_x + b\,u)\,dx. \qquad (3.4.6)$$

Expression (3.4.6) represents the classical Riemann formula. Consequently, by means of the initial value on the curve ADB we are able to determine the function $u\,(x, y)$. Let us apply Riemann's formula to (3.4.2), i. e., let $R\,(\mathfrak{z}, \eta;\, \mathfrak{z}_0, \eta_0)$ be the Riemann function of equation (3.4.2). Then

$$\Psi^*\,(\mathfrak{z}, \eta) = \Psi^*\,(\mathfrak{z}_0, \eta_0)\,R\,(\mathfrak{z}, \eta;\, \mathfrak{z}_0, \eta_0) + \int_{\eta_0}^{\eta} \Psi_y^*\,(\mathfrak{z}_0, y)\,R\,(\mathfrak{z}_0, y;\, \mathfrak{z}_0, \eta_0)\,dy +$$

$$+ \int_{\mathfrak{z}_0}^{\mathfrak{z}} \Psi_x^*\,(x, \eta_0)\,R\,(x, \eta_0;\, \mathfrak{z}_0, \eta_0)\,dx. \qquad (3.4.7)$$

Since for $\Psi^*\,(\mathfrak{z}_0, \eta_0)$ we may write an arbitrary prescribed constant A and for $\Psi^*\,(x, \eta_0)$ and $\Psi^*\,(\mathfrak{z}_0, y)$ we may write two arbitrary continuous functions $g\,(x)$ and $h\,(y)$ with the condition that $g\,(\mathfrak{z}_0) = h\,(\eta_0) = A$, the formula

$$\Psi^*\,(\mathfrak{z}, \eta) = A\,R\,(\mathfrak{z}, \eta;\, \mathfrak{z}_0, \eta_0) + \int_{\mathfrak{z}_0}^{\mathfrak{z}} g_x\,(x)\,R\,(x, \eta_0;\, \mathfrak{z}_0, \eta_0)\,dx +$$

$$+ \int_{\eta_0}^{\eta} h_y\,(y)\,R\,(\mathfrak{z}_0, y;\, \mathfrak{z}_0, \eta_0)\,dy, \qquad (3.4.8)$$

represents an integral operator which transforms two arbitrary functions g and h of a real variable into a solution of equation (3.4.2). Thus according to (3.4.8) the values of $\Psi^*\,(\mathfrak{z}, \eta)$ are determined within a rectangle having vertices at $\mathfrak{z}_0, \eta_0;\ \mathfrak{z}, \eta_0;\ \mathfrak{z}_0, \eta;\ \mathfrak{z}, \eta$, by prescribing its values along the line segments: $y = \eta_0$, $\mathfrak{z}_0 \leqq x \leqq \mathfrak{z}$, and $x = \mathfrak{z}_0$, $\eta_0 \leqq y \leqq \eta$. It is possible to show that the Riemann function of equation (3.4.2) can be written in the form

$$R\,(\mathfrak{z}, \eta;\, \mathfrak{z}_0, \eta_0) = \sum_0^{\infty} (-1)^n\,R^{(n)}\,(\mathfrak{z}, \eta;\, \mathfrak{z}_0, \eta_0), \qquad (3.4.9)$$

where

$$R^{(0)}\,(\mathfrak{z}, \eta;\, \mathfrak{z}_0, \eta_0) = 1, \qquad (3.4.10\,\mathrm{a})$$

$$R^{(1)}\,(\mathfrak{z}, \eta;\, \mathfrak{z}_0, \eta_0) = \int_{\mathfrak{z}_0}^{\mathfrak{z}} \int_{\eta_0}^{\eta} F_1\,(x + y)\,dx\,dy, \qquad (3.4.10\,\mathrm{b})$$

$$R^{(2)}(\mathfrak{z}, \eta; \mathfrak{z}_0, \eta_0) = \int\limits_{\mathfrak{z}_0}^{\mathfrak{z}} \int\limits_{\eta_0}^{\eta} F_1(x+y)\, R^{(1)}(x, y; \mathfrak{z}_0, \eta_0)\, dx\, dy, \qquad (3.4.10\,\text{c})$$

$$R^{(n)}(\mathfrak{z}, \eta; \mathfrak{z}_0, \eta_0) = \int\limits_{\mathfrak{z}_0}^{\mathfrak{z}} \int\limits_{\eta_0}^{\eta} F_1(x+y)\, R^{(n-1)}(x, y; \mathfrak{z}_0, \eta_0)\, dx\, dy. \qquad (3.4.10\,\text{d})$$

The tabulation of $R = R(\mathfrak{z}, \eta; \mathfrak{z}_0, \eta_0)$ is rather complicated since R is a function of four variables. Below, for the illustrative purpose, the second term $R^{(1)}$ of the expansion (3.4.9) is computed.

5. Evaluation of the Function $R^{(1)}(\mathfrak{z}, \eta; \mathfrak{z}_0, \eta_0)$

In this section, the second term $R^{(1)}(\mathfrak{z}, \eta; \mathfrak{z}_0, \eta_0)$ in the series for the Riemann function will be evaluated. The function F_1 is a function of the variable $\Lambda = \dfrac{\mathfrak{z}+\eta}{2}$ and therefore

$$\int\limits_{\mathfrak{z}_0}^{\mathfrak{z}} F_1(x+y)\, dx = \Gamma^{(1)}(\mathfrak{z}+y) - \Gamma^{(1)}(\mathfrak{z}_0+y), \qquad (3.5.1)$$

where

$$\int\limits_{t_0}^{t_1} F_1(\tau)\, d\tau = \Gamma^{(1)}(t_1) - \Gamma^{(1)}(t_0), \qquad (3.5.1\,\text{a})$$

and

$$\int\limits_{\eta_0}^{\eta} dy \int\limits_{\mathfrak{z}_0}^{\mathfrak{z}} F_1(x+y)\, dx = \int\limits_{\eta_0}^{\eta} [\Gamma^{(1)}(\mathfrak{z}+y) - \Gamma^{(1)}(\mathfrak{z}_0+y)]\, dy =$$
$$= \Gamma^{(2)}(\mathfrak{z}+\eta) - \Gamma^{(2)}(\mathfrak{z}+\eta_0) - \Gamma^{(2)}(\mathfrak{z}_0+\eta) + \Gamma^{(2)}(\mathfrak{z}_0+\eta_0), \qquad (3.5.2)$$

where

$$\int\limits_{t_0}^{t_1} \Gamma^{(1)}(\tau)\, d\tau = \Gamma^{(2)}(t_1) - \Gamma^{(2)}(t_0). \qquad (3.5.2\,\text{a})$$

For $\Lambda = 0$, i.e., $M = 1$ [see (3.2.8)], the quantities $N_1(M) = N_1(1)$, (3.2.9a), $F_1(B) = F_1(0)$, (3.4.3), or $F_1(\Lambda) = F_1(0)$ tend to infinity and this fact causes a certain amount of inconvenience in the tabulation of the functions $\Gamma^{(1)}$ and $\Gamma^{(2)}$. According to (3.4.3), for $k = 1{\cdot}4$,

$$F_1(B) = 0{\cdot}45\, B^{-6} + 0{\cdot}63\, B^{-4} - 0{\cdot}21\, B^{-2} - 0{\cdot}51 - 0{\cdot}12\, B^2, \qquad (3.5.3)$$

and

$$\Lambda = \sqrt{6}\, \arctan\!\left[(\sqrt{6})^{-1} B\right] - \arctan B. \qquad (3.5.4)$$

The values of M, B, F_1, $\Gamma^{(1)}(2\,\Lambda)$, $\Gamma^{(2)}(2\,\Lambda)$ are tabulated in [1], where an ordinary numerical procedure was applied.

6. Second Type of Integral Operator

In analogy to the subsonic case, where formula (1.9.2) yields a representation for solutions of equation (1.9.1) in terms of an arbitrary analytic function of one complex variable, there is derived in this section a representation for solutions of equation (3.2.9) in terms of two arbitrary differentiable functions of a real variable. Obviously, instead of equation (3.2.9), equation (3.4.2) for the reduced stream-function Ψ^* may be considered. The desired representation can be obtained as a consequence of the following theorem:

Let $E^{(n)}(\tau)$, $\tau = 2\,\Lambda$, denote functions defined by the following recurrence relations

$$E_\tau^{(1)}(\tau) = -F_1(\tau), \quad \text{or} \quad E^{(1)}(\tau) = -\int\limits_0^{\tau} F_1(\tau_1)\, d\tau_1, \qquad (3.6.1)$$

$$E_\tau^{(n+1)}(\tau) = -\,E_{\tau\tau}^{(n)}(\tau) - F_1(\tau)\,E^{(n)}(\tau), \qquad (3.6.1\,a)$$

or

$$E^{(n+1)}(\tau) = -\int_0^\tau [E_{\tau\tau}^{(n)}(\tau_1) + F_1(\tau_1)\,E^{(n)}(\tau_1)]\,d\tau_1, \qquad (3.6.1\,b)$$

where $F_1(\tau)$ is the given function which must satisfy the inequality

$$|F_1(\tau)| \leq \frac{2\,\mu\,\tau_{00}^2}{|\tau_{00} - \tau|^2}, \quad \text{for}\quad 0 \leq \tau \leq \tau_0, \qquad (3.6.2)$$

where μ and τ_{00} are conveniently chosen constants, and the derivatives of F_1 must satisfy the inequalities

$$\left|\frac{d^n F_1(\tau)}{d\tau^n}\right| \leq \frac{2\,(n+1)!\,\mu\,\tau_{00}^2}{|\tau_{00} - \tau|^{n+2}}, \quad \text{for}\ 0 \leq \tau \leq \tau_0, \qquad (3.6.2\,a)$$

i. e., for τ real and positive. Then the series

$$V^{(1)}(\mathfrak{z}, \eta) = f(\mathfrak{z}) + \sum_1^\infty E^{(n)}(\mathfrak{z} + \eta)\,f^{(n)}(\mathfrak{z}), \qquad (3.6.3)$$

and

$$V^{(2)}(\mathfrak{z}, \eta) = g(\eta) + \sum_1^\infty E^{(n)}(\mathfrak{z} + \eta)\,g^{(n)}(\eta), \qquad (3.6.4)$$

where f and g are two arbitrary differentiable functions of a real variable and

$$f^{(n+1)}(\mathfrak{z}) = \int_0^{\mathfrak{z}} f^{(n)}(\mathfrak{z}_1)\,d\mathfrak{z}_1, \quad f^{(0)}(\mathfrak{z}) = f,$$
$$g^{(n+1)}(\eta) = \int_0^\eta g^{(n)}(\eta_1)\,d\eta_1, \quad g^{(0)}(\eta) = g, \qquad (3.6.5)$$

represent the solutions of (3.4.2). The proof of this theorem is carried out under the assumption that the series (3.6.3) and (3.6.4), as well as those for their first derivatives, converge uniformly. This assumption permits one to interchange the operations of summation and differentiation. A formal computation yields

$$V_{\mathfrak{z}}^{(1)} = f_{\mathfrak{z}} + \sum_1^\infty \left(E_{\mathfrak{z}}^{(n)} f^{(n)} + E^{(n)} f_{\mathfrak{z}}^{(n)}\right), \qquad (3.6.6)$$

or with the use of (3.6.5), i. e., $f_{\mathfrak{z}}^{(n)} = f^{(n-1)}$, we obtain

$$V_{\mathfrak{z}}^{(1)} = f_{\mathfrak{z}} + f^{(0)} E^{(1)} + \sum_1^\infty f^{(n)} \left(E_{\mathfrak{z}}^{(n)} + E^{(n+1)}\right). \qquad (3.6.6\,a)$$

The second differentiation (with respect to η) gives

$$V_{\mathfrak{z}\eta}^{(1)} = f\,E_\eta^{(1)} + \sum_1^\infty f^{(n)} \left(E_{\mathfrak{z}\eta}^{(n)} + E_\eta^{(n+1)}\right). \qquad (3.6.7)$$

Also

$$F_1\,V^{(1)} = f\,F_1 + \sum_1^\infty f^{(n)}\,F_1\,E^{(n)}. \qquad (3.6.8)$$

Substituting (3.6.6a) and (3.6.8) into (3.4.2) yields:

$$V_{\mathfrak{z}\eta}^{(1)} + F_1\,V^{(1)} = f\left[E_\eta^{(1)} + F_1\right] + \sum_1^\infty f^{(n)}\left[E_{\mathfrak{z}\eta}^{(n)} + E_\eta^{(n+1)} + F_1\,E^{(n)}\right]. \qquad (3.6.9)$$

Since $\tau = 2\,\Lambda = \mathfrak{z} + \eta$, we have $E_{\mathfrak{z}}^{(n)}(2\,\Lambda) = E_{\mathfrak{z}}^{(n)}(\tau) = E_\tau^{(n)}$, $E_{\mathfrak{z}\eta}^{(n)} = E_{\tau\tau}^{(n)}$. Thus the first bracket on the right-hand side of (3.6.9) vanishes due to (3.6.1) and the second one vanishes due to (3.6.1a).

7. The Proof of the Uniform Convergence

In section III.6, the assumption was made that the series (3.6.3) and (3.6.4), as well as their first derivatives, converge uniformly. Of course, it is sufficient to prove that the series and both their formal (i. e., term-by-term) derivatives converge uniformly. Assume that in a certain interval I there hold for certain functions $A(\tau)$ and $\overline{A}(\tau)$, and all their derivatives the inequalities

$$|A(\tau)| < \overline{A}(\tau), \quad \left|\frac{d^n A(\tau)}{d\tau^n}\right| \leq \frac{d^n \overline{A}(\tau)}{d\tau^n}, \tag{3.7.1}$$

for τ contained in the interval I (which symbolically will be denoted by $\tau \,\epsilon\, I$), then, as is already known from the section I., the function $\overline{A}(\tau)$ is a dominant of $A(\tau)$, which fact is symbolized by writing $\overline{A} \gg A$, $\tau \,\epsilon\, I$. If $\overline{E}^{(1)}(\tau)$ is given by

$$\overline{E}^{(1)}(\tau) = \int_0^\tau \overline{F}_1(\tau_1)\, d\tau_1, \quad \overline{F}_1(\tau) \gg F_1(\tau), \tag{3.7.2}$$

then

$$|E^{(1)}(\tau)| \leq \overline{E}^{(1)}(\tau), \; \tau \,\epsilon\, I, \tag{3.7.3}$$

and also

$$\left|\frac{d^n E^{(1)}(\tau)}{d\tau^n}\right| \leq \frac{d^n \overline{E}^{(1)}(\tau)}{d\tau^n}. \tag{3.7.4}$$

Thus $\overline{E}^{(1)}(\tau)$ is a dominant of $E^{(1)}(\tau)$, i. e., $\overline{E}^{(1)} \gg E^{(1)}$. Suppose now, that

$$E^{(n+1)}(\tau) = \int_0^\tau \left[\overline{E}^{(n)}_{\tau_1 \tau_1}(\tau_1) + F_1(\tau_1)\, \overline{E}^{(n)}(\tau_1)\right] d\tau_1 + \overline{H}^{(n+1)}(\tau), \tag{3.7.5}$$

where $E^{(n)}(\tau) \ll \overline{E}^{(n)}(\tau), \quad 0 \ll \overline{H}^{(n)}(\tau)$. Then it follows immediately from (3.6.1b) and (3.7.5) that

$$|E^{(n+1)}(\tau)| \leq \overline{E}^{(n+1)}(\tau), \tag{3.7.6}$$

$$\left|\frac{dE^{(n+1)}}{d\tau}\right| \leq \frac{d\overline{E}^{(n+1)}}{d\tau}; \tag{3.7.6a}$$

and by considering the corresponding derivatives of $\left[\overline{E}^{(n)}_{\tau\tau} + \overline{F}_1 \overline{E}^{(n)}\right]$ in comparison with $\left[E^{(n)}_{\tau\tau} + F_1 E^{(n)}\right]$ it follows that

$$E^{(n+1)}(\tau) \ll \overline{E}^{(n+1)}(\tau). \tag{3.7.7}$$

Thus starting with the assumption that $E^{(n)} \ll \overline{E}^{(n)}$, we have proved by induction that $E^{(n+1)} \ll \overline{E}^{(n+1)}$. Now by expression (3.6.2) we may assume that F_1 is dominated by the right-hand side of (3.6.2), i. e.,

$$F_1(\tau) \ll \frac{2\mu\tau_{00}^2}{|\tau_{00} - \tau|^2} = \overline{F}_1(\tau). \tag{3.7.8}$$

Suppose now that $\overline{H}^{(n)}(\tau)$ is given by

$$\overline{H}^{(n)} = \frac{c_n}{\tau_{00}^n} \gg 0, \tag{3.7.9}$$

where the c_n's are some conveniently chosen positive constants, which will be determined later. Then, an explicit expression may be obtained for $\overline{E}^{(n)}$,[1] and namely with $\overline{E}^{(0)} = 1$ and $c_1 = 2\mu\tau_{00}^2$, (3.7.9), we have from (3.7.5)

$$\overline{E}^{(1)} = \int_0^\tau \overline{F}_1(\tau_1)\, d\tau_1 + \overline{H}^{(1)} = 2\mu\tau_{00}^2 (\tau_{00} - \tau)^{-1}. \tag{3.7.10}$$

Assume that

$$\overline{E}^{(n)}(\tau) = c_n \, (\tau_{00} - \tau)^{-n}, \qquad (3.7.10\,a)$$

and in order to obtain the recurrence relation holding between the c_n, write with the use of (3.7.9)

$$\overline{E}^{(n+1)}(\tau) = \int_0^\tau c_n \left[\frac{n(n+1)}{(\tau_{00} - \tau_1)^{n+2}} + \frac{2\mu \tau_{00}^2}{(\tau_{00} - \tau_1)^{n+2}} \right] d\tau_1 + \overline{H}^{(n+1)} = c_{n+1} \, (\tau_{00} - \tau)^{-(n+1)},$$

$$(3.7.11)$$

$$c_{n+1} = c_n \left[n + 2\mu \tau_{00}^2 \, (n+1)^{-1} \right]. \qquad (3.7.12)$$

Relation (3.7.12) supplies the quantity $c_1 = 2\mu \tau_{00}^2$ if the assumption is made that $c_0 = 1$. Consider large values of n, which permit us to write

$$2\mu \tau_{00}^2 \, (n+1)^{-1} \leqq 1, \qquad (3.7.13)$$

or

$$c_{n+1} \leqq c_n \, (n+1) \leqq c_{n-1} \, n \, (n+1) \qquad (3.7.14)$$

$$\leqq c_{n-2} \, (n-1) \, n \, (n+1) \leqq \ldots \leqq c_0 \, 1.2.3 \ldots n \, (n+1).$$

Therefore we see that for every value of μ we can put down

$$c_{n+1} \leqq \mu^* \, (n+1)!, \qquad (3.7.15)$$

where $\mu^* \geqq c_0$ is a conveniently chosen constant.

The function $f(\mathfrak{z})$ is assumed to be differentiable and finite, which means that for all values of $\mathfrak{z}$ less than a certain value $\mathfrak{z}_2$, say, the values of the function $f(\mathfrak{z})$ are less than a certain constant value μ_2, say, which can be written symbolically in the form

$$|f(\mathfrak{z})| < \mu_2 \quad \text{for} \quad |\mathfrak{z}| \leqq \mathfrak{z}_2. \qquad (3.7.16)$$

From the expression (3.6.5) in conjunction with (3.7.16) we obtain

$$|f^{(1)}| \leqq \mu_2 \, \mathfrak{z}, \quad |f^{(2)}| \leqq \mu_2 \frac{\mathfrak{z}^2}{2}, \quad |f^{(3)}| \leqq \mu_2 \frac{\mathfrak{z}^3}{2 \cdot 3}, \ldots \qquad (3.7.17)$$

or in general:

$$|f^{(n)}(\mathfrak{z})| \leqq \frac{\mu_2 \, \mathfrak{z}^n}{n!} \quad \text{for} \quad |\mathfrak{z}| \leqq \mathfrak{z}_2. \qquad (3.7.17\,a)$$

Consequently, substituting (3.7.10a), (3.7.15) and (3.7.17) into (3.6.3) we obtain

$$V^{(1)}(\mathfrak{z}, \eta) \ll \mu_2 \left[1 + \frac{2! \, \mu^* \, |\mathfrak{z}|}{1! \, (\tau_{00} - \tau)} + \ldots + \frac{(n+1)! \, \mu^* \, |\mathfrak{z}|^n}{n! \, (\tau_{00} - \tau)^n} + \ldots \right]. \qquad (3.7.18)$$

The series inside the square bracket itself has a dominant, namely the series

$$\left[1 + \frac{2! \, \mu^* \, |\mathfrak{z}|}{1! \, |\tau_{00} - \tau|} + \ldots + \frac{(n+1)! \, \mu^* \, |\mathfrak{z}|^n}{n! \, |\tau_{00} - \tau|^n} + \ldots \right], \qquad (3.7.18')$$

which converges if

$$\frac{|\mathfrak{z}|}{|\tau_{00} - \tau|} < 1, \quad \text{or} \quad -1 < \frac{\mathfrak{z}}{|\tau_{00} - \tau|} < 1. \qquad (3.7.19)$$

The series inside the square bracket of equation (3.7.18) converges if

$$-1 < \frac{\Lambda + \theta}{\tau_{00} - 2\Lambda} < 1, \quad \text{for} \quad 2\Lambda < \tau_0 = 2\Lambda_0. \qquad (3.7.20)$$

Thus we proved that the series $V^{(1)}$ has a dominant which converges and has all terms positive. According to the fundamental concepts of analysis, the series $V^{(1)}$ is uniformly convergent.

8. The Domain of Convergence

The series $V^{(1)}$ converges in the domain given by (3.7.20). Assume that the quantity τ_{00} is negative: $\tau_{00} < 0$. Then the domain of convergence is bounded by the following two straight lines in the $\tau \, (= 2\,\varLambda)$, θ-system of coordinates: $\frac{3}{2}\tau + + \theta - \tau_{00} = 0$, $\frac{1}{2}\tau - \theta - \tau_{00} = 0$. Similarly, we may prove that the series $V^{(2)}$, given by (3.6.4), converges uniformly in the domain

$$\frac{|\eta|}{|\tau_{00} - \tau|} < 1, \tag{3.8.1}$$

or

$$-1 < \frac{\varLambda - \theta}{|\tau_{00} - \tau|} < 1, \tag{3.8.1a}$$

which is bounded by two straight lines: $\frac{3}{2}\tau - \theta - \tau_{00} = 0$, $\frac{1}{2}\tau + \theta - \tau_{00} = 0$. The intersection of both domains determines the domain in which both series converge. The upper limit on the positive part of the $(2\,\varLambda)$-axis is given by the quantity $\tau_0 = 2\,\varLambda_0$, (i. e., vertical line). In a particular case we may assume $\tau_{00} = 0$, which implies that the domain of convergence is a triangle bounded by the following three lines: $\frac{1}{2}\tau \pm \theta = 0$, and $\tau = \tau_0$.

The last question which should be mentioned is the investigation of the behavior of the function $F_1\,(2\,\varLambda)$. That is, we should show whether and where its derivatives satisfy the inequalities (3.6.2a). This question is too complicated to be discussed here, and the reader who is more advanced mathematically is referred to the original literature[1], where proof is presented that the derivatives of F_1 satisfy the inequalities (3.6.2a).

9. Solution Regular at $\varLambda = \tau = 0$

The solution (3.6.3) and (3.6.4) and the theorem presented in section III. 6, cannot be applied directly to equation (3.4.2), as F_1 given by (3.4.3) has a pole for $\varLambda = 0$. That is, if M varies from 1 to ∞, the quantity $B = (M^2 - 1)^{1/2}$ varies from 0 to ∞, and $\varLambda$, given by (3.2.8) varies from 0 to $\frac{1}{2}\,(h^{-1} - 1)\,\pi$. Thus for $\varLambda = 0$, the quantity $B = 0$ and the function $F_1 \to \infty$. In some cases, however, it is possible to overcome this difficulty by shifting the origin. Let a be a positive number a and

$$\mathfrak{z}^* = \mathfrak{z} - \frac{1}{2}\,a, \qquad \eta^* = \eta - \frac{1}{2}\,a, \tag{3.9.1}$$

so that $\varLambda^*$ will mean $\varLambda^* = \varLambda - \frac{a}{2}$, $\tau^* = \tau - a$. Equation (3.4.2) then assumes the form

$$W_{\mathfrak{z}^* \eta^*} + F_2\,(\mathfrak{z}^*, \eta^*)\,W = 0, \tag{3.9.2}$$

where $W\,(\mathfrak{z}^*, \eta^*) = \varPsi^*\,(\mathfrak{z}^* + \frac{1}{2}\,a, \; \eta^* + \frac{1}{2}\,a)$, $F_2\,(\tau^*) = F_1\,(\tau^* + a)$, so that F_2 is analytic for $\tau^* = 0$. The corresponding changes in $E^{(n)}$ will be indicated by writing $E^{(n)}\,(\tau)$ thus:

$$E^{(1)}\,(\tau^*) = -\int_0^{\tau^*} F_2\,(\tau_1^*)\,d\tau_1^* = -\int_a^{\tau} F_1\,(\tau_1)\,d\tau_1 = E_1\,(\tau), \tag{3.9.3}$$

$$E^{(n+1)}\,(\tau^*) = -\int_0^{\tau^*} \left[E^{(n)}_{\tau_1^* \tau_1^*}\,(\tau_1^*) + F_2\,(\tau_1^*)\,E^{(n)}\,(\tau_1^*)\right] d\tau_1^* =$$
$$\tag{3.9.3a}$$
$$= -\int_a^{\tau} \left[E^{(n)}_{\tau_1 \tau_1}\,(\tau_1) + F_1\,(\tau)\,E^{(n)}(\tau_1)\right] d\tau_1 = E^{(n+1)}\,(\tau).$$

The domain of convergence, remaining the same, referred to the τ^* system, is merely a shift of the original domain. It may be desirable to have explicit formulas for the $E^{(n)}$ and their derivatives, as these may be employed in computation; if n is given the values $n = 1, 2, 3, \ldots$ in (3.9.3a), then

$$\underline{E}^{(1)}(\tau) = - \int_a^{\tau} F_1(\tau_1)\, d\tau_1, \tag{3.9.4}$$

$$\underline{E}^{(2)}(\tau) = F_1(\tau) - F_1(a) + 1/2\, [\underline{E}^{(1)}(\tau)]^2, \tag{3.9.5}$$

$$\underline{E}^{(3)}(\tau) = F_1(\tau)\,\underline{E}^{(1)}(\tau) - F_{1\tau}(\tau) + F_{1\tau}(a) -$$

$$- \int_a^{\tau} F_1^2(\tau_1)\, d\tau_1 - F_1(a)\,\underline{E}^{(1)}(\tau) + 1/6\, [\underline{E}^{(1)}(\tau)]^3, \tag{3.9.6}$$

where

$$\underline{E}_\tau^{(1)}(\tau) = - F_1(\tau), \tag{3.9.7}$$

$$\underline{E}_\tau^{(2)}(\tau) = F_{1\tau}(\tau) - F_1(\tau)\,\underline{E}^{(1)}(\tau), \text{ etc.} \tag{3.9.8}$$

The expressions for $E^{(4)}$ and $E_\tau^{(3)}(\tau)$, $E_\tau^{(4)}(\tau)$, are cited in [3]. Attention must be called to the fact that in those formulas instead of $\varLambda$, the quantity τ should be substituted. In the same report Bergman cites the expressions for $\underline{E}^{(1)}(\varLambda)$, $F_{1\varLambda}(\varLambda)$ and $\int_a^{\varLambda} F_1^2(\varLambda_1)\, d\varLambda_1$.

Assume for the quantity a in (3.9.1) the value $a = (h^{-1} - 1)\,\pi$, so that

$$\tau^* = 2\,\varLambda^* = 2\,\varLambda - (h^{-1} - 1)\,\pi. \tag{3.9.9}$$

This means that the function $F_2(\tau^*)$ has a singularity for $\varLambda = 0$ or $\tau^* = 2\,\varLambda^* = = - (h^{-1} - 1)\,\pi$, corresponding to an extreme case $M \to \infty$. This transformation may be very convenient in some cases and a more thorough discussion of it may be found in [1]. Let us mention that because the quantity $\varLambda = 1/2\,(h^{-1} - 1)\,\pi$ represents an upper bound for $\varLambda$, it is obvious that for the upper bound of the variable τ, the quantity $\tau_0 = (h^{-1} - 1)\,\pi$ may be chosen. That quantity τ_0 should be used in the considerations presented in section III.6. It is obvious that $F_2(\tau^*) = F_2(2\,\varLambda^*) = = F_2[2\,\varLambda - \pi\,(h^{-1} - 1)] = F_1(2\,\varLambda)$ is an analytic function of $\varLambda$ which is certainly regular in the circle of radius $\dfrac{1}{2}\,\pi\,(h^{-1} - 1)$.

10. Transformation to the Physical Plane

In order to carry out the transition to the physical plane, the values of $\varPsi_q^*$ and $\varPsi_\theta^*$ are needed. Without going into too many details which may be found in the original literature[3], we shall present here the final results with $\overline{H}$ given by (3.2.3):

$$\varPsi_q^* = B\overline{H}\,(q - 1)\left[- \frac{(k + 1)}{4}\,\frac{(B^2 + 1)}{B^3}\,V + M_1 + 2\,M_2 + 3\,M_3 + M_4 \right], \tag{3.10.1}$$

$$\varPsi_\theta^* = \overline{H}\,[M_1 - M_4], \tag{3.10.2}$$

where $V = \sum_i{}' W^{(i)}(\mathfrak{z}^*, \eta^*)$, $i = 1, 2$, represents the solution of (3.9.2) and the values of M_i's are given in the original papers of Bergman. (See the section discussing the tables available.) If the integration is carried out along a streamline, $\varPsi^* = \text{const.}$,

---

[3] S. Bergman: On supersonic and partially supersonic flows. N. A. C. A., T. N. No. 1096 (1946).

then the corresponding values of x and y, that is, the image in the physical plane of the hodograph flow, are given by the formulas[3]

$$x = \int (\varrho \, q^2 \, \Psi_\theta{}^*)^{-1} \, [B^2 \, \Psi_\theta{}^{*2} - q^2 \, \Psi_q{}^{*2}] \cos \theta \, dq, \qquad (3.10.3)$$

$$y = \int (\varrho \, q^2 \, \Psi_\theta{}^*)^{-1} \, [B^2 \, \Psi_\theta{}^{*2} - q^2 \, \Psi_q{}^{*2}] \sin \theta \, dq. \qquad (3.10.4)$$

11. Alternative Formulas

Sometimes the range of variability of the speed is comparatively small. In these instances, it is useful to replace F_1 by a constant, say F_0. Equation (3.4.2) then becomes

$$\Psi^*_{\mathfrak{z}\,\eta} + F_0 \, \Psi^* = 0, \qquad (3.11.1)$$

and its solution can be written in the form

$$\Psi^* \, (\mathfrak{z}, \eta) = \int\limits_{t=-1}^{t=+1} \left\{ \cos\left[\frac{2\,t\,(\mathfrak{z}\,\eta)^{1/2}}{F_0}\right] \right\} \left\{ f\left[\frac{\mathfrak{z}}{F_0}\,\frac{(1-t^2)}{2}\right] + g\left[\frac{\eta}{F_0}\,\frac{(1-t^2)}{2}\right] \right\} \frac{dt}{(1-t^2)^{1/2}},$$

$$(3.11.2)$$

where f and g are two arbitrary, twice-continuously differentiable functions of one variable. When applying formula (3.11.2), the independent variables of functions f and g should be substituted by the expressions $\left[\frac{1}{2}\,\mathfrak{z}\,F_0{}^{-1}\,(1-t^2)\right]$ and $\left[\frac{1}{2}\,\eta\,F_0{}^{-1}\,(1-t^2)\right]$, respectively.

In many instances, it is necessary to have representations of the solutions which are valid in the whole strip $-\infty < \theta < \infty$, $0 < 2\,\Lambda < 2\,\pi\,(h^{-1}-1)$, say. One possibility of obtaining such formulas consists in approximating $F_2\,(2\,\Lambda')$, $2\,\Lambda' = 2\,\Lambda - (h^{-1}-1)\,\pi$, by a sequence of functions $F_2{}^{(m)}\,(2\,\Lambda')$, each of which satisfies the inequalities (3.6.2) and (3.6.2a), and which are chosen in such a manner that

$$\lim_{m \to \infty} F_2{}^{(m)} \, (2\,\Lambda') = F_2 \, (2\,\Lambda'), \qquad (3.11.3)$$

in the interval $0 < 2\,\Lambda' < 2\,(h^{-1}-1)\,\pi$. Let $\Psi^{*(m)}\,(\mathfrak{z}',\eta')$ denote the solutions of

$$\Psi^{*\,(m)}_{\mathfrak{z}'\,\eta'} + F_2{}^{(m)}\,(\mathfrak{z}'+\eta')\,\Psi^{*(m)} = 0, \qquad m = 1, 2, \ldots, \qquad (3.11.4)$$

where $\mathfrak{z}'$ and η' are taken in the interval $-\mathfrak{z}^{(0)} < \mathfrak{z}' < \mathfrak{z}^{(0)}$, $-\eta^{(0)} < \eta' < \eta^{(0)}$, and $\mathfrak{z}^{(0)}$ and $\eta^{(0)}$ are values, corresponding to the quantity $\tau^{(0)} = 2\,(h^{-1}-1)\,\pi$, equivalent to $\tau_0 = 2\,\Lambda_0$ in the section III.7. The quantity $\tau^{(0)}$ is the upper bound for the variable $\tau' = 2\,\Lambda' < \tau^{(0)} = 2\,(h^{-1}-1)\,\pi$. The functions $\Psi^{*(m)}$ are also to satisfy the conditions:

$$\Psi^{*(m)}\,(\mathfrak{z}',0) = \Psi^*\,(\mathfrak{z}',0), \qquad \Psi^{*(m)}\,(0,\eta') = \Psi^*\,(0,\eta'). \qquad (3.11.5)$$

Then, in a manner similar to that in Part I, it can easily be shown that

$$\lim_{m \to \infty} \Psi^{*\,(m)}\,(\mathfrak{z}',\eta') = \Psi^*\,(\mathfrak{z}',\eta'), \qquad (3.11.6)$$

in the given interval. A sequence of functions $F_2{}^{(m)}\,(2\,\Lambda')$ can be obtained in the following manner: Let

$$F_2\,(2\,\Lambda') = \alpha_0 + \alpha_1\,\Lambda' + \alpha_2\,\Lambda'^2 + \ldots, \qquad (3.11.7)$$

be the series development of F_2 around the point $2\,\Lambda' = 0$, i. e. $2\,\Lambda = (h^{-1}-1)\,\pi$. This series obviously converges in the circle of radius $\Lambda = \frac{1}{2}\,(h^{-1}-1)\,\pi$. Then, it is relatively easy to show that the series (3.6.3) and (3.6.4), obtained as solutions of equation (3.11.4), with $F_2{}^{(m)}$ given by equation

$$F_2{}^{(m)}\,(2\,\Lambda') = \sum_{n=0}^{\infty} \frac{\alpha_n}{\Gamma\,(1+n/m)}\,\Lambda'^{\,n}, \tag{3.11.8}$$

will converge for every finite positive number $m < \infty$, in the strip $-\infty < \theta < \infty$, $0 < 2\,\Lambda < \pi\,(h^{-1}-1)$. On the other hand, according to a classical theorem of the theory of functions, the relation (3.11.3) holds in this case, so that for the functions $\Psi^{*(m)}\,(\mathfrak{z}',\eta')$ obtained in the foregoing manner, the relation (3.11.6) holds in the whole interval $-\mathfrak{z}^{(0)} < \mathfrak{z}' < \mathfrak{z}^{(0)}$, $-\eta^{(0)} < \eta' < \eta^{(0)}$.

Below, we shall give another form of the solution of equation (3.4.2). Suppose that F_m is a function which possesses a continuous first derivative. Let $E_1^*\,(\mathfrak{z},\eta,t)$ and $E_2^*\,(\mathfrak{z},\eta,t)$ be solutions of

$$(1-t^2)\,E^*{}_{1\,\eta t} - t^{-1}\,E^*{}_{1\,\eta} + 2\,t\,\mathfrak{z}\,[E^*{}_{1\,\mathfrak{z}\,\eta} + F_m\,E_1{}^*] = 0, \tag{3.11.9}$$

and

$$(1-t^2)\,E^*{}_{2\,\mathfrak{z}\,t} - t^{-1}\,E^*{}_{2\,\mathfrak{z}} + 2\,t\,\eta\,[E^*{}_{2\,\mathfrak{z}\,\eta} + F_m\,E_2{}^*] = 0, \tag{3.11.10}$$

respectively. Let E_1 and E_2 possess continuous second derivatives, and let $E^*{}_{1\,\mathfrak{z}}\,(\eta\,t)^{-1}$ and $E^*{}_{2\,\eta}\,(\mathfrak{z}\,t)^{-1}$ be finite for $t = 0$. Then

$$\Psi^*\,(\mathfrak{z},\eta) = \int_{-1}^{+1} \left\{ E_1^*\,(\mathfrak{z},\eta,t)\,f_1\!\left[\frac{1}{2}\,\mathfrak{z}\,(1-t^2)\right] + E_2^*\,(\mathfrak{z},\eta,t)\,f_2\!\left[\frac{1}{2}\,\eta\,(1-t^2)\right] \right\} (1-t^2)^{-1/2}\,dt, \tag{3.11.11}$$

where f_k, $k = 1, 2$, are two arbitrary, twice-continuously differentiable functions of their respective arguments, is a solution of the equation

$$\Psi^*{}_{\mathfrak{z}\,\eta} + F_m\,\Psi^* = 0. \tag{3.11.12}$$

The proof of this theorem is given in[4]. Let $F_m\,(\beta)$ possess derivatives of all orders in the interval $\beta_0 \leqq \beta \leqq \beta_1$, $0 < \beta_0 < \beta_1 < \infty$. If a constant c exists such that the inequalities

$$\left| \frac{d^K F_m}{d\beta^K} \right| \leqq \frac{c\,(K+1)!}{\beta^{K+2}}, \qquad K = 0, 1, 2, \ldots n, \tag{3.11.13}$$

can be obtained, then there exist solutions E_1 and E_2 of (3.11.9) and (3.11.10), respectively, satisfying the conditions of the theorem cited above. Thus, we obtain a solution for Ψ^* in terms of two arbitrary, twice-differentiable functions. In[5] Bergman shows that (1.9.2) and (3.11.11) are different forms of the same operator, the former valid in the subsonic case, while the latter holds in the supersonic range.

Similarly, as in the subsonic range, if $\Psi_\nu\,(\Lambda,\theta)$, $\nu = 1, 2, 3, \ldots$, represents a set of particular solutions of equation (3.2.9), and A_ν, arbitrary constants, any linear combination

$$\sum_{\nu=1}^{N} A_\nu\,\Psi_\nu\,(\Lambda,\theta), \tag{3.11.14}$$

is also a solution of equation (3.2.9). This is so since the equations are linear and therefore the principle of superposition of solutions holds. By varying the constants A_ν, $\nu = 1, 2, \ldots$, flows around various shapes can be obtained. On the other hand, it is often necessary to determine constants A_ν to yield a flow which approximates

[4] S. Bergman: The approximation of function satisfying a linear partial differential equation. Duke Mathematical Journal, Vol. 6 (1940), pp. 537—561.

[5] S. Bergman: Methods for determination and computation of flow patterns of a compressible fluid. N. A. C. A., T. N. No. 1018 (1946).

that about a prescribed boundary curve, the equation of which is, say, $F(x, y) = 0$. In this case, the problem of how to determine the constants A_v is somewhat complicated and unsolved up to the present. Some remarks on this subject are in[1].

Part IV

Transonic Flow

1. General Remarks

The problem of transonic flow is perhaps the most difficult. We have to find solutions which must partly cover the subsonic, and partly the supersonic regions. The complexity of the problem is the origin of several methods of solutions. Below, we shall try to outline and discuss a few of them. In the second half of this part a more general discussion of the problem of transonic flow will be presented without deriving the proofs which a reader who is more advanced mathematically may find in Bergman's original papers.

First we shall present a continuation of the subsonic flow solution through the sonic line, according to the proposition of von Mises[6].

2. Behavior of the G_n at $m = 1$

The equation (1.5.3a), discussed above, provides a recursion formula for the determination of the G_n. One may write it in the form:

$$G_{n+1} = \frac{dG_n}{d\lambda} + \int^{\lambda} f G_n \, d\lambda, \qquad G_0 = 1. \tag{4.2.1}$$

If we compute this last integral, we obtain an integration constant each time, since the lower limit is arbitrary, i. e., G_n would depend on n constants of integration. By means of reasoning identical with that in section I.9, we choose as the lower limit of the integrals the value $\lambda = -\infty$, corresponding to $f = m = 0$ (1.3.11) and (1.3.13). However, at $\lambda = 0$ or $m = 1$, all G_n become infinite. It is therefore proper to select other functions r_n in the place of the G_n which remain bounded (i. e., finite). To find such r_n one has to study the behavior of the G_n near $m = 1$, or, what is the same, the behavior of $f(T)$ and $G_n(T)$ in a neighborhood of $T = 0$. Suppose that

$$G_n = c_n T^{-3n} (1 + b_n T^2 + \ldots) = c_n T^{-3n} r_n(T); \tag{4.2.2}$$

we insert into the recursion formula for the G_n the function $r_n(T)$ given by

$$r_n(T) = 1 + b_n T^2 + \ldots \; . \tag{4.2.3}$$

Since $G_0 = 1$, one has $c_0 = 1$, $r_0 = 1$. From (4.2.2) one obtains:

$$\frac{dG_n}{dT} = c_n T^{-(3n+1)} (-3n\, r_n + T\, r_n'), \tag{4.2.4}$$

where r_n' denotes the derivative of r_n with respect to T. Following (1.3.13) we have, after a few transformations,

$$\frac{d\lambda}{dT} = -2 T^2 [(k+1) - 2k T^2 + (k-1) T^4]^{-1}. \tag{4.2.5}$$

By multiplying the right-hand sides of (1.3.14a) and (4.2.5), we have:

$$B = f \frac{d\lambda}{dT} = -\frac{1}{8} (1 - T^2) T^{-4} (1 - h^2 T^2)^{-1} [5(k+1) + 2(5-k) T^2 - (3k-1) T^4] =$$
$$= -\frac{5}{8} (k+1) T^{-4} [1 - 2k \{5(k+1)\}^{-1} T^2 \ldots]. \tag{4.2.6}$$

---

[6] R. von Mises and M. Schiffer: On Bergman's integration method in two-dimensional compressible fluid flow. Advances in Applied Mechanics, Vol. I. New York: Academic Press Inc. (1948).

But from the relation:

$$G_{n+1} = \frac{dG_n}{d\lambda} = \int_{-\infty}^{\lambda} f G_n \, d\lambda = \frac{dG_n}{dT} A + \int_{1}^{T} B G_n \, dT, \tag{4.2.7}$$

where the used symbol denotes:

$$A = \frac{dT}{d\lambda} = -\frac{1}{2} \left[(k+1) - 2 k T^2 + (k-1) T^4 \right] T^{-2}, \tag{4.2.7a}$$

we get with the use of (4.2.3) and (4.2.6):

$$c_{n+1} T^{-(3n+3)} (1 + b_{n+1} T^2 + \ldots) = -\frac{1}{2} c_n T^{-(3n+3)} \cdot [-3n(1 + b_n T^2 + \ldots) +$$

$$+ 2 b_n T^2 + \ldots] [(k+1) - 2 k T^2 + (k-1) T^4] - \frac{5}{8} (k+1) c_n \int_{1}^{T} x^{-4} \{1 -$$

$$- 2 k [5(k+1)]^{-1} x^2 + \ldots\} (1 + b_n x^2 + \ldots) x^{-3n} \, dx; \tag{4.2.8}$$

this furnishes:

$$\frac{c_{n+1}}{c_n} [T^{-(3n+3)} + b_{n+1} T^{-(3n+1)} + \ldots] =$$

$$= T^{-(3n+3)} \left[\frac{1}{2} \cdot 3n(k+1) + \frac{5}{8} (3n+3)^{-1} (k+1) \right] + \tag{4.2.9}$$

$$+ T^{-(3n+1)} \left\{ (k+1) \left[\frac{1}{2} (3n-2) + \frac{5}{8} (3n+1)^{-1} \right] b_n - k \left[3n + \frac{1}{4} (3n+1)^{-1} \right] \right\} + \ldots$$

Comparing the factors of $T^{-(3n+3)}$ on both sides of (4.2.9), we get the relation:

$$\frac{c_{n+1}}{c_n} = \frac{1}{2} 3n(k+1) + \frac{1}{8} 5(k+1)(3n+3)^{-1} =$$

$$= \frac{1}{2} 3n(k+1) \{1 + 5 [36 n(n+1)]^{-1}\}; \tag{4.2.10}$$

this is a recursion formula for the c_n; this gives finally:

$$c_{n+1} = \left[\frac{1}{2} 3(k+1) \right]^n n! \prod_{\nu=1}^{n} \{1 + 5 [36 \nu (\nu+1)]^{-1}\} c_1. \tag{4.2.11}$$

Since $c_0 = 1$, we obtain further from equation (4.2.10):

$$c_1 = \frac{5}{24} (k+1), \tag{4.2.12}$$

which determines all c_n. Let us compare now the coefficients of $T^{-(3n+1)}$; using (4.2.10), we obtain the relation:

$$b_{n+1} \{1 + 5 [36 n(n+1)]^{-1}\} = b_n \{1 - 2(3n)^{-1} + 5 [12 n(3n-1)]^{-1}\} -$$

$$- \{2 k(k-1)^{-1} + k [6 n(3n+1)(k+1)]^{-1}\}, \tag{4.2.13}$$

which together with $b_0 = 1$ determines all b_n; this justifies our specific choice of the functions $r_n(T)$. If n is large, we may simplify relation (4.2.10):

$$c_{n+1} c_n^{-1} \approx \frac{1}{2} 3n(k+1), \tag{4.2.14}$$

and also

$$b_n \approx -2 n k (k+1)^{-1}, \quad \lim_{n \to \infty} \frac{b_{n+1}}{b_n} = 1. \tag{4.2.15}$$

Choosing $k = 1{\cdot}4$, we obtain from (4.2.13): $b_1 = -0{\cdot}7$, $b_2 = -1{\cdot}4$, $b_3 = -2{\cdot}1$, $b_4 = -2{\cdot}8$, etc.

3. Recursion Formula for the Functions r_n

By substituting (4.2.2) into (4.2.7) we get the relation:

$$c_{n+1} r_{n+1}(T) \, T^{-(3n+3)} = A c_n (T^{-3n} r_n' - 3n \, T^{-(3n+1)} r_n) + c_n \int_1^T B \, x^{-3n} r_n(x) \, dx. \tag{4.3.1}$$

Representing the function A (4.2.7a) by a polynomial $A = T^{-2} \sum_{\nu=0}^{4} \alpha_\nu T^\nu$, and the function B (4.2.6) by a power series $B = T^{-4} \sum_{\nu=0}^{\infty} \beta_\nu T^\nu$, and substituting them into (4.3.1), we obtain

$$c_{n+1} c_n^{-1} r_{n+1} = \sum_{\nu=0}^{4} \alpha_\nu T^\nu (T r_n' - 3n r_n) + T^{3n+3} \int_1^T x^{-(3n+4)} \sum_{\nu=0}^{\infty} \beta_\nu x^\nu r_n(x) \, dx. \tag{4.3.2}$$

To compute the coefficient $r_n(T)$ from this recursion formula, one has to keep in mind that $r_0 = 1$ and that one obtains from (4.2.10) for $k = 1 \cdot 4$, the values $c_1/c_0 = 1/2$, $c_2/c_1 = 77/20$, etc. This furnishes the relation:

$$\frac{1}{2} r_1(T) = T^3 \int_1^T x^{-4} \sum_{\nu=0}^{\infty} \beta_\nu x^\nu \, dx = T^3 \int_1^T B(x) \, dx. \tag{4.3.3}$$

Applying this result in the next recursive step, we get the relation:

$$\frac{77}{20} r_2(T) = \sum_{\nu=0}^{4} \alpha_\nu T^\nu (T r_1' - 3 r_1) + T^6 \left(\int_1^T B(\tau) \, d\tau \right)^2, \tag{4.3.4}$$

where the second term on the right-hand side originates from the expression

$$T^6 \int_1^T x^{-3} B(x) r_1(x) \, dx = T^6 \int_1^T 2 B(x) \left(\int_1^x B(\tau) \, d\tau \right) dx, \tag{4.3.4a}$$

by means of integrating by parts. The results, (4.3.3) and (4.3.4), have an interesting feature. $B(T)$, taken in its original form (4.2.6) without developing into a power series, contains only even powers of T; it appears that $r_1(T)$ and $r_2(T)$ are free from logarithmic terms and can be developed into power series at $T = 0$. In general, this is not always true. The power series for $r_2(T)$, when introduced into the integral in (4.3.2) in order to determine $r_3(T)$, produces a logarithmic term. But it is multiplied by T^9, so that $r_3(T)$ has a regular development at the point $T = 0$ up to the ninth order. Obviously, further application of the recursion formula will preserve this logarithmic term and will produce new additional terms of the type $T^\nu (\log T)^\mu$. One can easily verify that in general one has:

$$r_n(T) = \sum_{\nu=0}^{n-2} P_\nu^{(n)}(T) \, (\log T)^\nu, \qquad n > 2, \tag{4.3.5}$$

where the $P_\nu^{(n)}(T)$ are regular series of T. It is relatively easy to deduce recursion formulas for the $P_\nu^{(n)}$ by means of (4.3.2). One may point out that each $P_\nu^{(n)}(T)$ with $\nu > 0$ has just those powers of T which are larger than 8. We may determine the first coefficients of $P_0^{(n)}(T)$. One may apply the development:

$$P_0^{(n)}(T) = \sum_{\nu=0}^{\infty} \varrho_\nu^{(n)} T^\nu, \qquad \varrho_0^{(n)} = 1. \tag{4.3.6}$$

Relation (4.3.2) gives for the coefficients $\varrho_\nu^{(n)}$ the following recursion formula as long as $\nu \leqq 8$, $\nu < 3n + 3$:

$$c_{n+1}\, c_n^{-1}\, \varrho_\nu^{(n+1)} = \sum_{\mu=0}^{\nu} \{(\mu - 3n)\, \alpha_{\nu-\mu} + \beta_{\nu-\mu}\, [\nu - (3n+3)]^{-1}\}\, \varrho_\mu^{(n)}. \qquad (4.3.7)$$

Certain numerical values of the coefficients $\varrho_\mu^{(n)}$ are given in[6], p. 269.

Let us note that it is possible to apply another expansion, since in the neighborhood of $T = 0$ (or $M = 1$), T and T^{-1} can be developed in a series of $\varepsilon = [-3(k+1)(\lambda/2)]^{1/3}$ or $\varepsilon = (-3\cdot6\,\lambda)^{1/3}$ for $k = 1\cdot4$. This choice yields $T = \varepsilon + \ldots$ If these series are substituted in (1.3.14a) for T and T^{-1}, a development of f in the power series of ε, which holds for $M < 1$, is obtained in the neighborhood of $M = 1$.

4. Methods of Analytic Continuation

In the region $(M \leqq M_0 < 1)$, where M_0 is near 1, the series (1.9.2) converges very slowly, and it is therefore necessary to employ a large number of terms in order to obtain a good approximation for Ψ^*. When applying equation (1.7.7), the number m must be chosen rather large. If this is the case, it is then expedient to replace the expansion (1.9.2) by (1.9.9). Theoretically, this is, however, not the only way of overcoming this difficulty, and in the following other means of doing so will be indicated: this alternative approach employs the method of analytic continuation.

Let $\Psi(q, \theta)$ be determined in a domain, say H, and let Ψ_n, $n = 1, 2, \ldots$, be a complete system of particular solutions, each Ψ_n being determined in a domain G. Suppose that H and G actually do overlap, and denote their common part by I. Further, let $\sum\limits_{n=1}^{\infty} a_n \Psi_n$ be the series expansion of Ψ in I. Frequently, $\sum\limits_{n=1}^{\infty} a_n \Psi_n$ will converge outside of I, say in the domain $H_2 - I$, where H_2 is G or some part of it. If, in addition, $\sum\limits_{n=1}^{\infty} a_n \Psi_n$ can be termwise differentiated twice in H_2, it represents the analytic continuation of Ψ in $H_2 - I$. Frequently, the domain H_2 in which the stream function can be represented in the form $\sum\limits_{n=1}^{\infty} a_n \Psi_n$ covers a supersonic region as well, and consequently this method will then yield the flow in this latter region. In this manner, a method for determining a mixed flow may be obtained. Below, we shall present some such methods and at first let us cite an auxiliary lemma:

Let $p(q, \theta)$, $q_0 \leqq q \leqq q_1$, $-L \leqq \theta \leqq L$, be an analytic function of two real variables, q and θ, and let

$$\sum_{\nu=0}^{\infty} \left[a_\nu(q) \cos\left(\frac{\pi\nu\theta}{L}\right) + b_\nu(q) \sin\left(\frac{\pi\nu\theta}{L}\right) \right], \qquad (4.4.1)$$

$$a_0(q) = \frac{1}{2} L^{-1} \int_{-L}^{L} p(q, \theta)\, d\theta, \qquad (4.4.1a)$$

$$a_\nu(q) = L^{-1} \int_{-L}^{L} p(q, \theta) \cos\left(\frac{\pi\nu\theta}{L}\right) d\theta, \qquad (4.4.1b)$$

$$\nu = 1, 2, \ldots,$$

$$b_\nu(q) = L^{-1} \int_{-L}^{L} p(q, \theta) \sin\left(\frac{\pi\nu\theta}{L}\right) d\theta, \qquad (4.4.1c)$$

be its Fourier development. The series (4.4.1) converges uniformly and can be differentiated termwise any finite number of times with respect to both q and θ. The proof of this lemma can be found in any book on Fourier series, and is also cited in [7]. Since every solution of an elliptic equation with analytic coefficients is an analytic function of two real variables, the result obtained can be applied to the case where $p\,(q,\theta)$ is the stream function $\Psi\,(q,\theta)$ of a subsonic flow. Thus

$$\sum_{\nu=0}^{\infty}\left[a_{\nu}\,(q)\cos\left(\frac{\pi\,\nu\,\theta}{L}\right)+b_{\nu}\,(q)\sin\left(\frac{\pi\,\nu\,\theta}{L}\right)\right], \qquad (4.4.2)$$

$$a_{0}\,(q)=\frac{1}{2}\,L^{-1}\int_{-L}^{L}\Psi\,(q,\theta)\,d\theta, \qquad (4.4.2a)$$

$$a_{\nu}\,(q)=L^{-1}\int_{-L}^{L}\Psi\,(q,\theta)\cos\left(\frac{\pi\,\nu\,\theta}{L}\right)d\theta, \qquad (4.4.2b)$$

$$\nu=1,2,\ldots,$$

$$b_{\nu}\,(q)=L^{-1}\int_{-L}^{L}\Psi\,(q,\theta)\sin\left(\frac{\pi\,\nu\,\theta}{L}\right)d\theta, \qquad (4.4.2c)$$

can be differentiated termwise. If, now, following Chaplygin, we introduce instead of q, the variable

$$\tau=q^2\,(2\,a_0{}^2)^{-1}, \qquad (4.4.3)$$

then the equation (1.1.5) for Ψ assumes the form

$$[2\,\tau\,(1-\tau)^{-\beta}\,\Psi_{\tau}]_{\tau}+[1-(2\,\beta+1)]\,[2\,\tau\,(1-\tau)]^{-1}\,(1-\tau)^{-\beta}\,\Psi_{\theta\theta}=0, \qquad (4.4.4)$$

where $\beta=(k-1)^{-1}$. Using the representation of $\Psi\,(q,\theta)$ by means of the Fourier series, given above, and differentiating termwise, gives the result:

$$\sum_{\nu=0}^{\infty}\left\{\left[\frac{d}{d\tau}\left(\tau\,(1-\tau)^{-\beta}\,\frac{da_{\nu}}{d\tau}\right)-[1-(2\,\beta+1)\,\tau]\cdot[\tau\,(1-\tau)]^{-1}\,(1-\tau)^{-\beta}\cdot\right.\right.$$

$$\left.\cdot\,\frac{\nu^2\,\pi^2\,a_{\nu}}{4\,L^2}\right]\cos\frac{\nu\,\pi\,\theta}{L}+\left[\frac{d}{d\tau}\left(\tau\,(1-\tau)^{-\beta}\,\frac{db_{\nu}}{d\tau}\right)-[1-(2\,\beta+1)\,\tau]\,[\tau\,(1-\tau)]^{-1}\cdot\right.$$

$$\left.\left.\cdot\,(1-\tau)^{-\beta}\cdot\frac{\nu^2\,\pi^2\,b_{\nu}}{4\,L^2}\right]\sin\frac{\nu\,\pi\,\theta}{L}\right\}=0. \qquad (4.4.5)$$

In this equation, the coefficients of $\cos\left(\frac{\nu\,\pi\,\theta}{L}\right)$ and $\sin\left(\frac{\nu\,\pi\,\theta}{L}\right)$, for $\nu=1,2,\ldots$, must be equal to zero separately, and therefore the a_{ν} and b_{ν} are each solutions of equations

$$\frac{d}{d\tau}\left\{\tau\,(1-\tau)^{-\beta}\,\frac{da_{\nu}}{d\tau}\right\}-\{1-(2\,\beta+1)\,\tau\}\,\{\tau\,(1-\tau)\}^{-1}\cdot(1-\tau)^{-\beta}\,\frac{\nu^2\,\pi^2\,a_{\nu}}{4\,L^2}=0, \qquad (4.4.6)$$

$$\frac{d}{d\tau}\left\{\tau\,(1-\tau)^{-\beta}\,\frac{db_{\nu}}{d\tau}\right\}-\{1-(2\,\beta+1)\,\tau\}\,\{\tau\,(1-\tau)\}^{-1}\cdot(1-\tau)^{-\beta}\,\frac{\nu^2\,\pi^2\,b_{\nu}}{4\,L^2}=0. \qquad (4.4.7)$$

Each of these equations can easily be transformed into a hypergeometric equation and thus every solution of them may be written in the form

$$(A_{\nu}F_{\nu}+B_{\nu}F_{\nu}{}^{*})\,\tau^{\nu/(2\,L)}, \qquad (4.4.8)$$

---

[7] S. Bergman: On two-dimensional flows of compressible fluids. N. A. C. A., T. N. No. 972 (1945).

where A_ν and B_ν are constants, and

$$F_\nu = F(\alpha_\nu, \beta_\nu; -\beta; 1 - \tau), \qquad (4.4.9)$$

$$F_\nu^* = (1 - \tau)^{-\beta + 1} F(\gamma_\nu - \alpha_\nu, \gamma_\nu - \beta_\nu; 2 + \beta; 1 - \tau), \qquad (4.4.10)$$

$F(\alpha, \beta, \gamma, \tau)$ being the hypergeometric series. The symbols used are given by:

$$\gamma_\nu = (\nu L^{-1} + 1), \quad \alpha_\nu = \frac{1}{2}\left[\left(\frac{\nu}{L} - \beta\right) + \Delta_\nu\right], \qquad (4.4.11)$$

$$\beta_\nu = \frac{1}{2}\left[\left(\frac{\nu}{L} - \beta\right) - \Delta_\nu\right], \quad \Delta_\nu = \left[\left(\frac{\nu}{2L}\right)^2(2\beta + 1) + \beta^2\right]^{1/2}.$$

In order to determine the constants A_ν, B_ν, the following theorem is employed: Let $\Psi^{(1)}(q, \theta)$ and $\Psi^{(2)}(q, \theta)$, $[q_0 \leq q \leq q_1, -L \leq \theta \leq L]$ be solutions of an equation of elliptic type. If, along a line, say $q = q_0$,

$$\Psi^{(1)}(q_0, \theta) = \Psi^{(2)}(q_0, \theta), \qquad (4.4.12)$$

and

$$\Psi_q^{(1)}(q, \theta)\big|_{q=q_0} = \Psi_q^{(2)}(q, \theta)\big|_{q=q_0}, \qquad (4.4.12\text{a})$$

then in the whole domain $[q_0 \leq q \leq q_1, -L \leq \theta \leq L]$,

$$\Psi^{(1)}(q, \theta) = \Psi^{(2)}(q, \theta). \qquad (4.4.13)$$

Now, suppose that the function $g(\overline{Z})$, (1.9.2a), is regular in some domain $H_1 + H_2$,

$$H_1 = [q_0 \leq q \leq q_1, \quad -L \leq \theta \leq L], \qquad (4.4.14\text{a})$$

$$H_2 = [q_1 \leq q \leq q_2, \quad -L \leq \theta \leq L], \qquad (4.4.14\text{b})$$

which domains lie in $[\theta^2 < 3\lambda^2, \lambda < 0]$. Then, by the formula (1.9.2), it follows that Ψ is also regular in $H_1 + H_2$. Suppose that Ψ has been evaluated in the domain H_1, but it is desired to avoid the evaluation of Ψ in H_2 by means of (1.9.2), since this series converges very slowly in H_2. To achieve that, represent the function Ψ in H_1 by a Fourier series, which is always possible since we evaluated Ψ in H_1:

$$\Psi^{(1)} = \sum_{\nu=0}^{\infty} \left[a_\nu(q) \cos\left(\frac{\pi \nu \theta}{L}\right) + b_\nu(q) \sin\left(\frac{\pi \nu \theta}{L}\right)\right]. \qquad (4.4.15)$$

Assume that in H_2 we wish to represent Ψ by another Fourier series

$$\Psi^{(2)} = \sum_{\nu=0}^{\infty} \left[S_\nu^{(1)}(\tau) \cos\left(\frac{\pi \nu \theta}{L}\right) + S_\nu^{(2)}(\tau) \sin\left(\frac{\pi \nu \theta}{L}\right)\right], \qquad (4.4.16)$$

$$S_\nu^{(k)}(\tau) = \left[A_\nu^{(k)} F_\nu + B_\nu^{(k)} F_\nu^*\right]\tau^{\frac{\nu L}{2}}, \quad k = 1, 2. \qquad (4.4.16\text{a})$$

In order that this series will be the solution Ψ under consideration, the constants $A_\nu^{(k)}$ and $B_\nu^{(k)}$ must be determined so that

$$S_\nu^{(1)}(\tau)\big|_{q = q_0} = a_\nu(q_0), \qquad \nu = 1, 2, \ldots, \qquad (4.4.17\text{a})$$

$$S_\nu^{(2)}(\tau)\big|_{q = q_0} = b_\nu(q_0), \qquad \nu = 0, 1, 2, \ldots, \qquad (4.4.17\text{b})$$

$$\frac{dS_\nu^{(1)}(\tau)}{dq}\bigg|_{q = q_0} = \frac{da_\nu(q)}{dq}\bigg|_{q = q_0}, \qquad (4.4.17\text{c})$$

$$\frac{dS_\nu^{(2)}(\tau)}{dq}\bigg|_{q = q_0} = \frac{db_\nu(q)}{dq}\bigg|_{q = q_0}. \qquad (4.4.17\text{d})$$

Since in $[q_1 \leq q \leq q_2, -L \leq \theta \leq L]$, $\Psi(q, \theta)$ is an analytic function, the series (4.4.16) and its derivatives converge uniformly and absolutely in this domain, and this series represent the solution Ψ under consideration in the region H_2.

Moreover, this series and its derivatives may also converge outside of H_2, say in $H_3 = [q_2 \leqq q \leqq q_3, \; -L \leqq \theta \leqq L]$. If H_3 lies partially outside of the domain $[\theta^2 < 3\,\lambda^2, \; \lambda < 0]$, then the expression obtained gives the analytic continuation of the solution outside the domain of representation by the integral formulas $(1.9.2)$. In particular, H_3 may include some region which lies in $M > 1$. Of course, in practical application the summation will include only a finite number of terms N, say. Thus the procedure of continuation consists of determining a linear combination $\sum A_\nu \, \chi_\nu \, (M, \theta)$, so that along a line $M = M_0 < 1$, this latter function and its derivative with respect to M coincide with the corresponding quantities for the stream function of the subsonic flow pattern which is to be continued.

Another similar method is proposed by Bergman in [7] and [8].

Unfortunately, the power series for the functions of $(4.4.6)$ and $(4.4.7)$ converge so slowly that in most cases this approach must be abandoned for computational purposes. On the other hand, if certain changes in this approach are made, numerical results may be obtained. Two such procedures will be discussed in the following[3]:

(a) The solutions of $(4.4.6)$ and $(4.4.7)$ are expanded in the form of a power series, not, however, around $\tau = 0$, which is the case of the hypergeometric series, but rather around some conveniently chosen value of τ which lies inside the interval under consideration $(-L \leqq \theta \leqq L)$. For example, it will often be convenient to introduce for τ the variable $\sigma = \tau - 0{\cdot}15$, say. Since the particular solutions are often considered only for a small range of variation of τ, say $0{\cdot}13 \leqq \tau \leqq 0{\cdot}20$, the coefficients of $(4.4.6)$ and $(4.4.7)$ may be approximated by polynomials in σ, so that if, say, $L = \pi/2$, the equation approximating $(4.4.6)$ and $(4.4.7)$ is (for $\beta = (k-1)^{-1} = 5/2$):

$$(-\sigma^3 + 0{\cdot}5500\,\sigma^2 + 0{\cdot}2325\,\sigma + 0{\cdot}0191)\frac{d^2 U}{d\sigma^2} + (1{\cdot}5000\,\sigma^2 + 1{\cdot}4500\,\sigma + 0{\cdot}1838)\frac{dU}{d\sigma} +$$

$$+ (6{\cdot}000\,\sigma - 0{\cdot}1000)\,\nu^2\,U = 0, \qquad (4.4.18)$$

where U denotes a function approximating the actual solution of $(4.4.6)$ and $(4.4.7)$. As is known, every solution of $(4.4.18)$ can be represented as a linear combination of two independent solutions, each of which can be written in the form of an infinite series in σ, which at least converges for $|\sigma| < 0{\cdot}15$, a range of convergence which will be sufficient for most purposes. Solutions $U_1\,(\sigma; \nu)$, $U_2\,(\sigma; \nu)$ are chosen so that

$$U_1\,(0; \nu) = 0, \qquad U_1{}'\,(0; \nu) = 1, \qquad (4.4.19\,\mathrm{a})$$

$$U_2\,(0; \nu) = 1, \qquad U_2{}'\,(0; \nu) = 0. \qquad (4.4.19\,\mathrm{b})$$

Thus

$$U_1\,(\sigma; \nu) = \sigma + a_2\,\sigma^2 + a_3\,\sigma^3 + \ldots, \qquad (4.4.20)$$

$$U_2\,(\sigma; \nu) = 1 + b_2\,\sigma^2 + b_3\,\sigma^3 + \ldots, \qquad (4.4.21)$$

a_i, b_i, ..., $i = 2, 3, \ldots$, being functions of ν. Substituting $(4.4.20)$ and $(4.4.21)$ into $(4.4.18)$, we obtain a set of equations which give the values of a_i and b_i as polynomials in ν^2 $(^3, p.\,35)$. Since it is often sufficient to consider the series $(4.4.20)$ and $(4.4.21)$ in a small interval, say $|\sigma| < 0{\cdot}05$, it will suffice to employ only a few terms of these series, and hence it is necessary to compute only a small number of a_i, b_i.

(b) To employ the second method mentioned, equations $(4.4.6)$ and $(4.4.7)$ are approximated by the equations

$$(0{\cdot}15 + \sigma)\,(0{\cdot}85 - \sigma)\frac{d^2 u}{d\sigma^2} + (1{\cdot}225 + 1{\cdot}5\,\sigma)\frac{du}{d\sigma} - \nu^2\left(\frac{0{\cdot}1 - 6\,\sigma}{0{\cdot}15 + \sigma}\right)u = 0, \quad (4.4.22)$$

<hr>

[8] S. Bergman: The hodograph method in the theory of compressible fluids. Supplement to „Fluid Dynamics" by R. v. Mises und K. O. Friedrichs, Brown University (1941—1942).

and two independent solutions of this equation, $u_1(\sigma; \nu)$, $u_2(\sigma; \nu)$, are determined not by the power series (4.4.20) and (4.4.21), but by an approximation method. That is, we write

$$\frac{du}{d\sigma} = v, \quad \text{or} \quad \Delta u = v\,\Delta\sigma. \tag{4.4.23}$$

Then

$$\Delta v = \left\{ -\left[\frac{(1\cdot225 + 1\cdot5\,\sigma)}{(0\cdot15 + \sigma)(0\cdot85 - \sigma)} \right] v + \nu^2 \left[\frac{(0\cdot1 - 6\,\sigma)}{(0\cdot15 + \sigma)^2(0\cdot85 - \sigma)} \right] u \right\} \Delta\sigma. \tag{4.4.24}$$

Assume that $\Delta\sigma$ is contained in a small interval, say $0\cdot001$, and that for $\sigma = 0$, we have $u = 0$ and $v = 1$. Then, if (4.4.23) and (4.4.24) are employed, $\Delta u(\sigma)$ and $\Delta v(\sigma)$ may be determined at $\sigma = 0$, and hence

$$u(0\cdot001) = u(0) + \Delta u(0), \tag{4.4.25a}$$

$$v(0\cdot001) = v(0) + \Delta v(0). \tag{4.4.25b}$$

From these values of u and v, $\Delta u(0\cdot001)$ and $\Delta v(0\cdot001)$, and hence $u(0\cdot002)$, $v(0\cdot002)$ may be found and the entire procedure continued until an approximate curve is found for $u(\sigma)$, $0 \leq \sigma \leq 0\cdot15$. In this manner, the desired integration is performed. Both methods described above are appropriate for the purpose of computation, but they are insufficient to give an insight into the behavior of the particular solutions when $\nu \to \infty$.

5. Application of Integral Operators to Transonic Flow. Basic Equations

Below, the application of integral operators to transonic flow will be discussed. In general, the possible method of continuation, which will be discussed, will consist of the determination of the limit values of the subsonic stream function and of its derivatives with respect to M on the line $M = 1$, and then of the consideration of the initial value problem for the compressibility equation in the supersonic portion of the flow, $M > 1$. Since it is advisable to use different variables in the region of transonic flow, we have to begin with the basic equations. Assuming that the thermo-dynamical conditions follow the isentropic law, and introducing as new variables θ and

$$H = \int_{q_1}^{q} \varrho\,(dq/q) = \int_{q_1}^{q} q^{-1} \left[1 - \frac{1}{2}(k-1)q^2 \right]^{\frac{1}{(k-1)}} dq, \tag{4.5.1}$$

we obtain the following linear equation for the stream function:

$$\Psi_{HH} + l(H)\,\Psi_{\theta\theta} = 0, \quad l(H) = (1 - M^2)\,\varrho^{-2}, \tag{4.5.2}$$

where

$$M = q \left[1 - \frac{1}{2}(k-1)q^2 \right]^{-\frac{1}{2}}. \tag{4.5.3}$$

It is noted that in considering purely subsonic flow in previous sections, the variable λ (1.2.1) was used; and for purely supersonic flows, the variable $\Lambda = i\,\lambda$. In studying transonic flow patterns, it is found to be more convenient to use the variables H, θ. A formal computation shows that the Taylor development of $l(H)$ in the neighborhood of $H = 0$ is

$$l(H) = [2/(k-1)]^{(2-k)/(k-1)} \{ (-2H) - [(2k+5)/(2k+2)] \cdot$$
$$\cdot [(k+1)/2]^{2k/(k-1)} (-2H)^2 + \dots . \tag{4.5.4}$$

In purely subsonic flow, we applied formula (1.9.1) in previous sections. Introducing the complex variables $Z = \lambda + i\,\theta$, $\overline{Z} = \lambda - i\,\theta$, equation (1.9.1) transforms into

$$4\,\Psi_{z\bar{z}} + 4\,N\,(\Psi_z + \Psi_{\bar{z}}) = 0, \qquad (4.5.5)$$

where the following transformation formulas may be helpful:

$$\frac{\partial}{\partial Z} = \frac{1}{2}\left(\frac{\partial}{\partial \lambda} - i\,\frac{\partial}{\partial \theta}\right), \qquad \frac{\partial}{\partial \overline{Z}} = \frac{1}{2}\left(\frac{\partial}{\partial \lambda} + i\,\frac{\partial}{\partial \theta}\right), \qquad (4.5.5\,\text{a})$$

$$\frac{\partial}{\partial \lambda} = \frac{\partial}{\partial Z} + \frac{\partial}{\partial \overline{Z}}, \qquad \frac{\partial}{\partial \theta} = i\left(\frac{\partial}{\partial Z} - \frac{\partial}{\partial \overline{Z}}\right). \qquad (4.5.5\,\text{b})$$

It should be noted that the interval $-\infty < \lambda < 0$ corresponds to the interval $-\infty < H < 0$. In the supersonic case, we applied equation (3.2.9). Equations (1.9.1) and (3.2.9) can be transformed. If Ψ is replaced by

$$\Psi^* = \Psi\,R^{-1}, \qquad \partial R/\partial \overline{Z} = N, \qquad (4.5.6)$$

then Ψ^* satisfies the equations

$$\Psi^*_{\lambda\lambda} + \Psi^*_{\theta\theta} + 4\,F\,\Psi^* = 0, \qquad \text{and} \qquad \Psi^*_{\Lambda\Lambda} - \Psi^*_{\theta\theta} - 4\,F_1\,\Psi^* = 0, \qquad (4.5.7)$$

where

$$F = F_1 = \frac{(k+1)\,M^4}{64}\left[\frac{-(3\,k-1)\,M^4 - 4\,(3-2\,k)\,M^2 + 16}{(1-M^2)^3}\right]. \qquad (4.5.8)$$

The expansions of N and F in the neighborhood of $\lambda = 0$ are([9], p. 861):

$$N = (1/12\,\lambda)\left\{1 - \frac{1}{4}\,(k+1)^{1/2}\left[\left(2 + \frac{3}{5}\,2^{1/2}\right)k + 5 \cdot 2^{1/2} - 2\right] \cdot \right.$$
$$\left. \cdot\,[2^{-1/6}\,3^{2/3}\,(k+1)^{-5/6}\,(-\lambda)^{2/3}] + \ldots\right\}, \qquad (4.5.9)$$

$$F = 5/[36\,(-2\,\lambda)^2] + \ldots \quad . \qquad (4.5.10)$$

In the vicinity of $H = 0$, $l(H)$ may be replaced by the first term in its expansion. Using this value of $l(H)$ in (4.5.2), we obtain the so-called "simplified" compressibility equation:

$$-\,CH\,\Psi_{\theta\theta} + \Psi_{HH} = 0, \qquad (4.5.11)$$

where C is a constant. In considering transonic flows, the solutions of the simplified equation will give a fair approximation of the exact stream function. The expressions for N, F and H will, in this case, reduce to

$$N = N^+ = 1/(12\,\lambda), \qquad (4.5.12\,\text{a})$$

$$F = F^+ = 5/36\,[1/(-2\,\lambda)^2], \qquad (4.5.12\,\text{b})$$

$$H = H^+ = (3^{2/3}/2)\,[2/(k-1)]^{(k-2)/(3\,k-3)}\,(-\lambda)^{2/3}. \qquad (4.5.12\,\text{c})$$

In his important investigation, Tricomi[10] studied the boundary value problem of equation (4.5.11) and showed that if we consider a finite domain located partly in the subsonic and partly in the supersonic region, then under certain conditions the boundary value problem has a unique solution. Frankl[11] considered questions allied with Tricomi's investigations in the case of the exact compressibility equation.

[9] S. Bergman: Two-dimensional transonic flow patterns. Amer. J. Math. 70, pp. 856—891 (1948).

[10] F. Tricomi: Sulle equazioni lineari alle derivato parziali di 2° ordine di tipo misto. Atti R. Accad. Naz. Linzei, Mem. della Cl. Sci. Fis. Mat. e. Nat., Ser. 5, Vol. 14, 133—247 (1923).

[11] F. I. Frankl: On the problems of Chaplygin for mixed sub- and supersonic flows. Bull. Acad. Sci. URSS, 8, 195—245 (1944) (in Russian). — On the problems of Chaplygin for mixed sub- and supersonic flows. Bull. Acad. Sci. URSS. Vol. 9, 121—143 (1945) (in Russian). Also: N. A. C. A., T. N. No. 1155 (1947).

We shall discuss below the integral operators (of two kinds). The integral operator of the first kind may be applied to the subsonic flow. It is useful to shift the origin to the point λ_0, θ_0 (point of reference of the operator). Every real solution of equation (4.5.5) can be represented in a neighborhood of the origin $Z' = 0$ as the imaginary part of the right-hand side of (4.5.13) —

$$p(g) = R(Z', \bar{Z}')\left[g(Z') + \sum_{n=1}^{\infty} 2^{-2n}\, \frac{\Gamma(2n+1)}{\Gamma(n+1)}\, p^{(n)}(Z', \bar{Z}')\, g^{[n]}(Z')\right], \quad (4.5.13)$$

with

$$g^{[n]}(Z') = \int_0^{Z'}\int_0^{Z_1}\!\!\ldots\int_0^{Z_{n-1}} g(Z_n)\,dZ_n \ldots dZ_1 = \frac{(-1)^{n-1}}{(n-1)!}\int_{\mathfrak{z}=0}^{\mathfrak{z}=Z'}(Z' - \mathfrak{z})^{n-1} g(\mathfrak{z})\,d\mathfrak{z}, \quad (4.5.13a)$$

where

$$g(Z') = \int_{t=-1}^{+1} f\left[\frac{1}{2} Z'(1 - t^2)\right] dt/(1 - t^2)^{1/2}, \quad (4.5.13b)$$

with a suitably chosen function f (or g). The function $g(Z')$ is called the associate of $p(g)$ with regard to the integral operator of the first kind, p. The function R is defined by the formula

$$R(Z', \bar{Z}') = \exp\left[-\int_0^{\bar{Z}'} N(Z' + \bar{Z}_1')\,d\bar{Z}_1'\right], \quad (4.5.14)$$

and the symbols Z' and $\bar{Z}'$ denote:

$$Z' = \lambda' + i\theta'; \quad \bar{Z}' = \lambda' - i\theta', \quad \lambda' = \lambda - \lambda_0, \quad \theta' = \theta - \theta_0, \quad \lambda_0 < 0. \quad (4.5.15)$$

Another form of the operator of the first kind is

$$P_1(f) = \int_{-1}^{+1} E_1(Z', \bar{Z}', t)\, f\left[\frac{1}{2} Z'(1 - t^2)\right] dt/(1 - t^2)^{1/2}, \quad (4.5.16)$$

where f is a suitably chosen associate function, called the generating function of the operator P_1, and E_1 is equal to

$$E_1(Z', \bar{Z}', t) = R(Z', \bar{Z}')\, E_1^*(Z', \bar{Z}', t), \quad (4.5.17)$$

where E_1^* has the development

$$E_1^* = 1 + \sum_1^{\infty} Z'^n\, t^{2n}\, P^{(n)}(Z', \bar{Z}'). \quad (4.5.18)$$

The functions $P^{(n)}$ satisfy the following recurrence relations:

$$P_{\bar{Z}'}^{(1)} + 2F = 0, \quad (2n+1)\, P_{\bar{Z}'}^{(n+1)} + 2\, P_{Z'\bar{Z}'}^{(n)} + 2F P^{(n)} = 0, \quad n = 1, 2, 3, \ldots, \quad (4.5.19)$$

with

$$P^{(n)}(Z', 0) = 0, \quad n = 1, 2, 3, \ldots$$

By the above requirement, the $P^{(n)}$ and hence the generating functions E_1 (of the first kind) are uniquely determined. All the proofs—for example, the proof that the series (4.5.18) converges absolutely and uniformly in a sufficiently small neighborhood of the origin $Z' = 0$, $\bar{Z}' = 0$—can be found in [9]. Thus, assuming that the associate function is regular in a sufficiently large domain, we may prove that by applying to it the integral operator of the first kind, we obtain a solution of (4.5.5) defined in a sufficiently small neighborhood of the origin. As the next step, one can show that if f is regular in the entire domain $M < 1$, this solution can be continued throughout the whole subsonic region, i. e., the representation holds for $E[M < 1, -\infty < \theta < \infty)$, where the symbol $E[..]$ denotes a domain. There exists still another form of the integral operator of the first kind, namely

$$p(g) = R(Z', \overline{Z}')\left\{g_1(Z') - \int\limits_0^{Z}\int\limits_0^{\overline{Z}'} F g_1 \, dZ_2 \, d\overline{Z}_1 + \right.$$

$$\left. + \int\limits_0^{Z'}\int\limits_0^{\overline{Z}'} F\left[\int\limits_0^{Z_1}\int\limits_0^{\overline{Z}_1} F g_1 Z_2 \, dZ_2\right] dZ_1 \, d\overline{Z}_1 + \ldots\right\}. \tag{4.5.20}$$

Making some modifications, we can extend the definition of the operator P_1 so that it can be applied to functions of one variable $\varLambda + \theta$ and $\varLambda - \theta$, respectively, thus generating solutions of equation (3.2.9) in the almost entire supersonic region. The expression

$$P_1[f_1(\varLambda + \theta)] + P_2[f_2(\varLambda - \theta)], \tag{4.5.21}$$

where f_1 and f_2 are two linearly independent functions, will represent a (possible) stream function of a supersonic flow pattern, i. e., for $E[M > 1, -\infty < \theta < \infty]$. The integral operator of the first kind, convenient though it is for many purposes, has the disadvantage that it does not represent solutions of the compressibility equation in the neighborhood of the sonic line. Furthermore, it has the disadvantage from the practical point of view that the $P^{(n)}(Z', \overline{Z}')$ are functions of two variables Z' and $\overline{Z}'$, or λ and θ, which makes tabulation of the values of $P^{(n)}$ very time-consuming.

6. Integral Operator of the Second Kind in the Case of the Simplified Compressibility Equation

The two disadvantages of the operator of the first kind can be removed by the use of another operator—to be termed "operator of the second kind"—for which the $P^{(n)}$ are functions of one variable only, and which yields a representation of the stream function in the neighborhood of the sonic line. This operator has a number of other distinctive features, which will best be elucidated by the detailed discussion of the so-called "simplified" compressibility equation, i. e., where $N = N^+ = (12\,\lambda)^{-1}$, in equation (1.9.1), or, alternatively, where $F = F^+ = 5/(144\,\lambda^2)$ in equation (4.5.8). At first, let us represent the generating function in the case of the simplified compressibility equation. It is of the form

$$E = HE^*, \tag{4.6.1}$$

where H is given by (1.9.7); it can be given by the formula

$$H(2\,\lambda) = S_0(-2\,\lambda)^{-1/6}\, p[(-2\,\lambda)^{2/3}], \tag{4.6.2}$$

with

$$p[(-2\,\lambda)^{2/3}] = 1 + S_1(-2\,\lambda)^{2/3} + S_2(-2\,\lambda)^{4/3} + \ldots, \tag{4.6.3}$$

$$S_0 = 2^{(2\,k+1)/(6\,k-6)}\, 3^{-1/6}(k+1)^{(2-k)/(6\,k+6)}, \tag{4.6.3a}$$

$$S_1 = (1/10)(3/4)^{2/3}(k+1)^{-1/3}(2\,k+5), \tag{4.6.3b}$$

$$S_2 = -(1/1400)(3/4)^{4/3}(k+1)^{-2/3}(64\,k^2 + 70\,k + 75), \text{ etc.} \tag{4.6.3c}$$

In the case of the simplified equation, we have $N^+ = (12\,\lambda)^{-1}$, $p[(-2\,\lambda)^{2/3}] = 1$, $H^+(2\,\lambda) = S_0(-2\,\lambda)^{-1/6}$, $F^+ = 5/(144\,\lambda^2)$. One can show that in this case, the generating function E can be given in the form

$$E^+(Z', \overline{Z}', t) = E^+(\lambda, \theta, t) = A_1 S_0(-2\,\lambda)^{-1/6} F(1/6, 5/6, 1/2, X) +$$

$$+ B_1 S_0(-2\,\lambda)^{-2/3}[-t^2(\lambda + i\,\theta)]^{1/2} F(2/3, 4/3, 3/2, X),$$

$$\text{for } |X| = |- t^2 (\lambda + i\,\theta)/(- 2\,\lambda)| < 1, \qquad (4.6.4a)$$

and

$$E^+ (Z', \overline{Z'}, t) = E^+ (\lambda, \theta, t) = A_2 S_0 [(- t^2) (\lambda + i\,\theta)]^{-1/6} F (1/6, 2/3, 1/3, X) +$$
$$+ B_2 S_0 (- 2\,\lambda)^{-2/3} [(- t^2) (\lambda + i\,\theta)]^{-5/6} F (5/6, 4/3, 5/3, X),$$
$$\text{for } |X| = |- 2\,\lambda/- t^2 (\lambda + i\,\theta)| < 1. \qquad (4.6.4b)$$

One can prove in this case of the simplified equation that the imaginary part of $P_2^+ (f)$,

$$P_2^+ (f) = \int_{-1}^{+1} E^+ (\lambda, \theta, t)\, f \left[\frac{1}{2} Z (1 - t^2) \right] dt/(1 - t^2)^{1/2}, \qquad (4.6.5)$$

$$Z = \lambda + i\,\theta, \qquad (4.6.5a)$$

is a solution of $(1.9.1)$ with $N = N^+ = 1/(12\,\lambda)$, i. e.,

$$\Psi = I\,m\,[P_2^+ (f)]. \qquad (4.6.6)$$

The solution obtained is valid in the large, i. e., in a finite domain. In particular, if the function

$$g (Z) = \int_{t=-1}^{+1} f \left[\frac{1}{2} Z (1 - t^2) \right] dt/(1 - t^2)^{1/2}, \qquad (4.6.7)$$

is regular in a region B which possesses in its interior a number of branch points, each of finite order, then the obtained solution $(4.6.6)$ is defined in B and possesses branch points at the same points and of the same order as $(4.6.7)$. We assume here that the only singularities of g in B are branch points. In applying the integral operator method in the case where g has poles or logarithmic singularities, certain modifications are needed. If the function $(4.6.7)$ is defined for all values $\lambda < 0$, $- \infty < \theta < \infty$, then the solution Ψ will be defined in the same region.

The expressions $(4.6.4)$ obtained by means of the explained procedure, will not necessarily be analytical continuations of each other. Since, however, the hypergeometric equation has only two linearly independent solutions, the constants, A_1, B_1, and A_2, B_2 can always be so adjusted as to make these two solutions analytical continuations of each other.

The integral representation $(4.6.5)$ can immediately be generalized to the supersonic case where it will produce, in an analogous manner, solutions of $(3.2.9)$ with $N_1 = N^+ (4.5.12a)$. Indeed, replacing λ by the variable $i\,\Lambda$, it may be verified relatively easily that $\Psi (i\,\Lambda, \theta)$ satisfies equation $(3.2.9)$ with $N_1 = 1/(12\,\Lambda)$. Repeating the procedure which led to the generating function $(4.6.4)$ in the subsonic case, we now obtain the generating function

$$E^+ = [a_1 S_0/(2\,\Lambda)^{1/6}] F (1/6, 5/6, 1/2, X) + [b_1 S_0 (\Lambda + \theta)^{1/2} t/(2\,\Lambda)^{2/3}] F (2/3, 4/3, 3/2, X),$$
$$\text{for } |X| = |t^2 (\Lambda + \theta)/(2\,\Lambda)| < 1. \qquad (4.6.8)$$

As is known, a hypergeometric series in z defines a function which is analytic when $|z| < 1$. This function has a branch point at $z = 1$ and if a cut (i. e., an impassable barrier) is made from $+ 1$ to $+ \infty$ along the real axis, the function is analytic and one-valued throughout the cut plane. In our case, where $|X|$ is substituted for z, if $F (\alpha_1, \beta_1, \gamma_1, X)$ denotes only the hypergeometric series and not the hypergeometric function, $(4.6.8)$ has to be replaced for $|X| = |t^2 (\Lambda + \theta)/(2\,\Lambda)| > 1$, by

$$E^+ = [a_2 S_0/(t^2 (\Lambda + \theta))^{1/6}] F (1/6, 2/3, 1/3, X) +$$
$$+ [b_2 S_0 (2\,\Lambda)^{2/3}/(t^2 (\Lambda + \theta))^{5/6}] F (5/6, 4/3, 5/3, X), \qquad (4.6.9)$$

where the constants a_2, b_2 are expressable in terms of a_1, b_1.

If $\theta = \Lambda$ and $t^2 = 1$, there arise certain difficulties, since the hypergeometric functions in (4.6.8) will then become singular. By the transformation formulas of the hypergeometric function, (4.6.8) may be written in the neighborhood of $[(\Lambda + \theta)(2\,\Lambda)^{-1}]\,t^2 = 1$, in the form

$$E^+ = [a_3\,S_0/(2\,\Lambda)^{1/6}]\,F\,(1/6,\,5/6,\,3/2,\,X)\,+$$
$$+\,b_3\,S_0\,[2\,\Lambda/(2\,\Lambda - t^2\,(\Lambda + \theta))]^{1/2}\,F\,(1/3,\,-\,1/3,\,1/2,\,X), \qquad (4.6.10)$$

with now

$$X = 1 - t^2\,(\Lambda + \theta)/(2\,\Lambda).$$

In order to avoid the complications which arise from the fact that the second term of (4.6.10) has a singularity for $\Lambda = \theta$, $t^2 = 1$, we shall therefore take $b_3 = 0$. The function E^+ will accordingly be of the form

$$E^+ = [a_3\,S_0/(2\,\Lambda)^{1/6}]\,F\,(1/6,\,5/6,\,3/2,\,X). \qquad (4.6.11)$$

We note further that all these considerations can be repeated with Λ replaced by $-\,\Lambda$. Our operator will therefore yield two independent types of solutions, depending on whether the argument of the associate function is taken as $\Lambda + \theta$, or $\Lambda - \theta$. The exact conditions under which our operator can generate solutions of the compressibility equation in the supersonic case are given in the following theorem:

Suppose $f_s\,(\beta)$, $s = 1, 2$, are real functions of the real variable and everywhere differentiable with the possible exception of $\beta = 0$; suppose further that in a fixed neighborhood of $\beta = 0$, f_s can be approximated to any prescribed degree of accuracy by the expressions of the form

$$\sum_{n=1}^{N} A_n^{(N)}\,\beta^{K_n}, \qquad K_n \geqq 1, \qquad (4.6.12)$$

then

$$\Psi\,(\Lambda,\,\theta) = R_1^+\,(f_1) + R_2^+\,(f_2), \qquad (4.6.13)$$

$$R_s^+\,(f_s) = \int_{t=-1}^{+1} E^+\,(\Lambda,\,-\,(-\,1)^s\,\theta,\,t)\,f_s\left(\frac{1}{2}\,(\Lambda - (-\,1)^s\,\theta)\,(1 - t^2)\right)\,dt/(1 - t^2)^{1/2}, \qquad (4.6.14)$$

$$s = 1, 2,$$

represents a solution of the compressibility equation, which is defined for any $\Lambda > 0$ and can be interpreted as a stream function of a (possible) supersonic flow pattern. The proof of this theorem is given in [9]. In (4.6.14), the function E^+ is defined by (4.6.8) or (4.6.9) or (4.6.11), depending upon the value of $|X|$.

Before we proceed to investigate the behavior of our solutions on the sonic line, we note that (4.6.4a) is not the only solution which depends on one variable. The generating function (4.6.4) for the operator of the second kind may be replaced by

$$E^+\,(\lambda,\,\theta,\,t) = A_1\,S_0\,(-\,2\,\lambda)^{-1/6}\,F\,(1/6,\,5/6,\,1/2,\,X), \qquad (4.6.15)$$
$$X = t^2\,(\lambda - \lambda_0 + i\,\theta)/(2\,\lambda).$$

Besides its greater generality, the generating function (4.6.15) (with $\lambda_0 \neq 0$) has a number of additional features, which make it superior to (4.6.4) in many cases. In (4.6.4), the point $\lambda = 0$, $\theta = 0$, is a singularity since by approaching this point in a suitable manner, the argument X of the hypergeometric function can be given an arbitrary value. In (4.6.15), such a singularity does not exist, since λ and $\lambda - \lambda_0 + i\,\theta$ cannot vanish simultaneously if $\lambda_0 \neq 0$. Another singularity, which can be removed by using (4.6.15) with $\lambda_0 \neq 0$, occurs in the supersonic case. For $t^2 = 1$ and $\Lambda = \theta$, the second term of the right-hand side of (4.6.10) becomes singular. In order to allow for this case, we had to assume that $b_3 = 0$, thus somewhat restricting the generality of the solutions we could obtain. Setting $\lambda = i\,\Lambda$, it is seen that in the supersonic case the argument of the hypergeometric function in (4.6.15) becomes

$$X = t^2\,(\Lambda + i\,\lambda_0 + \theta)/(2\,\Lambda). \qquad (4.6.16)$$

For $t = \mp 1$, we have $X \neq 1$ for real values of Λ, θ; the singularity in question is therefore removed and the constant b_3 in (4.6.10) may now take any arbitrary value.

Thus, by the use of integral operators of the second kind with the generating functions (4.6.4), (4.6.8), (4.6.9), as well as (4.6.15), we obtain solutions $\Psi = I\,m\,[P_2^+\,(f)]$ which are defined in the subsonic and supersonic regions respectively. Let us repeat that if g (4.6.7) is defined in a domain G_1 in the subsonic region and has, as its only singularities in G_1, branch points of finite order (but not poles or logarithmic singularities), then the generated function is again defined in G_1. In particular, it has branch points at the same points and of the same order as g. If g is a twice-differentiable function of one real variable in a domain G_2 in the supersonic region, the generated function will be a solution of (4.5.5). Thus, by this procedure we can generate solutions of (4.5.2) which are defined in adjacent domains, one in the subsonic, the other in the supersonic region. Our object is to show that these solutions can be continued across the sonic line. This fact is seen immediately if we introduce a new variable $\bar{s}$

$$\begin{aligned} \bar{s} &= (-\lambda)^{2/3}, \quad &\text{for} \quad \lambda < 0, \\ \bar{s} &= -\Lambda^{2/3}, \quad &\text{for} \quad \Lambda > 0. \end{aligned} \qquad (4.6.17)$$

We note that in the simplified case we have:

$$\bar{s} = -2\cdot 3^{-2/3}\,[2/(k-1)]^{(2-k)/(3k-3)}\,H^+. \qquad (4.6.18)$$

The variable $\bar{s} = \bar{s}\,(M^+)$ (where M^+ denotes the fact that the simplified case is taken into account) considered as a function of Mach number M^+, possesses the property that $\bar{s}\,(1) = 0$, and that

$$d\bar{s}\,(M^+)/dM^+ = -2^{5/3}\,(3\,k+3)^{-2/3}\,M^+ + 0\,(1 - M^{+2}), \qquad (4.6.19)$$

is non-vanishing and bounded (finite) in a sufficiently small neighborhood of $M^+ = 1$. In (4.6.19), the symbol $0\,(1 - M^{+2})$ denotes the terms of order $(1 - M^{+2})$. More details concerning this particular item may be found in the original paper[9]. Thus our solutions can be continued across the sonic line, since by expressing both of them (one on each side of the sonic line) in terms of the one variable $\bar{s}$, we may easily show that the generated function $\Psi\,(\theta, \bar{s})$ is an analytic function of θ and $\bar{s}$ (and therefore of θ and M^+) at the point $\theta = 0$, $\bar{s} = 0$.

7. Integral Operator of the Second Kind in the Case of the "Exact" Compressibility Equation

As was shown in Part I, in the subsonic region, equation (4.5.5) has a solution in the form

$$\Psi = I\,m\,[P\,(f)], \qquad (4.7.1)$$

where

$$P\,(f) = \int_{-1}^{+1} E\,(Z, \overline{Z}, t)\,f\left(\tfrac{1}{2}\,Z\,(1 - t^2)\right)\,dt/(1 - t^2)^{1/2}. \qquad (4.7.2)$$

The generating function of the integral operator $P\,(f)$, which produces solutions of this equation, is

$$E = H\,(2\,\lambda)\,E^*, \qquad (4.7.3)$$

and the function E^* can be represented by a series

$$E^* = 1 + \sum_{n=1}^{\infty} (t^2 Z)^n Q^{(n)} (2\,\lambda), \tag{4.7.4}$$

where the $Q^{(n)}$'s are solutions of the system:

$$(2\,n + 1)\, Q_\lambda^{(n+1)} + Q_{\lambda\lambda}^{(n)} + 4\,F\,Q^{(n)} = 0, \tag{4.7.5}$$

$$Q^{(1)} = -\,4 \int_{-\infty}^{\lambda} F\,d\lambda, \tag{4.7.5a}$$

[for relation between λ, Λ and M see (1.3.13) and (3.2.8)] the function F being given by equation (4.5.8). The above series converges for $|Z| < 2\,|\lambda|$, or for $\theta^2 < 3\,\lambda^2$ and was thoroughly discussed in Part I. It is therefore desirable to obtain solutions which are defined in $|Z| > 2\,|\lambda|$ or $\theta^2 > 3\,\lambda^2$. Similarly, in the supersonic region a solution was presented which is convergent for $-\,3\,\Lambda < \theta < \Lambda$, if only the series $V^{(1)}$ is taken into account, or for $-\,\Lambda < \theta < \Lambda$, if the sum of the series $V^{(1)} + V^{(2)}$ is considered (see Section III.8). It is desirable, therefore, to obtain solutions which are defined for $-\,3\,\Lambda > \theta$ and for $\theta > \Lambda$. At first, we shall discuss the subsonic region. We shall present the following theorem without the proof which the reader may find in [9]:

Let

$$q^{(n, \varkappa)} = \sum_{\nu=0}^{\infty} C_\nu^{(n, \varkappa)} (-\,\lambda)^{n - \frac{1}{2} + (2/3)\,(\varkappa + \nu)}, \qquad \varkappa = 1, 2, \tag{4.7.6}$$

be a set of functions which are connected by the relations

$$q_{\lambda\lambda}^{(0, \varkappa)} + 4\,F\,q^{(0, \varkappa)} = 0, \tag{4.7.7a}$$

$$2 \left(n + \frac{2}{3}\,\varkappa\right) q_\lambda^{(n, \varkappa)} + q_{\lambda\lambda}^{(n+1, \varkappa)} + 4\,F\,q^{(n+1, \varkappa)} = 0, \tag{4.7.7b}$$

$$n = 1, 2, \ldots, \qquad \varkappa = 1, 2.$$

Then each of the functions

$$E^{*(\varkappa)} = \sum_{n=0}^{\infty} q^{(n, \varkappa)} / (-\,t^2 Z)^{n - \frac{1}{2} + (2/3)\,\varkappa}, \tag{4.7.8}$$

multiplied by H is a generating function of the integral operator $P\,(f)$ and each of these series converges in $E\,[2\,|\lambda| < |Z|]$.

In order to obtain a continuation of a given stream function to the supersonic region, replace λ by the variable $\bar{s} = (-\,\lambda)^{2/3}$. Then the generating function in the subsonic case may be written as

$$E^{(\varkappa)}\,(\lambda, \theta, t) = H\,(2\,\lambda)\,E^{*(\varkappa)}\,(\lambda, \theta, t) =$$

$$= 2^{-1/6} S_0\,p\,(\bar{s})\,\bar{s}^{\varkappa - 1} \sum_{n=0}^{\infty} \sum_{\nu=0}^{\infty} C_\nu^{(n, \varkappa)}\,\bar{s}^{(3/2)\,n}\,\bar{s}^\nu / (-\,t^2 Z)^{n - \frac{1}{2} + (2/3)\,\varkappa}, \tag{4.7.9}$$

for

$$|\bar{s}| < \bar{s}_0, \qquad 2\,|\bar{s}|^{3/2} < |Z|, \qquad \varkappa = 1, 2,$$

where

$$Z = (-\,\bar{s})^{3/2} + i\,\theta, \tag{4.7.9a}$$

and the domain of regularity is a circle of radius $\bar{s}_0$, with the center at the origin. By replacing λ by $i\,\Lambda$ we see that $(-\,\lambda)^{2/3}$ is changed to $-\,\Lambda^{2/3}$. Thus we obtain as the generating function in the supersonic case,

$$E^{(\varkappa)}\,(\Lambda, \theta, t) = 2^{-1/6} S_0\,p\,(\bar{s})\,\bar{s}^{\varkappa - 1} \sum_{n=0}^{\infty} \sum_{\nu=0}^{\infty} i^n \cdot C_\nu^{(n, \varkappa)}\,(-\,\bar{s})^{(3/2)\,n}\,\bar{s}^\nu / (-\,t^2 Z)^{n - \frac{1}{2} + (2/3)\,\varkappa},$$

$$\tag{4.7.10}$$

for

$$|\bar{s}| < \bar{s}_0, \qquad 2\,|\bar{s}|^{3/2} < |Z|, \qquad \varkappa = 1, 2,$$

where

$$Z = i\,(-\bar{s})^{3/2} + i\,\theta. \qquad\qquad (4.7.10a)$$

As mentioned above, the proof is given in [9].

Let us now restrict ourselves to a neighborhood of the point $\lambda = 0$, $\theta = \theta_0$, $\theta_0 \neq 0$, lying entirely in the domain $D = E\,[|\bar{s}| < \bar{s}_0,\ |\bar{s}| < 3^{-1/3}\,|\theta|^{2/3}]$. We may show that if f is regular in D, then both the generating function and the associate function, and therefore the expression $P\,[f\,(-\bar{s}^{3/2} + i\,\theta)]$ in the subsonic case, may be expanded in integral powers of $(\theta - \theta_0)$ and $\bar{s}^{1/2}$. Similarly, in the supersonic case, the expression $P\,[f\,(i\,(-\bar{s})^{3/2} + i\,\theta)]$ may be expanded in integral powers of $(\theta - \theta_0)$ and $(-\bar{s})^{1/2}$. One may also show that both functions P (in sub- and supersonic regions) are determined and equal to each other for $\bar{s} = 0$. In the next step, we may show that if f is regular in D, then the solution $P\,[f\,(-\bar{s}^{3/2} + i\,\theta)]$ is an analytic function of $\bar{s}$ and $(\theta - \theta_0)$ at $(0, \theta_0)$. Thus

$$P\,[f\,(-\bar{s}^{3/2} + i\,\theta)] = \sum_{n=0}^{\infty} f_n\,(\theta - \theta_0)\,\bar{s}^n. \qquad\qquad (4.7.11)$$

Since $\bar{s}$ is unchanged by the result of putting $i\,\Lambda$ for λ, we obtain

$$P\,[f\,(i\,(-\bar{s}^{3/2} + i\,\theta)] = \sum_{n=0}^{\infty} f_n\,(\theta - \theta_0)\,\bar{s}^n. \qquad\qquad (4.7.12)$$

The expressions $P\,[f\,(-\bar{s}^{3/2} + i\,\theta)]$ and $P\,[f\,(i\,(-\bar{s})^{3/2} + i\,\theta)]$, equal functions of $\bar{s}$ and $(\theta - \theta_0)$, are analytic continuations of each other across the sonic line. Thus, assuming that the associate function is regular in a sufficiently large domain, and applying the integral operator of the second kind, we obtain solutions of a compressibility equation defined in four adjacent domains, namely

$$D_1 = E\,[M < 1,\quad \theta > 3^{1/2}\,|\lambda\,(M)|] + E\,[M > 1,\quad \theta > \Lambda\,(M)],$$

$$D_2 = E\,[M < 1,\quad \theta < 3^{1/2}\,|\lambda\,(M)|],$$

$$D_3 = E\,[M < 1,\quad \theta < -\,3^{1/2}\,|\lambda\,(M)|] + E\,[M > 1,\quad \theta < -\,3\,\Lambda\,(M)],$$

$$D_4 = E\,[M > 1,\quad -\,3\,\Lambda\,(M) < \theta < \Lambda\,(M)].$$

The solutions defined in D_1 and D_3 were presented in the present section, while those defined in D_2 and D_4 were derived in Parts I and III, respectively.

In the simplified case, using the theory of hypergeometric equations, it was possible to combine these representations into one, which yields solutions of (4.5.2) defined in the whole $(M, -\,\theta)$-plane. In the exact compressibility equation, the problem remains of combining these four representations into one. This problem can be attacked by using the integral operator of the first kind in addition to that of the second kind, and, in analogy to the simplified case, developing a theory of differential equations with singular coefficients, which would furnish us with information corresponding to that used in the simplified case. More information concerning this item may be found in [9].

We should like to add here a remark regarding the general question of analytic continuation of a solution $\Psi\,(Z, \bar{Z})$ of a linear partial differential equation. If Ψ is given in two different domains, say B_1 and B_2, by different representations, say in B_1 by the integral operator of the first kind in the form

$$\Psi = \Psi_2 \equiv p\,[g\,(Z)] + \bar{p}\,[h\,(\bar{Z})], \qquad\qquad (4.7.13)$$

[see (4.5.20)], and in B_2 by another operator

$$\Psi = \Psi_2 \equiv \int_{t=-1}^{+1} E\left(Z, \overline{Z}, t\right) f\left(\frac{1}{2} Z \left(1 - t^2\right)\right) \left(dt/(1 - t^2)^{1/2}\right), \qquad (4.7.14)$$

[see (4.5.16)], (not necessarily of the first kind), and the origin lies in the intersection of both domains B_1 and B_2, which may be denoted mathematically by a symbol $B_1 \cap B_2$, then the problem of the analytic continuation of Ψ_2 into the domain B_1 is equivalent to the determination of g and h from a given $f = \sum\limits_{n=0}^{\infty} \alpha_n \, Z^n$. One may show[12] that setting $Z = 0$, and $\overline{Z} = 0$, respectively, in the relation

$$p \left[g\left(Z\right)\right] + \overline{p} \left[h\left(\overline{Z}\right)\right] = \int_{t=-1}^{+1} E\left(Z, \overline{Z}, t\right) f\left(\frac{1}{2} Z \left(1 - t^2\right)\right) dt/(1 - t^2)^{1/2}, \qquad (4.7.15)$$

we obtain the identities

$$\sum_{n=0}^{\infty} Z^n \sum_{\nu=0}^{n} \tau_{n\nu}^{(1)} \, \alpha_\nu = g\left(Z\right) + \overline{R}\left(0, Z\right) h\left(0\right), \qquad (4.7.16)$$

$$\alpha_{00} \sum_{n=0}^{\infty} \overline{Z}^n \tau_{nn}^{(2)} = R\left(0, \overline{Z}\right) g\left(0\right) + h\left(\overline{Z}\right), \qquad (4.7.17)$$

where

$$\tau_{n\nu}^{(1)} = \int_{t=-1}^{1} E_{n-\nu}^{(1)} \left(1 - t^2\right)^{\nu - 1/2} 2^{-\nu} \, dt, \qquad (4.7.18)$$

$$\tau_{nn}^{(2)} = \int_{t=-1}^{1} E_n^{(2)} \left(t\right) \left(1 - t^2\right)^{-1/2} \, dt, \qquad (4.7.19)$$

$$E\left(Z, 0, t\right) = \sum_{n=0}^{\infty} E_n^{(1)} \left(t\right) Z^n, \qquad (4.7.20)$$

$$E\left(0, \overline{Z}, t\right) = \sum_{n=0}^{\infty} E_n^{(2)} \left(t\right) \overline{Z}^n. \qquad (4.7.21)$$

It is thus seen that the analytic continuation of Ψ_2 into B_1 and all the problems arising from it, such as the determination of the singularities, etc., are reduced to similar problems in the theory of functions of one complex variable which are given by their power series expansions.

One more item may be mentioned briefly, and this is the determination of the associate function in terms of the given stream function. There are possible cases in which the stream function is given in the form of a power series, say, and we want to have the associate function. In many instances, the stream function Ψ is given in some different form, say the values of Ψ and $\partial\Psi/\partial M$ are given on a line $M = $ const. If these quantities are analytic functions of θ, then from these data it is possible to determine the associate functions. Problems of this type are sometimes of importance. In [9] one may find the procedure for the derivation of the formula expressing f in terms of the values $\Psi = \chi_1\left(\theta\right)$ and $\partial\Psi/\partial M = \chi_2\left(\theta\right)$ on the sonic line, $M = 1$.

8. Combination of the Integral Operator and Chaplygin's Solution

We now want to indicate another procedure which aims at constructing "mixed" flows. In applying this method, we use in addition to the integral operator,

---

[12] S. Bergman: Certain classes of analytic functions of two real variables and their properties. Trans. Amer. Math. Soc. 57, Vol. 299—331 (1945).

Chaplygin's solutions. Suppose that $g(Z^4)$ is a complex potential which in the (incompressible) physical plane yields a flow past a closed curve. For the sake of simplicity, let us assume that the flow is symmetric, and the complex potential (in the logarithmic plane) is a two-valued function, possessing as its only singularity a branch point of the second order at the point λ_∞ corresponding to the flow at infinity (see the section on singularities for more details on this subject). Employing the operator of the first kind, we obtain a solution of equation $\Psi^{(1)} = I\,m\,[p\,(g)]$ in the subsonic region. For the illustrative purposes, let us assume that this solution will not be determined at first in its entire domain of definition, but only in a portion, say D, of this domain, namely in the portion situated in $[\lambda \leq \lambda_0]$, $\lambda_0 < 0$, say, so that, in general, the boundary curve $\Psi^{(1)} = 0$ is not closed in D. Let us assume that $\lambda_\infty < \lambda_0$ holds. We shall now describe a procedure to determine the analytic continuation of the solution $\Psi^{(1)}$ outside of D. Since in the following we have also to operate in the supersonic region, and as we shall employ Chaplygin's solutions, it is convenient to introduce, instead of λ, the variable $\tau = (k - 1)\,q^2/(2\,a_0{}^2)$, where a_0 is the speed of sound at a stagnation point. Let $\tau_\infty = \tau\,(\lambda_\infty)$, and let us introduce Chaplygin's solution:

$$(A_\nu\,F_\nu + B_\nu\,F_\nu{}^*)\,\tau^{\nu/(2L)}\,\exp\,[i\,(\pi\,\nu\,\theta)/L], \tag{4.8.1}$$

where

$$F_\nu = F\,(\alpha_\nu,\ \beta_\nu,\ -\beta,\ 1 - \tau), \tag{4.8.2}$$

$$F_\nu{}^* = (1 - \tau)^{\beta+1}\,F\,(\gamma_\nu - \alpha_\nu,\ \gamma_\nu - \beta_\nu,\ 2 + \beta,\ 1 - \tau), \tag{4.8.3}$$

F being the hypergeometric series. Here

$$\gamma_\nu = \left(\frac{\nu}{L} + 1\right), \quad \beta = (k - 1)^{-1}, \quad \alpha_\nu = \frac{1}{2}\left[\left(\frac{\nu}{L} - \beta\right) + \Delta_\nu\right], \tag{4.8.4a}$$

$$\beta_\nu = \frac{1}{2}\left[\left(\frac{\nu}{L} - \beta\right) - \Delta_\nu\right], \quad \Delta_\nu = \left[\left(\frac{\nu}{2L}\right)^2 (2\beta + 1) + \frac{1}{4}\,\beta^2\right]^{1/2}. \tag{4.8.4b}$$

We assumed, above, that we know the solution $\Psi^{(1)}\,(\tau,\theta) = \Psi^{(1)}\,(\lambda\,(\tau),\theta)$ in D. In order to compute the continuation $\Psi^{(2)}$ of $\Psi^{(1)}$, obtained in D by the use of the integral operator, into the domain B_1, say, we write $\Psi^{(2)}$ in the form:

$$\Psi^{(2)} = \sum_{\nu=1}^\infty \left[(A_\nu^{(1)}\,F_\nu + B_\nu^{(1)}\,F_\nu{}^*)\,\tau^{\nu/(2L)}\,\cos\left(\frac{\pi\,\nu\,\theta}{L}\right) + (A_\nu^{(2)}\,F_\nu + B_\nu^{(2)}\,F_\nu{}^*)\,\tau^{\nu/(2L)}\,\sin\left(\frac{\pi\,\nu\,\theta}{L}\right)\right],$$
$$\tag{4.8.5}$$

and determine the coefficients $A_\nu^{(\varkappa)}$, $B_\nu^{(\varkappa)}$, $\varkappa = 1, 2$, $\nu = 0, 1, 2, \ldots$, so that on $\tau = \tau_0$, $-L \leq \theta \leq L$ (where $\mp L$ denote the lower and upper bounds of θ),

$$\Psi^{(1)}\,(\tau_0,\theta) = \Psi^{(2)}\,(\tau_0,\theta), \tag{4.8.6}$$

$$\Psi_\tau^{(1)}\,(\tau_0,\theta) = \Psi_\tau^{(2)}\,(\tau_0,\theta), \tag{4.8.7}$$

holds. It can be shown that the series (4.8.5) converges in a certain strip, say $[\tau_0 \leq \tau \leq \tau_1,\ -L_1 \leq \theta \leq L_1]$. By replacing τ_0 by τ_1, L by L_1, and repeating the same procedure, we may continue the function to a range of values $[\tau_1 \leq \tau \leq \tau_2,\ -L_2 \leq \theta \leq L_2]$, and so on. If we employ the integral operator of the first kind, under the assumption that the function g in the expression $\Psi^{(1)} = I\,m\,[p\,(g)]$ is regular in a sufficiently large domain, we can prove the existence of Ψ in the whole subsonic region, as well as the fact that by this procedure the function Ψ can be determined in the subsonic portion of the domain of definition of the flow. If the flow can be continued to the supersonic region, then the procedure described also yields the continuation to that region. If, on the other hand, we replace the infinite series (4.8.5) by a finite one (in doing this, we may require that

$$|\Psi^{(1)}(\tau_0,\theta) - \Psi^{(2)}(\tau_0,\theta)| \leqq \varepsilon, \qquad (4.8.8\,\mathrm{a})$$

$$|\Psi_\tau^{(1)}(\tau_0,\theta) - \Psi_\tau^{(2)}(\tau_0,\theta)| \leqq \varepsilon, \qquad (4.8.8\,\mathrm{b})$$

where ε is any preassigned quantity, however small), then the finite sum is defined everywhere, and if $\Psi = 0$ is a closed curve, we obtain in this way a "mixed" solution of the exact equation, which together with its normal derivative has a jump (of the value $\leqq \varepsilon$) on the line $\tau = \tau_0$. This jump, however, can be made arbitrarily small.

9. Tables for the Determination of Transonic Flow Patterns. Alternative Formulas

The preparation of tables for the determination of transonic flow patterns needs some explanation, which will be given below. For a better understanding, let us begin with an incompressible fluid. A plane is introduced whose Cartesian coordinates are the variables θ and

$$\tilde{\eta} = \int_1^q \tilde{\varrho}_0\, q^{-1}\, dq = \log q, \qquad (4.9.1)$$

with $\tilde{\varrho}_0 = 1$ for an incompressible fluid. The complex potential for a large class of symmetric obstacles in these variables can be written in the form

$$g(\tilde{z}) = \sum_{\nu=0}^{\infty} \tilde{b}_\nu\, (\tilde{z} - \eta_0)^{\nu - 1/2}, \qquad (4.9.2)$$

with $\tilde{z} = \tilde{\eta} + i\,\theta$ and η_0 equal to the logarithm of the speed at infinity. In the case of symmetric obstacles, the $\tilde{b}_\nu$'s are purely imaginary. The representation (4.9.2) holds only in a circle with origin at $\tilde{\eta} = \eta_0$ and of radius equal to the distance from $\eta = \eta_0$ to the nearest singularity. However, by employing suitable summation methods, e. g.,

$$g(\tilde{z}) = \lim_{s \to \infty} \sum [\tilde{b}_\nu / \Gamma(1 + s_\nu)]\, [\tilde{z} - \eta_0]^{\nu - 1/2}, \qquad (4.9.3)$$

where the symbol Γ denotes the Gamma function, a representation is obtained for $g(\tilde{z})$ in a much larger domain (for details, see [12a]). With the constants b_ν purely imaginary, the stream function $\widetilde{\Psi}$ can be written

$$\widetilde{\Psi} = Im\,[g(\tilde{z})] = \sum_{\nu=0}^{\infty} i\,\tilde{b}_\nu\, \widetilde{\Psi}^{(\nu)}(\tilde{\eta}, \theta;\, \eta_0), \qquad (4.9.4)$$

$$\widetilde{\Psi}^{(\nu)}(\tilde{\eta}, \theta;\, \eta_0) = Re\,[\tilde{z} - \eta_0]^{\nu - 1/2}. \qquad (4.9.4\,\mathrm{a})$$

The streamlines of the flow patterns corresponding to $g(\tilde{z})$ in the physical plane are given by

$$x = \sum_{\nu=0}^{\infty} i\,b_\nu\, \widetilde{X}^{(\nu)}(\tilde{\eta}, \theta;\, \eta_0), \qquad (4.9.5\,\mathrm{a})$$

$$y = \sum_{\nu=0}^{\infty} i\,b_\nu\, \widetilde{Y}^{(\nu)}(\tilde{\eta}, \theta;\, \eta_0), \qquad (4.9.5\,\mathrm{b})$$

where

$$\widetilde{X}^{(\nu)}(\tilde{\eta}, \theta;\, \eta_0) = \int \tilde{\varrho}_0^{-1}\, e^{\eta_0}\, \{[- \widetilde{\Psi}_\theta^{(\nu)} \cos\theta -$$
$$- \widetilde{\Psi}_{\tilde{\eta}}^{(\nu)} \sin\theta]\, d\tilde{\eta} + [\widetilde{\Psi}_{\tilde{\eta}}^{(\nu)} \cos\theta - \widetilde{\Psi}_\theta^{(\nu)} \sin\theta]\, d\theta\}, \qquad (4.9.6\,\mathrm{a})$$

$$\widetilde{Y}^{(\nu)}(\tilde{\eta}, \theta;\, \eta_0) = \int \tilde{\varrho}_0^{-1}\, e^{\eta_0}\, \{[- \widetilde{\Psi}_\theta^{(\nu)} \sin\theta +$$
$$+ \widetilde{\Psi}_{\tilde{\eta}}^{(\nu)} \cos\theta]\, d\tilde{\eta} + [\widetilde{\Psi}_{\tilde{\eta}}^{(\nu)} \sin\theta + \widetilde{\Psi}_\theta^{(\nu)} \cos\theta]\, d\theta\}. \qquad (4.9.6\,\mathrm{b})$$

[12a] F. Lindeloef: Calcul des residues. Paris (1905).

The coefficients b_ν depend on the boundary curve of the profile. In employing the hodograph method in the incompressible fluid case, it is convenient to prepare a set of tables for

$$\widetilde{\Psi}^{(\nu)}\,(\tilde{\eta},\,\theta;\,\eta_0),\quad \widetilde{X}^{(\nu)}\,(\tilde{\eta},\,\theta;\,\eta_0),\quad \widetilde{Y}^{(\nu)}\,(\tilde{\eta},\,\theta;\,\eta_0),\tag{4.9.7}$$

for a number of values of the parameter η_0. For, by using a set of b_ν's appropriate to the given obstacle, $\widetilde{\Psi}\,(\tilde{\eta},\,\theta;\,\eta_0)$ could immediately be determined by making use of equations (4.9.4) and (4.9.7). To find the image of a streamline,

$$\widetilde{\Psi}\,(\tilde{\eta},\,\theta;\,\eta_0) = s = \text{const.},\tag{4.9.8}$$

in the physical plane, a sufficiently dense set of points $(\tilde{\eta}_s,\,\theta_s) = 1, 2, \ldots$, is first found, corresponding to the streamline $\widetilde{\Psi}\,(\tilde{\eta},\,\theta;\,\eta_0) = s$. From the tables, the values are read for

$$\widetilde{X}^{(\nu)} = \widetilde{X}^{(\nu)}\,(\tilde{\eta}_s,\,\theta_s;\,\eta_0),\quad \widetilde{Y}^{(\nu)} = \widetilde{Y}^{(\nu)}\,(\tilde{\eta}_s,\,\theta_s;\,\eta_0),\tag{4.9.9}$$

and by using (4.9.5), the image $\widetilde{\Psi}\,(x,\,y) = s$ of $\widetilde{\Psi}\,(\tilde{\eta},\,\theta;\,\eta_0) = s$ is found in the physical plane.

The essence of this method is (as explained in previous sections) the separation of the quantities $\widetilde{X}^{(\nu)}$, $\widetilde{Y}^{(\nu)}$, $\widetilde{\Psi}^{(\nu)}$, that are independent of the profile from the b_ν's which depend on it. For the former, a set of tables can be prepared once and for all. The problem of determining the latter recurs with each profile. This usually is a relatively simple matter, in the case of incompressible flow. In the compressible case, the problem is much more involved, but in most cases of interest, the b_ν's found for the incompressible case may be used in the compressible case as a first approximation to the exact b_ν's. Since the published tables were calculated for solutions having forms a little different from those cited above, we shall present these alternative formulas. The equation corresponding to (4.5.11) now has the form

$$\Psi_{\eta\eta} + C\,\eta\,\Psi_{\theta\theta} = 0,\quad C = -\,2\,[2/(k+1)]^{(2-k)/(k-1)},\tag{4.9.10}$$

with $\eta \equiv H$, and the complex solutions of (4.9.10) are given in the form:

$$P^*\,[f\,(z)] = \int_{-1}^{+1} E^*\,(\eta,\,\theta;\,\eta_0)\,f\,[(z/2)\,(1-t^2)]\,[dt/(1-t^2)^{1/2}],\tag{4.9.11a}$$

or

$$p^*\,[g\,(z)] = S_0\,(-\eta)^{-1/4}\,\{g\,(z) + \sum (1/2^{2n})\,[\Gamma(2n+1)/\Gamma(n+1)]\,P^{(n)}\,(z)\,g^{(n)}\,(z)\},\tag{4.9.11b}$$

where $z = \eta + i\,\theta$, and $g\,(z)$ is an analytic function of z. The function E^* may be taken in the form

$$E^*\,(\eta,\,\theta;\,\eta_0) = S_0\,(-\eta)^{-1/4}\,\overline{F}\{1/6,\,5/6,\,1/2,\,X\},\tag{4.9.12}$$

where $\overline{F}\,(\alpha,\,\beta,\,\gamma,\,X)$ is the hypergeometric function and the other symbols used are

$$X = t^2\,[c\,(-\eta)^{3/2} + i\,\theta - \eta_0]/2\,c\,(-\eta)^{3/2},\tag{4.9.13}$$

$$c = (2^{3/2}/3)\,[(k-1)/2]^{(k-2)/[2\,(k-1)]}.\tag{4.9.13a}$$

The quantity S_0 is given by (4.6.3a), and the polynomials $P^{(n)}\,(z)$ are defined by the recursion formula and initial conditions:

$$P_z^{(1)} + 2\,F^* = 0,\quad (2n+1)\,P_z^{(n+1)} + 2\,P_{z\bar{z}}^{(n)} + 2\,F^*\,P^{(n)} = 0,\tag{4.9.14}$$

$$P^{(n)}\,(0) = 0,\tag{4.9.14a}$$

with

$$F^* = (5/36)\,(1/4\,c^2)\,(-\eta)^{-3}.\tag{4.9.15}$$

The functions $g^{(n)}(z)$ may be written as

$$g^{(n)}(z) = \int\limits_0^z \int\limits_0^{z_1} \ldots \int\limits_0^{z_{n-1}} g(z_n)\,dz = [(-1)^n/(n-1)!] \int\limits_{t=0}^{t=z} (z-t)^{n-1} g(t)\,dt. \qquad (4.9.16)$$

The stream function is defined as the imaginary part, i. e., $\Psi = I\,m\,[p^*(g)]$. A form similar to that one in (4.9.4) may be given, namely

$$p^*[g(z)] = \sum_{\nu=0}^{\infty} b_\nu\, p^* \{(z - \eta_0)^{\nu - 1/2}\}, \qquad (4.9.17)$$

where

$$p^* \{(z - \eta_0)^{\nu - 1/2}\} = B^{(\nu)}[p^*(z), \overline{p^*(z)}; \eta_0]\,[p^*(z) - \eta_0]^{\nu - 1/2}, \qquad (4.9.18)$$

$$p^*(z) = p^*(\eta + i\,\theta) = -c\,(-\eta)^{3/2} + i\,\theta = \mathfrak{z}, \qquad (4.9.18a)$$

$$B^{(\nu)}[p^*(z), \overline{p^*(z)}; \eta_0] = B^{(\nu)}[\mathfrak{z}, \bar{\mathfrak{z}}; \eta_0] = S_0\,(-\eta)^{-1/4}\,F\,[1/6, 5/6, \nu + 1/2, X], \qquad (4.9.18b)$$

$$X = (\mathfrak{z} - \eta_0)/[(-2\,c)\,(-\eta)^{3/2}]. \qquad (4.9.18c)$$

It is seen, then, that in the case of the stream function, the transition from the incompressible to the compressible case is made by multiplying each term of its power series expansion by a "compressibility factor" $B^{(\nu)}(\mathfrak{z}, \bar{\mathfrak{z}}; \eta_0)$ and by replacing the variable $z = \eta + i\,\theta$ by the variable $\mathfrak{z} = -c\,(-\eta)^{3/2} + i\,\theta$. This latter change implies a distortion in the η direction mentioned above. Corresponding to expression (4.9.4) for the incompressible case, the stream function for a symmetric profile can be written in the form

$$\Psi(\eta,\,\theta;\,\eta_0) = \sum i\,b_\nu\,\Psi^{(\nu)}(\eta,\,\theta;\,\eta_0), \qquad (4.9.19)$$

where

$$\Psi^{(\nu)} = Re\,[B^{(\nu)}(\mathfrak{z}, \bar{\mathfrak{z}};\,\eta_0)\,(\mathfrak{z} - \eta_0)^{\nu - 1/2}]. \qquad (4.9.20)$$

The implicit representation of the streamline $\Psi(\eta,\,\theta;\,\eta_0) = s$ in the physical plane is given by

$$x = \sum i\,b_\nu\,X^{(\nu)}(\eta_s,\,\theta_s;\,\eta_0), \quad y = \sum i\,b_\nu\,Y^{(\nu)}(\eta_s,\,\theta_s;\,\eta_0), \qquad (4.9.21)$$

with

$$X^{(\nu)} = \int A_1\,[(-B_1\,\Psi_\theta^{(\nu)}\cos\theta - \Psi_\eta^{(\nu)}\sin\theta)\,d\eta + (C_1\,\Psi_\eta^{(\nu)}\cos\theta - \Psi_\theta^{(\nu)}\sin\theta)\,d\theta], \qquad (4.9.22)$$

$$Y^{(\nu)} = \int A_1\,[(-B_1\,\Psi_\theta^{(\nu)}\sin\theta - \Psi_\eta^{(\nu)}\cos\theta)\,d\eta + (C_1\,\Psi_\eta^{(\nu)}\sin\theta + \Psi_\theta^{(\nu)}\cos\theta)\,d\theta], \qquad (4.9.23)$$

$$A_1 = [\varrho\,\exp\,(\int \sqrt{-l^*(\eta)}\,d\eta)]^{-1}, \qquad (4.9.24a)$$

$$B_1 = (1 - M^2)^{1/2}\,[-l^*(\eta)]^{1/2}, \qquad (4.9.24b)$$

$$C_1 = (1 - M^2)^{1/2}\,[\sqrt{-l^*(\eta)}]^{-1}, \qquad (4.9.24c)$$

$$l^*(\eta) = C\,\eta. \qquad (4.9.24d)$$

In order to evaluate the "compressibility factors" $B^{(\nu)}(\eta,\,\theta;\,\eta_0)$, suitable power series representations must be found for the hypergeometric function appearing in (4.9.18b). It is well known that for $|X| < 1$, $\overline{F}\,(\alpha, \beta, \gamma, X)$ may be represented by a hypergeometric series $F\,(\alpha, \beta, \gamma, X)$ and for $|X| > 1$ a combination of two hypergeometric series with arguments and parameters different from those appearing in the hypergeometric function may be used. In the case of the hypergeometric function appearing in (4.9.18b), the boundary separating the domains of convergence is

$$|[- c (- \eta)^{3/2} - \eta_0 + i \theta]/[- 2 c (- \eta)^{3/2}]| = 1. \qquad (4.9.25)$$

Condition (4.9.25) may be presented in the form:

$$3 c^2 \eta^3 + \theta^2 + 2 c (- \eta)^{3/2} \eta_0 + \eta_0{}^2 = 0, \quad \text{for } \eta < 0, \qquad (4.9.26a)$$

$$3 c^2 \eta^3 - \theta^2 - 2 c (- \eta)^{3/2} \eta_0 - \eta_0{}^2 = 0, \quad \text{for } \eta > 0. \qquad (4.9.26b)$$

In these domains, the following representations are used respectively:

$$\overline{F} (1/6, 5/6, \nu + 1/2, X) = F (1/6, 5/6, \nu + 1/2, X), \qquad (4.9.27)$$

$$\overline{F} (1/6, 5/6, \nu + 1/2, X) = A_\nu (1/X)^{1/6} F (1/6, 2/3 - \nu, 1/3, 1/X) +$$
$$+ B_\nu (1/X)^{5/6} F (5/6, 4/3 - \nu, 5/3, 1/X), \qquad (4.9.28)$$

where the quantity X is given by (4.9.18c). The A_ν's and B_ν's denote the quantities:

$$A_\nu = \Gamma (- 2/3) \Gamma (\nu + 1/2)/[\Gamma (- 1/3 - \nu) \Gamma (5/6)], \qquad (4.9.29a)$$

$$B_\nu = \Gamma (\nu + 1/2) \Gamma (2/3)/[\Gamma (1/3 - \nu) \Gamma (1/6)]. \qquad (4.9.29b)$$

It has been shown above how it is possible to extend the methods used in the case of an incompressible fluid to the case of a compressible one. The analogues of equations (4.9.2), (4.9.4), (4.9.5), and (4.9.6) of the incompressible flow were obtained, namely equations (4.9.17), (4.9.19), (4.9.21), (4.9.22) and (4.9.23). These permit the setting up of tables for a compressible fluid of the type that was outlined for an incompressible fluid.

In the literature [13] one may find two types of tables. Type I depends upon a parameter η_0 and has to be prepared for a number of values of η_0, say $\eta_{0,\nu}$ $(\nu = 1, \ldots, n)$. In this case, one must resort to interpolation methods in order to obtain streamlines for intermediary values of η_0: $\eta_{0,\nu} < \eta_0 < \eta_{0,\nu+1}$. A table of $\Psi^{(\nu)} (\eta, \theta; \eta_0)$ (4.9.20) for a specific value of η_0 $(\eta_0 = - 0{\cdot}02)$ is given in [13]. In the same paper, a table is also included for

$$\varphi^{(\nu)} (\eta, \theta; - 0{\cdot}02) = I m \{p^* [(z - 0{\cdot}02)^{\nu - 1/2}]\}, \qquad (4.9.30)$$

in view of possible later applications, as well as a conversion table from the variable η to the local Mach number M.

The second type of tables which may be calculated are the tables containing the values of $\overline{F} (1/6, 5/6, \nu + 1/2, \alpha + i \beta)$. These values may be computed once and for all, as well as derivatives of $\overline{F}$ with respect to the argument for a number of values of α and β, where

$$\alpha = 1/2 - \eta_0/[- 2 c (- \eta)^{3/2}], \quad \beta = \theta/[- 2 c (- \eta)^{3/2}]. \qquad (4.9.31)$$

By the use of tables of fractional powers and trigonometric functions, and the tables outlined above, the computation can now be made of (4.9.20), namely:

$$\Psi^{(\nu)} = Re \{S_0 (- \eta)^{-1/4} [- c (- \eta)^{3/2} - \eta_0 + i \theta]^{\nu - 1/2} \cdot \overline{F}, \qquad (4.9.32)$$

$$\overline{F} = \overline{F} (1/6, 5/6, \nu + 1/2, \alpha + i \beta). \qquad (4.9.32a)$$

Similarly, the derivatives $\Psi_\eta^{(\nu)} (\eta, \theta; \eta_0)$ and $\Psi_\theta^{(\nu)} (\eta, \theta; \eta_0)$ can be calculated. The stream functions in the hodograph plane are obtained by forming $\Psi (\eta, \theta; \eta_0) = \sum i b_\nu \Psi^{(\nu)} (\eta, \theta; \eta_0)$. The integrands of equations (4.9.22) and (4.9.23) may be evaluated now by using the tables for the derivatives of $\overline{F}$. Since the integrands of (4.9.22) and (4.9.23) are perfect differentials, the value of the integral is independent of the path and it may conveniently be taken along a streamline. The transition to the physical plane is carried out by numerical integration of (4.9.22) and (4.9.23).

In [13] one may find the application of the tables, mentioned above, to the calculation of the flow around an oval-shaped obstacle in the compressible flow domain corresponding to the flow around an ellipse in the incompressible flow domain. The example shows clearly the entire procedure.

Part V

Axially Symmetric Flow

1. General Equations, Explanation of the Method of Infinite Approximation by a Sequence of Linear Equations

Perhaps the most basic, and consequently the most important, particular case of a spatial flow is an axially symmetric spatial flow, as for example flow past bodies of revolution. Let x, r, χ denote the cylindrical coordinates with the x-axis coincident with the axis of symmetry of the flow. The velocity vector then no longer depends on the rotation angle χ, but rather entirely on x and r, and always lies in a meridian plane (plane through the x-axis). As a result, the flow in all meridian planes is the same, and needs to be studied only in a given meridian plane. Below we shall apply the hodograph transformation. Thus, x, r denote the cylindrical coordinates in the physical plane in which the flow actually takes place, whereas the corresponding velocity components, u, v, determine the cylindrical coordinates in the hodograph plane. The velocity potential equation of an axially symmetric flow in the physical plane has the form:

$$\varphi_{xx}\left(1 - \varphi_x^2/a^2\right) + \varphi_{rr}\left(1 - \varphi_r^2/a^2\right) - 2\,\varphi_{xr}\,\frac{\varphi_x\,\varphi_r}{a^2} + \varphi_r/r = 0, \qquad (5.1.0)$$

where φ is the velocity potential and the subscripts denote partial differentiation. As is known, the following relations hold: $u = \varphi_x$, $v = \varphi_r$. When transforming this equation into the hodograph plane, we introduce a new function Φ connected with φ by means of the formula:

$$\Phi = x\,\varphi_x + r\,\varphi_r - \varphi. \qquad (5.1.1)$$

Then in the hodograph plane (u, v), the equation for the function Φ is of the form

$$L(\Phi) - N(\Phi) = 0, \quad \text{or} \quad L(\Phi) = N(\Phi), \qquad (5.1.2)$$

where

$$L(\Phi) = \left(1 - \frac{v^2}{a^2}\right)\Phi_{uu} + \left(1 - \frac{u^2}{a^2}\right)\Phi_{vv} + 2\,\frac{uv}{a^2}\,\Phi_{uv}, \qquad (5.1.2a)$$

$$N(\Phi) = -\frac{v}{\Phi_v}\left(\Phi_{uu}\Phi_{vv} - \Phi_{uv}^2\right) = -\frac{v}{\Phi_v}\,\frac{\partial(\Phi_u, \Phi_v)}{\partial(u, v)}, \qquad (5.1.2b)$$

$$a^2 = \gamma\left(\hat{q}^2 - u^2 - v^2\right), \qquad \gamma = \frac{1}{2}(k - 1). \qquad (5.1.2c)$$

The quantities γ and $\hat{q} = q_{max}$ are constants. We note that in polar coordinates we have $q = (u^2 + v^2)^{1/2}$, $\theta = \arctan(v/u)$ and equations (5.1.2a) and (5.1.2b) assume the form

---

[13] S. Bergman: On tables for the determination of transonic flow patterns. Hans Reissner Anniversary Volume: Contributions to Applied Mechanics. J. W. Edwards. 1949.

$$L (\Phi) = \left[q^2\, \Phi_{q\,q} + q \left(1 - \frac{q^2}{a^2}\right) \Phi_q + \left(1 - \frac{q^2}{a^2}\right) \Phi_{\theta\,\theta}\right] q^{-2}, \qquad (5.1.2\,\mathrm{d})$$

$$N (\Phi) = \Phi_v{}^{-1} \{q^{-1} \sin\theta\, [\Phi_{q\,q}\, (q\, \Phi_q + \Phi_{\theta\,\theta}) - (\Phi_{q\,\theta} - q^{-1}\, \Phi_\theta)^2]\}, \qquad (5.1.2\,\mathrm{e})$$

with

$$\Phi_v = \Phi_q \sin\theta + q^{-1}\, \Phi_\theta \cos\theta.$$

In the solution of the problem of an axially symmetric flow, we shall restrict our attention at first to the case of analytic solutions (for the definition of analytic functions, see Part 0), and then we shall consider these solutions for complex values of the arguments. The method of solution of equation (5.1.2) by means of successive approximations will be explained.

If we should have a plane flow, then, as one may easily verify, the equation in the hodograph plane would have the form

$$L (\Phi^{(0)}) = 0. \qquad (5.1.3)$$

Denote a solution of (5.1.3) by $\Phi^{(0)}$. Then:

$$\Phi^{(0)} = x\, \varphi_x^{(0)} + r\, \varphi_r^{(0)} - \varphi^{(0)}, \qquad (5.1.4)$$

where $\varphi^{(0)}$ is the velocity potential. Consider $\Phi^{(0)}$ as the first approximation to the solution of equation (5.1.2). The method to be discussed is based on the fact that for those $\varphi^{(0)}\, (x, r)$ which yield flow patterns past profiles of interest in applications, comparatively simple formulas can be obtained, using, for instance, the operator method. Hence, considering $\Phi^{(0)}$ as the initial approximation in that part of the plane where r (the distance of the flow under consideration from the rotation axis) is substantially larger than zero, the higher approximations can be obtained by solving the set of equations

$$L (\Phi^{(n)}) = N (\Phi^{(n-1)}), \qquad n = 1, 2, \ldots \qquad (5.1.5)$$

If the functions $\Phi^{(n)}$ converge to a limit Φ, then in many instances Φ will represent a solution of equation (5.1.2) with slightly distorted boundary conditions. The corresponding

$$\varphi = q\, \Phi_q - \Phi, \qquad (5.1.6)$$

will represent a potential in the hodograph plane of an axially symmetric flow past a profile which is obtained by a slight distortion from the boundary profile corresponding to $\varphi^{(0)}$. By some additional considerations, these distortions can be diminished considerably.

In the most general case, the equation $L (\Phi) = N (\Phi)$ is a non-linear one. Thus, applying the method of successive approximations (5.1.5), we replace the non-linear equation by a sequence of linear non-homogeneous equations. An important task which arises is the development of a procedure suitable for solving these equations, for practical purposes, at least for a small number of approximations, say one or two. As is known in the study of linear equations, say in the case $L (\Psi)$, the fact that if $\Psi_\nu, \nu = 1, 2, \ldots$, are solutions of $L = 0$, the linear combination $\Sigma_\nu\, c_\nu\, \Psi_\nu$, where c_ν are arbitrary constants, is also a solution of L, is of great advantage for numerical applications. Similarly, this fact permits us to prepare, once and for all, tables of a set of particular solutions Ψ_ν, which can then be used conveniently in determining a solution satisfying prescribed initial or boundary conditions. Below, we shall show which tables can be used in the initial approximations, in order to enable one to determine a complete set of particular solutions. Moreover, such tables can be computed for those values of the arguments in which we are particularly interested; in addition, this procedure gives us a certain insight into the variation of the solution if some parameters, for instance the boundary data, vary.

In our particular case, a considerable simplification may be achieved if the expression $N(\Phi^{(n-1)})$ in (5.1.5) is replaced by

$$N_0\left(\Phi^{(n)}\right) = -\frac{2\,v}{\Phi^{(n-1)}_{v\,\max} + \Phi^{(n-1)}_{v\,\min}}\;\frac{\partial\left(\Phi^{(n-1)}_u, \Phi^{(n-1)}_v\right)}{\partial(u,\,v)}, \tag{5.1.7}$$

where $\Phi^{(n-1)}_{v\,\max}$ and $\Phi^{(n-1)}_{v\,\min}$ are the maximum and minimum values $\Phi^{(n-1)}_v$ in the domain in which the function is considered. Below, we shall discuss the application of such a procedure in our case.

As mentioned above, we shall limit ourselves to the study of analytic solutions. The advantage in making this restriction and in considering the equation in the complex domain is that we can simultaneously consider the sub- and the supersonic regions and obtain solutions for the mixed case.

2. The Determination of $\Phi^{(0)}$

In this section, we shall determine the solution $\Phi^{(0)}$ of the equation $L\left(\Phi^{(0)}\right) = 0$, corresponding to flows past profiles of interest to us. As it has been shown in previous sections, the stream function $\Psi^{(0)}$ of a twodimensional flow pattern can be represented in the form

$$\Psi^{(0)} = \mathrm{Im}\left\{\int_{t=-1}^{t=+1} E\left(Z,\overline{Z},t\right) f\left(\tfrac{1}{2}\,Z\left(1-t^2\right)\right)\left(1-t^2\right)^{-1/2} dt\right\}, \tag{5.2.1}$$

$$Z = \lambda\left(q\right) + i\,\theta, \quad \overline{Z} = \lambda\left(q\right) - i\,\theta, \tag{5.2.1a}$$

where E is the generating function of the integral operator, and f is an arbitrary function of a complex variable. Here we may use alternatively $Z = \lambda - i\theta$, $\overline{Z} = \lambda + i\,\theta$, where

$$\lambda\left(q\right) = \int^{q} \left[a_0{}^2 - (\gamma+1)\,q^2\right]^{1/2}\left[a_0{}^2 - \gamma\,q^2\right]^{-1/2} q^{-1}\,dq. \tag{5.2.2}$$

If, instead of considering the "exact" compressibility equation, we consider the so-called "simplified" compressibility equation, then the generating function E becomes a hypergeometric function. This was indicated in Part IV. As is known, in this particular case, the quantities λ and H are connected by the relation:

$$H = c\left(-\lambda\right)^{2/3}, \tag{5.2.3}$$

where c is constant.

Considering plane flows of compressible fluid, the generating function E is of the form:

$$E = c\left(-2\,\lambda\right)^{1/6} F\left(1/6,\,5/6,\,1/2,\,X\right), \tag{5.2.4}$$

$$X = t^2\left[c\,H^{3/2}(q) + i\,\theta - \lambda_0\right]\left[2\,H^{3/2}(q)\right]^{-1}, \tag{5.2.4a}$$

where F is the hypergeometric function and λ_0 is a conveniently chosen constant. Let us keep in mind that in various parts of the plane, the hypergeometric function has to be represented by different power series developments. Concerning the function f, it may be chosen arbitrarily. For example, in order to obtain flow patterns past an elliptical shape, the function f should be of the form [1, pp. 217 and 218]:

$$f = \frac{1}{2}\,U\left\{\left[\frac{\exp Z - R^2\,U}{\exp Z - U}\right]^{1/2} + R^2\left[\frac{\exp Z - U}{\exp Z - R^2\,U}\right]^{1/2}\right\}, \tag{5.2.5}$$

where $R = \mathrm{const.}$ An analogous pattern will be preserved in the present case.

---

[1] S. Bergman and B. Epstein: Determination of a compressible fluid flow past an oval-shaped obstacle. J. Math. Physics **26**, 195—222 (1948).

Equation $L\,(\Phi^{(0)}) = 0$ may be solved by means of several methods. Applying the method of separating the variables, one obtains solutions of $L\,(\Phi^{(0)}) = 0$ in the form of an expression which represents an approximation

$$\sum_{\nu=-N}^{N} A_\nu\, q^{|\nu|} \exp\,(i\,\nu\,\theta)\, F\,(\overline{\alpha}, \overline{\beta},\, \overline{\gamma},\, q^2\,\hat{q}^{-2}), \tag{5.2.6}$$

$$\overline{\alpha} = \frac{1}{2}\,\nu\,\hat{q}^2 + (4\,\gamma)^{-1} + A, \quad \overline{\beta} = \frac{1}{2}\,\nu\,\hat{q}^2 + (4\,\gamma)^{-1} - A, \quad \overline{\gamma} = \nu + 1, \tag{5.2.6a}$$

$$4\,A^2 = \nu^2\,\hat{q}^4 + (\nu\,\hat{q}^2 + \nu^2 - \nu)\,\gamma^{-1} + (4\,\gamma^2)^{-1}. \tag{5.2.6b}$$

The symbol $F\,(\overline{\alpha}, \overline{\beta}, \ldots)$ denotes the hypergeometric function. It remains for us to calculate the coefficients A_ν, which correspond to the stream function (5.2.1) with f from (5.2.5), say. At first, let us note that equation (5.1.4) may be transformed into the form [analogous to equation (5.1.6)]:

$$\varphi^{(0)} = q\,\Phi_q^{(0)} - \Phi^{(0)}. \tag{5.2.7}$$

Our next step is to find a relation between $\Psi^{(0)}$ and $\varphi^{(0)}$. From the elementary relations between velocity potential and stream function, we have:

$$\varphi_\theta^{(0)} = \varrho_0\,\varrho^{-1}\,q\,\psi_q^{(0)}, \quad \psi^{(0)}\,\Psi^{(0)}, \tag{5.2.7a}$$

$$\varphi_q^{(0)} = -\,\varrho_0\,\varrho^{-1}\,(1 - M^2)\,q^{-1}\,\psi_\theta^{(0)}. \tag{5.2.7b}$$

From the relation (5.2.7), we have:

$$\varphi_\theta^{(0)} = q\,\Phi_{q\,\theta}^{(0)} - \Phi_\theta^{(0)}, \tag{5.2.7c}$$

$$\varphi_q^{(0)} = q\,\Phi_{qq}^{(0)} + \Phi_q^{(0)} - \Phi_q^{(0)} = q\,\Phi_{qq}^{(0)}. \tag{5.2.7d}$$

Combining (5.2.7a) through (5.2.7d), we get:

$$q\,\Phi_{qq}^{(0)} = -\,\varrho_0\,\varrho^{-1}\,(1 - M^2)\,q^{-1}\,\psi_\theta^{(0)}, \tag{5.2.8a}$$

$$q\,\Phi_{q\,\theta}^{(0)} - \Phi_\theta^{(0)} = \varrho_0\,\varrho^{-1}\,q\,\psi_q^{(0)}, \tag{5.2.8b}$$

where

$$\varrho = \varrho\,(q) = \varrho_0\,[1 - \gamma\,a_0^{-2}\,q^2]^{1/(2\,\gamma)}, \tag{5.2.8c}$$

and $a_0 = \text{constant}$, denotes the velocity of sound at rest. Suppose that we have found the values of the function $\psi^{(0)}$ by means of the integral operator method, for instance. Then, using (5.2.6), we express $\Phi^{(0)}$ by means of a polynomial or a series

$$\sum_{\nu=-\infty}^{\infty} B_\nu\, q^{|\nu|} \exp\,(i\,\nu\,\theta)\, F_{|\nu|}, \tag{5.2.9}$$

with N finite or $N \to \infty$. Assume also that $B_\nu = \overline{B}_{-\nu}$, i. e., that B_ν and $B_{-\nu}$ are conjugate. The coefficients B_ν can be found from (5.2.8b) with the use of (5.2.1):

$$q\,\Phi_{q\,\theta}^{(0)} - \Phi_\theta^{(0)} = \varrho_0\,\varrho^{-1}\,(q)\,q\,\frac{\partial}{\partial q}\left\{\mathrm{Im}\left[\int_{-1}^{+1} E^{(0)}\,(Z,\overline{Z},t)\cdot f\left(\frac{1}{2}\,Z\,(1-t^2)\right)(1-t^2)^{-1/2}\,dt\right]\right\}, \tag{5.2.10}$$

where $E^{(0)}$ and f are functions (5.2.4) and (5.2.5), say. The convergence of the series (5.2.9) is assured if the solutions represented by (5.2.9) are supposed to be analytic functions of q and θ in the interval $-\pi \le \theta \le \pi$. In particular, it is possible to insert for q a specific constant value, say $q = q_0$. This may be done since in the following, the expression (5.2.6) will always be considered in a comparatively small neighborhood, in which q varies only slightly. If, then, (5.2.9) is approximated, for practical purposes, by a polynomial in the sense that

$$\int_{\theta=-\pi}^{\pi}\left|\sum_{\nu=-\infty}^{\infty} B_\nu\, q_0^{|\nu|}\exp(i\,\nu\,\theta)\,F_{|\nu|} - \sum_{\nu=-N}^{N} B_\nu\, q_0^{|\nu|}\exp(i\,\nu\,\theta)\,F_{|\nu|}\right|^2 d\theta = \min., \qquad (5.2.11)$$

then we may show that

$$B_0^{(N)} = B_0 = \bar{A}_1\,(\bar{B}_1)^{-1}, \qquad B_\nu^{(N)} = B_\nu = \bar{A}_2\,(\nu)\,[\bar{B}_2\,(\nu)]^{-1}, \qquad \nu \neq 0, \qquad (5.2.12\,\mathrm{a})$$

$$\bar{A}_1 = -\,\hat{q}^4\,\varrho_0\,(1-q_0^2/a_0^2)\,\frac{\partial}{\partial\theta}\,\mathrm{Im}\left\{\int_{-1}^{+1} E\,(Z,\bar{Z},t)\cdot f\left(\frac{1}{2}Z\,(1-t^2)\right)(1-t^2)^{-1/2}\,dt\right\}\Bigg|_{\theta=0}, \qquad (5.2.12\,\mathrm{b})$$

$$\bar{B}_1 = 4\,\pi\,\varrho\,(q)\,q_0\,[2\,q_0^2\,F_0''\,(q_0^2/\hat{q}^2) + \hat{q}^2\,F_0'\,(q_0^2/\hat{q}^2) + q_0^2\,F_0\,(q_0^2/\hat{q}^2)], \qquad (5.2.12\,\mathrm{c})$$

$$\bar{A}_2 = \hat{q}^2\,\varrho_0\int_{-\pi}^{+\pi}\frac{\partial}{\partial q}\left[\mathrm{Im}\left\{\int_{-1}^{+1} E\,(Z,\bar{Z},t)\cdot\right.\right.$$

$$\left.\left.\cdot f\left(\frac{1}{2}Z\,(1-t^2)\right)(1-t^2)^{-1/2}\,dt\right\}\right]\exp(-\,i\,\nu\,\theta)\,d\theta\Bigg|_{q=q_0}, \qquad (5.2.12\,\mathrm{d})$$

$$\bar{B}_2 = 2\,\pi\,i\,\nu\,\varrho\,(q)\,q_0^{|\nu|-1}\,[\hat{q}^2\,F_{|\nu|}\,(q_0^2/\hat{q}^2) + 2\,q_0\,F'_{|\nu|}\,(q_0^2/\hat{q}^2)]. \qquad (5.2.12\,\mathrm{e})$$

In previous sections, we discussed widely the application of certain tables to the calculation of Ψ. These same tables can be used to calculate $\Psi^{(0)}$ (5.2.1), and, in turn, for the computation of further tables, for the evaluation of the right-hand side of (5.2.12).

3. Determination of the Solution of the Equation in the Transonic Case

As indicated above, it is convenient to approximate the set of equations (5.1.5) by the equations (5.1.7). As it will be explained in detail in the following, this permits, for certain values of n, the use of tables which can be computed once and for all. The set (5.1.5) can be approximated by (5.1.7) only in a domain where the variation of $\Phi_\nu^{(n-1)}$ is comparatively small. This item complicates the procedure a little. In the following, the determination of $\Phi^{(1)}$ will be discussed. The same procedure can then be repeated for larger n.

The domain B, in which $\Phi^{(0)}$ is considered, is divided into a number of sufficiently small domains N_μ, so that in every neighborhood N_μ the variation of $\Phi_\nu^{(0)}$ is sufficiently small. In order to determine $\Phi^{(1)}$, it will be necessary to use different procedures, depending upon whether N_μ lies in the sub-, super-, or transonic region. In this section, we shall consider an N_μ which lies in the transonic region.

In order to solve the equation

$$L\,(\Phi^{(1)}) = N_0, \qquad (5.3.1)$$

in a neighborhood N_μ of the sonic line, it is convenient to replace q by a new variable

$$Q = \gamma\,\hat{q}^2 - (\gamma+1)\,q^2. \qquad (5.3.2)$$

Notice, that $Q=0$ corresponds to points on the sonic line, i. e., we consider Q to be small in the transonic case. The expression $L\,(\Phi)$ assumes then the form

$$L\,(\Phi) = (\gamma+1)\,[\gamma\,(\gamma\,\hat{q}^2 - Q)\,(\hat{q}^2 + Q)]^{-1}\cdot\{4\,\gamma\,(\gamma\,\hat{q}^2 - Q)^2\,(\hat{q}^2 + Q)\,\Phi_{QQ} - 2\,(\gamma\,\hat{q}^2 - Q)\cdot$$

$$\cdot\,[\gamma\,(\hat{q}^2 + Q) + Q\,(\gamma+1)]\,\Phi_Q + Q\,(\gamma+1)\,\Phi_{\theta\theta}\}. \qquad (5.3.3)$$

We also write

$$\gamma\,(\gamma+1)^{-1}\,(\hat{q}^2+Q)\,(\gamma\,\hat{q}^2-Q)\,N_0 = \sum_{s=0}^{\infty}\ \sum_{\lambda=-(2N+1)}^{2N+1} c_{s,\lambda}\,Q^s \exp\,(i\,\lambda\,\theta), \qquad (5.3.4)$$

and now apply the method of power series to find $\Phi^{(1)}$. The result is

$$\Phi^{(1)} = A_0\,(\theta) + A_1\,(\theta)\,Q + \Phi_1^{(0)} + \sum_{n=-1}^{\infty}\left[(N_{n-3})^{-1}\sum_{\lambda=-(2N+1)}^{2N+1} M_{n+3,\lambda}\exp\,(i\,\lambda\,\theta)\right]Q^{n+3}, \ (5.3.5)$$

where $\Phi_1^{(0)}$ is an arbitrary solution of (5.1.3), and $A_0\,(\theta)$ and $A_1\,(\theta)$ are arbitrary analytic functions of θ;

$$a_{n,\lambda} = M_{n,\lambda}/N_n, \quad \text{and} \quad A_n\,(\theta) = \sum_{\lambda=-(2N+1)}^{2N+1} a_{n,\lambda}\cdot\exp\,(i\,\lambda\,\theta), \quad \text{for } n \ge 2. \qquad (5.3.6)$$

Here

$$N_m = \prod_{i=-1}^{m} S_i, \qquad (5.3.7)$$

and

$$M_{n+3,\lambda} =$$

$$\begin{array}{l|ccccccccccc}
c_{n+1,\lambda} & T_n & U_n & V_{n,\lambda} & 0 & 0 & 0 & \ldots\ldots & 0 & 0 & 0 \\
c_{n,\lambda} & S_{n-1} & T_{n-1} & U_{n-1} & V_{n-1,\lambda} & 0 & 0 & \ldots\ldots & 0 & 0 & 0 \\
c_{n-1,\lambda} & 0 & S_{n-2} & T_{n-2} & U_{n-2} & V_{n-1,\lambda} & 0 & \ldots\ldots & 0 & 0 & 0 \\
\hdashline
& \multicolumn{10}{c}{\ldots\ldots\ldots\ldots\ldots\ldots\ldots\ldots\ldots\ldots\ldots\ldots\ldots} \\
& \multicolumn{10}{c}{\ldots\ldots\ldots\ldots\ldots\ldots\ldots\ldots\ldots\ldots\ldots\ldots\ldots} \\
& \multicolumn{10}{c}{\ldots\ldots\ldots\ldots\ldots\ldots\ldots\ldots\ldots\ldots\ldots\ldots\ldots} \\
\hdashline
c_{3,\lambda} & 0 & 0 & 0 & \ldots\ldots & 0 & 0 & S_2 & T_2 & U_2 & V_{2,\lambda} \\
(c_{2,\lambda} - V_{1,\lambda}\,a_{1,\lambda}) & 0 & 0 & \ldots\ldots & 0 & 0 & 0 & S_1 & T_1 & U_1 & \\
(c_{1,\lambda} - V_{0,\lambda}\,a_{0,\lambda} - U_0\,a_{1,\lambda}) & 0 & 0 & \ldots\ldots & 0 & 0 & 0 & 0 & S_0 & T_0 & \\
(c_{0,\lambda} - T_{-1}\,a_{1,\lambda}) & 0 & 0 & 0 & \ldots\ldots & 0 & 0 & 0 & 0 & 0 & S_{-1}
\end{array} \qquad ,(5.3.8)$$

$$\text{for } n \ge 2.$$

Here

$$S_n = 4\,(n+3)\,(n+2)\,\gamma^3\,\hat{q}^6,$$
$$T_n = [4\,(n+1)\,\gamma^3\,\hat{q}^4 - (8\,n+10)\,\gamma^2\,\hat{q}^4]\,(n+2), \qquad (5.3.8\text{a})$$
$$U_n = [(4\,n-2)\,\gamma\,\hat{q}^2 - (8\,n+2)\,\gamma^2\,\hat{q}^2]\,(n+1),$$
$$V_{n,\lambda} = [((4\,n-6)\,\gamma-4)\,n - (\gamma+1)\,\lambda^2].$$

Similar formulas hold for $M_{2,\lambda}$, $M_{3,\lambda}$, and $M_{4,\lambda}$. In the computation of these constants, tables which can be prepared once and for all greatly facilitate the work. For the $c_{s,\lambda}$ we obtain the formulas

$$c_{s,\lambda} = (2\,i)^{-1}\sum_{\nu=-N}^{N} A_\nu\,(A_{\nu-\lambda-1}\,E^{(s)}_{\nu,\lambda-\nu-1} + A_{\nu-\lambda+1}\,E^{(s)}_{\nu,\lambda-\nu+1}) \quad \text{if } |\lambda| < 2\,N,$$

$$c_{s,\lambda} = (2\,i)^{-1}\sum_{\nu=-N}^{N} A_\nu\,\left(A_{\lambda-\nu-\delta_\lambda}\right)\cdot E^{(s)}_{\nu,\lambda-\nu-\delta_\lambda}$$

$$\text{if } \lambda = \delta_\lambda\,(2\,N) \quad \text{or} \quad \lambda = \delta_\lambda\,(2\,N+1), \quad \delta_\lambda = \pm 1. \qquad (5.3.9)$$

Here the $E^{(s)}_{\nu,\mu}$ are constants which can be determined once and for all since they are derived only from the coefficients of $F_{|\nu|}$. Therefore, to obtain a general solution $\Phi_\mu^{(1)}$ of (5.1.5) in an N_μ situated in a transonic region we write

$$\Phi_\mu^{(1)} = \Phi_\mu^{(0)} + \Phi^{(1)}, \qquad (5.3.10)$$

where $\Phi_\mu^{(0)} = \sum\limits_{\nu=-\infty}^{\infty} X_\nu^{(\mu)} \, q^{|\nu|} \, F_{|\nu|} \, \exp(i \, \nu \, \theta)$ is given by (5.2.9) with B_ν replaced by arbitrary constants $X_\nu^{(\mu)}$ which depend upon the specific solution and whose determination will be discussed below. On the other hand, in the solution $\Phi^{(0)}$ which does not involve any arbitrary constants depending upon the specific solution, the B_ν are completely determined (see 5.2.12), since we start from a known solution. $\Phi^{(1)}$ is determined as follows: Using the given B_ν, we compute $c_{s,\lambda}$ (see 5.3.9), which are substituted into (5.3.8) to obtain $M_{n,\lambda}$. Using these values, we finally, according to (5.3.5) calculate $\Phi^{(1)}$.

The coefficients $E_{\nu,\mu}^{(s)}$ are given by the formulas:

$$E_{\nu,\mu}^{(s)} = \gamma \, (\hat{q}^2 \, D_s^{(\nu,\mu)} + D_{s-1}^{(\nu,\mu)}), \qquad s > 1, \qquad (5.3.11\,\text{a})$$

where
$$E_{\nu,\mu}^{(0)} = \gamma \, \hat{q}^2 \, D_0^{(\nu,\mu)}, \qquad (5.3.11\,\text{b})$$

$$D_j^{(\nu,\mu)} = \sum_{j=0}^{\infty} \, \sum_{N=\max(j,\sigma-1)}^{\infty} B_{(N-\sigma+1)}^{(\nu,\mu)} \, \gamma^{(N-j)} \cdot (\gamma+1)^{-N} \left(\frac{N}{j}\right) \hat{q}^{(\sigma-1-2j)} \, Q^j, \qquad (5.3.12\,\text{a})$$

if $|\nu| + |\mu| = 2\,\sigma + 1$, σ integer,

$$D_j^{(\nu,\mu)} = \sum_{\mu=0}^{\infty} \left\{ \sum_{j=0}^{\infty} \frac{(2\,\mu - 2\,j)!}{2^{\mu-j} \, [(\mu-j)!]^2} \cdot \sum_{N=\max(j,\sigma-1)}^{\infty} B_{(N-\sigma+1)}^{(\nu,\mu)} \, \gamma^{(N-\mu)} \cdot \right.$$
$$\left. \cdot \, (\gamma+1)^{-N} \left(\frac{N}{j}\right) \hat{q}^{(\sigma-1-\mu-j)} \, Q^\mu \right\}, \qquad (5.3.12\,\text{b})$$

if $|\nu| + |\mu| = 2\,\sigma$, σ integer,

$$B_n^{(\nu,\mu)} = \sum_{j=0}^{n+3} \alpha_{\mu,j} \, \alpha_{\nu,\,n+3-j} \cdot (\nu^2 - |\nu|) \, (\mu^2 - |\mu|) + \nu \mu \, (|\nu \mu| + |\nu| + |\mu| + 1) +$$
$$+ \, 2 \, (n + 3 - j) \, (\nu \mu + |\mu|) \, (|\mu| - 1) + j \, (\mu \nu - 2 \, |\mu| - 1) \, (|\nu| - 1) +$$
$$+ \, 4 \, j \, (j - 1) \, (2 \, n + 4 - j - |\nu| \, (|\nu| - 1)), \qquad n = -1, 0, 1, \ldots, \qquad (5.3.13)$$

and
$$B_0^{(\nu,\mu)} = \alpha_{\nu 0} \, \alpha_{\mu 0} \, [(\nu^2 - |\nu|) \cdot (\mu^2 - |\mu|) + \nu \mu \, (|\nu \mu| + |\nu| + |\mu| + 1)], \qquad (5.3.14)$$

$$B_1^{(\nu,\mu)} = (\alpha_{\mu 0} \, \alpha_{\nu 1} + \alpha_{\mu 1} \, \alpha_{\nu 0}) \cdot [(\nu^2 - |\nu|) \, (\mu^2 - |\mu|) + \nu \mu \, (|\nu \mu| + |\nu| + |\mu| + 1)] +$$
$$+ \, 2 \, \alpha_{\mu 0} \, \alpha_{\nu 1} \, (\mu \nu + |\mu|) \, (|\mu| - 1) + 2 \, \alpha_{\mu 1} \, \alpha_{\nu 0} \, (\mu \nu - 2 \, |\mu| - 1) \, (|\nu| - 1), \qquad (5.3.15)$$

$$\alpha_{\nu,0} = 1, \qquad (5.3.16)$$

$$\alpha_{\nu,\,n+1} = [n! \, (\nu + 1 + n)!]^{-1} \left\{ \nu! \cdot \left[\left(\frac{1}{2}\right) \nu \, \hat{q}^2 + \left(\frac{1}{4\,\gamma}\right) + \overline{A}\right] \cdot \left[\left(\frac{1}{2}\right) \nu \, \hat{q}^2 + \left(\frac{1}{4\,\gamma}\right) + \overline{A} + 1\right] \ldots \right.$$
$$\left. \ldots \left[\left(\frac{1}{2}\right) \nu \, \hat{q}^2 + \left(\frac{1}{4\,\gamma}\right) + \overline{A} + n\right] \cdot \left[\left(\frac{1}{2}\right) \nu \, \hat{q}^2 + \left(\frac{1}{4\,\gamma}\right) - \overline{A}\right] \ldots \left[\left(\frac{1}{2}\right) \nu \, \hat{q}^2 + \left(\frac{1}{4\,\gamma}\right) - \overline{A} + n\right], \right.$$
$$\qquad (5.3.17)$$

where
$$\overline{A} = \left(\frac{1}{2}\right) \{\nu^2 \, \hat{q}^2 + \gamma^{-1} \, (\nu \, \hat{q}^2 + \nu^2 - \nu) + (4 \, \gamma^2)^{-1}\}^{1/2}. \qquad (5.3.18)$$

All the proofs are given in the Appendix to the version of Bergman's paper[2].

[2] S. Bergman: Determination of axially symmetric flow patterns of a compressible fluid. With an Appendix. Harvard University, Division of Engineering Sciences, Technical Report No. 16, Cambridge, Massachusetts (1951).

4. Determination of the Solution in the Subsonic and Supersonic Regions

In this and the next sections, we shall describe a method analogous to that in section V. 3, but where the subdomain N_μ lies completely in the subsonic or in the supersonic region. A difference which now arises is that we have to operate with the new variables, α, β:

$$\alpha = \theta + H(q), \qquad \beta = \theta - H(q), \tag{5.4.1}$$

(instead of the variables q and θ). Here,

$$\Lambda(q) = H(q) = - \arctan\left[\frac{q^2(1+\gamma) - \gamma\,\hat{q}^2}{\gamma(\hat{q}^2 - q^2)}\right]^{1/2} + \left(\frac{1+\gamma}{\gamma}\right)^{1/2} \arctan\left[\frac{q^2(1+\gamma) - \gamma\,\hat{q}^2}{(1+\gamma)(\hat{q}^2 - q^2)}\right]^{1/2},$$

in the supersonic case; and
$$\tag{5.4.2}$$

$$\lambda = H(q) = i\,\eta(q), \qquad \eta(q) = \left(\frac{1}{2}\right)\left(\log\frac{\varepsilon - 1}{\varepsilon + 1} - \sqrt{\frac{1+\gamma}{\gamma}}\,\log\frac{\sqrt{\gamma}\,\varepsilon - \sqrt{1+\gamma}}{\sqrt{\gamma}\,\varepsilon + \sqrt{1+\gamma}}\right),$$

$$\varepsilon = \left[\frac{\gamma\,\hat{q}^2 - q^2(1+\gamma)}{\gamma(\hat{q}^2 - q^2)}\right]^{1/2}, \tag{5.4.3}$$

in the subsonic case. The variables α and β are in the subsonic case conjugate imaginary quantities, while in the supersonic region they become two independent real variables.

Furthermore, in order to solve the non-homogeneous equation (5.1.5), instead of a direct attack such as discussed in section V. 3, we shall use the method of so-called fundamental solutions.

The expression (5.1.2a) assumes, in the variables α, β the form

$$L(\Phi) \equiv \left[2\left(1 - \frac{q^2}{a^2}\right) - 2\,[H'(q)]^2\,q^2\right]\Phi_{\alpha\beta}\,q^{-2} + \{H''(q) - [H'(q)]^3\,q\}\,(\Phi_\alpha - \Phi_\beta). \tag{5.4.4}$$

Here q and a are known functions of α and β.

Besides some minor differences, the considerations in both sub- and supersonic cases proceed in almost the same manner; therefore we carry out our discussions simultaneously.

The problem of solving the system (5.1.5) reduces to solving, for every n, a non-homogeneous equation. As we mentioned before, this will be done by using so-called fundamental solutions. We shall first explain this method in the general case of a linear equation.

Let a partial differential equation

$$M(\Phi) = \Phi_{\alpha\beta} + a\,\Phi_\alpha + b\,\Phi_\beta + c\,\Phi = 0, \tag{5.4.5}$$

be given, where a, b, and c are functions of α and β which are analytic in α and β in a sufficiently small domain, say B. A solution of the non-homogeneous equation

$$M(\Phi) = N(\alpha, \beta), \tag{5.4.6}$$

where $N(\alpha, \beta)$ is a given function which is analytic in B, can be obtained by using a fundamental solution

$$W(\alpha, \beta; A, B,) = U(\alpha, \beta; A, B)\,[\log(\alpha - A) + \log(\beta - B)] + V(\alpha, \beta; A, B), \tag{5.4.7}$$

of $M(\Phi) = 0$ (see [3]). Here U and V can be written in the form of infinite series which converge in B:

$$U = U_0 + U_1 + U_2 + \ldots, \tag{5.4.8}$$

where

$$U_0 = \left(\frac{1}{2}\right)\left[\exp\left(\int_\beta^B a(\alpha, Y)\,dY\right) + \exp\left(\int_\alpha^A b(X, \beta)\,dX\right)\right], \tag{5.4.9}$$

---

[3] R. Courant and D. Hilbert: Methoden der Mathematischen Physik, Vol. I and II. Berlin: Julius Springer. 1931 and 1937.

and U_{n+1} is given by

$$- U_{n+1} = \int\limits_{\alpha}^{A}\int\limits_{\beta}^{B} \left(a\,\frac{\partial U_n}{\partial \alpha_1} + b\,\frac{\partial U_n}{\partial \beta_1} + c\,U_n \right) d\alpha_1\,d\beta_1,$$

$$U_n \equiv U_n(\alpha_1, \beta_1;\, A,\, B). \tag{5.4.10}$$

Furthermore,

$$V = V_0 + V_1 + V_2 \ldots, \tag{5.4.11}$$

where V_0 and V_{n+1} are given by

$$V_0 = \frac{(a\,U - U_\beta)}{(A - \alpha)} + \frac{(b\,U - U_\alpha)}{(B - \beta)}, \tag{5.4.12}$$

$$- V_{n+1} = \int\limits_{\alpha}^{A}\int\limits_{\beta}^{B} \left(a\,\frac{\partial V_n}{\partial \alpha_1} + b\,\frac{\partial V_n}{\partial \beta_1} + c\,V_n \right) d\alpha_1\,d\beta_1,$$

$$V_n \equiv V_n(a_1, \beta_1;\, A,\, B). \tag{5.4.12a}$$

A fundamental solution of the equation (5.4.5) satisfies (5.4.5) and possesses the property that at the point (A, B), $U(\alpha, \beta;\, A,\, B) = 1$.

The method of a fundamental solution is usually applied to the equations of elliptic type (subsonic case), in which case α and β are conjugate.

In order to be able to apply it to the subsonic and the supersonic case, we continue the variables to complex values, and consider our functions in a four-dimensional domain B^4. Let

$$\gamma = \frac{\alpha + \beta}{2},$$

and

$$\delta = \frac{\alpha - \beta}{2\,i},$$

where $\gamma = \gamma_1 + i\,\gamma_2$, $\delta = \delta_1 + i\,\delta_2$. At a given point $\gamma_1^{(0)}$, $\gamma_2^{(0)}$, $\delta_1^{(0)}$, $\delta_2^{(0)}$, $\Phi\,(\gamma_1 + i\,\gamma_2,\, \delta_1 + i\,\delta_2)$ satisfies simultaneously the equations

$$\left(\frac{1}{4}\right)\left(\frac{\partial^2\Phi}{\partial\gamma_1{}^2} + \frac{\partial^2\Phi}{\partial\delta_1{}^2}\right) + \tilde{A}\,(\gamma_1 + i\,\gamma_2^{(0)},\, \delta_1 + i\,\delta_2^{(0)})\,\frac{\partial\Phi}{\partial\gamma_1} + \tilde{B}\,(\gamma_1 + i\,\gamma_2^{(0)},\, \delta_1 + i\,\delta_2^{(0)})\,\frac{\partial\Phi}{\partial\delta_1} +$$

$$+ \tilde{C}\,(\gamma_1 + i\,\gamma_2^{(0)},\, \delta_1 + i\,\delta_2^{(0)}) = N\,(\gamma_1 + i\,\gamma_2^{(0)},\, \delta_1 + i\,\delta_2^{(0)}), \tag{5.4.13}$$

as a function of γ_1 and δ_1, and

$$\left(\frac{1}{4}\right)\left(\frac{\partial^2\Phi}{\partial\gamma.^2} - \frac{\partial^2\Phi}{\partial\delta_2{}^2}\right) A\,(\gamma_1 + i\,\gamma_2^{(0)},\, \delta_1^{(0)} + i\,\delta_2)\,\frac{\partial\Phi}{\partial\gamma_1} + B\,\frac{(\gamma_1 + i\,\gamma_2^{(0)},\, \delta_1^{(0)} + i\,\delta_2)}{i}\,\frac{\partial\Phi}{\partial\delta_2} +$$

$$+ C\,(\gamma_1 + i\,\gamma_2^{(0)},\, \delta_1^{(0)} + i\,\delta_2) = N\,(\gamma_1 + i\,\gamma_2^{(0)},\, \delta_1^{(0)} + i\,\delta_2), \tag{5.4.14}$$

as a function of γ_1 and δ_2.

In the fundamental solution (5.4.7), the quantities α, β, A and B now have the following significance:

$$\alpha = \gamma_1 + i\,\gamma_2^{(0)} + i\,(\delta_1 + i\,\delta_2^{(0)}),$$

$$\beta = \gamma_1 + i\,\gamma_2^{(0)} - i\,(\delta_1 + i\,\delta_2^{(0)}),$$

$$A = \Gamma + i\,\Delta - \delta_2^{(0)} + i\,\gamma_2^{(0)}, \tag{5.4.15}$$

$$B = \Gamma + i\,\Delta + \delta_2^{(0)} + i\,\gamma_2^{(0)},$$

where Γ and Δ are real integration variables. We shall now denote it by $\tilde{W}\,(\gamma_1, \delta_1 \colon \gamma_2^{(0)}, \delta_2^{(0)};\, \Gamma, \Delta) \equiv W\,(\alpha, \beta;\, A,\, B)$.

According to classical results (see [3]),

$$\Phi\,(\gamma_1 + i\,\gamma_2^{(0)},\ \delta_1 + i\,\delta_2^{(0)}) =$$

$$= \iint\limits_{\Gamma^2 + \Delta^2 < 1} \tilde{W}\,(\gamma_1,\,\delta_1;\,\gamma_2^{(0)},\,\delta_2^{(0)};\,\Gamma,\,\Delta)\ N\,(\Gamma + i\,\gamma_2^{(0)},\,\Delta + i\,\delta_2^{(0)})\,d\Gamma\,d\Delta. \quad (5.4.16)$$

The integration has to be carried out over a circle $\Gamma^2 + \Delta^2 \leq 1$, in the plane $\gamma_2 = \gamma_2^{(0)}$, $\delta_2 = \delta_2^{(0)}$. The expression (5.4.16) can be used in the supersonic case. We wish, however, to stress that in the subsonic case, the integration domain K lies in the plane of the variables of the differential equation γ_1, δ_1, while in the supersonic case, K lies in the plane which is perpendicular to γ_2 and δ_1 of the equation.

5. Determination of $\Phi^{(1)}$ in N_μ Belonging entirely to the Subsonic or Supersonic Region

In this section, we shall apply the formulas derived in section V. 4 to our specific case. Since the explicit expressions for q and θ in terms of the variables α and β are rather involved, it is convenient to rewrite the formulas of V. 4, in terms of q and θ. This will form the main content of the present section.

In (5.4.2), (5.4.3), the function $\Lambda\,(q)$ has been introduced. In the following, we shall need the inverse function $\Omega\,(\Lambda)$,

$$q = \Omega\,(\Lambda), \quad\quad\quad (5.5.1)$$

or, even more,

$$q = \Omega\,(\Lambda_1), \quad \Lambda_1 = \Lambda + i\,\lambda. \quad\quad (5.5.1\,\mathrm{a})$$

The function Ω is a function of a complex variable Λ_1. In the present application, Λ_1 is real in the supersonic region (i. e., $\Lambda_1 \equiv \Lambda$), and purely imaginary in the subsonic region ($\Lambda_1 \equiv i\,\lambda$); i. e., $\lambda \equiv 0$, for $M > 1$, and $\Lambda \equiv 0$, for $M < 1$. We note, however, that in our treatment we need the values of Ω for complex values. $\Omega\,(\Lambda_1)$ can be represented in the form of power series. In the subsonic case, the power series of

$$T = [\gamma\,\hat{q}^2 - (1 + \gamma)\,q^2]^{1/2}\,[\gamma\,(\hat{q}^2 - q^2)]^{-1/2}, \quad (5.5.2)$$

and T^{-1} in terms of $\exp\,(2\,\lambda)$ have been given in [4, pp. 49, 75, 76].

We note that tables for $\Omega\,(\Lambda_1)$ are available for the subsonic region, i. e., for Λ_1 purely imaginary.

In the neighborhood of the point $\lambda = 0$, for $\gamma = 0.2$ ($k = 1.4$), a formal computation yields the representation

$$\lambda = c_1\,T + c_3\,T^3 + \ldots, \quad\quad (5.5.3)$$

$$T = d_1\,\mathfrak{H} + d_3\,\mathfrak{H}^3 + \ldots, \quad\quad (5.5.4)$$

Table 5.5.1

n	c_n	d_n
1	1	1
3	.23333	—.23333
5	.11619	+.047143
7	.07477	—.010329
9	.05435	+.002721
11	.04234	—.000669
13	.03448	+.000158
15	.02896	—.000038
17	.02488	+.000009
19	.02176	—.000002
21	.01930	
23	.01731	
25	.01568	

where $\mathfrak{H} = (-\,3.6\,\lambda)^{1/3}$ and the c_n and d_n are given in the table 5.5.1. We assume in the following that tables for the $\Omega\,(\Lambda_1)$ for complex values of the argument Λ_1 are available.

Let us substitute into the expressions (5.4.9), (5.4.10) and (5.4.12), the variables q and θ. If

$$G\,(q) = \left(\frac{1}{2}\right)\frac{\Lambda_1''\,(q)\,q^2 - [q\,\Lambda_1'\,(q)]^3}{1 - \dfrac{q^2}{a^2} - [\Lambda_1'\,(q)]^2\,q^2}, \quad\quad (5.5.5)$$

<hr>

[4] S. Bergman: On supersonic and partially supersonic flows. N. A. C. A.. T. N. No. 1096 (1946).

then we obtain

$$\tilde{U}_0\,(q,\,q';\,\theta,\,\Gamma;\,\lambda^{(0)},\,\theta_2^{(0)}) =$$

$$= \left(\frac{1}{2}\right)\left\{\exp\Big[2\int\limits_{\left(\sigma=\Omega\,\frac{i\Lambda+\theta_1-\Gamma+i\Delta-2\lambda^{(0)}}{2}\right)}^{\sigma=\Omega\,(i\Lambda-\lambda^{(0)})} G\,(\sigma)\Lambda_1{}'\,(\sigma)\,d\sigma\Big] + \right.$$

$$\left. + \exp\Big[2\int\limits_{\sigma=\Omega\,(i\Delta-\lambda^{(0)})}^{\sigma=\Omega\left(\frac{\Gamma+i\Delta+i\Lambda-\theta_1-2\lambda^{(0)}}{2}\right)} G\,(\sigma)\,\Lambda_1{}'\,(\sigma)\,d\sigma\Big]\right\}. \qquad (5.5.6)$$

Here and in the following, $\Lambda_1 \equiv \Lambda_1\,(q)$, $\Delta \equiv \Delta\,(q')$. Let

$$T\,(\Lambda,\,\lambda^{(0)},\,\theta_1,\,\theta_2^{(0)},\,\Gamma,\,\Delta) = \frac{i\left[\Lambda\left(\Gamma+i\,\theta_2^{(0)}\right) + i\,\lambda^{(0)},\,(\theta_1-\Gamma) - \Delta\left(\theta_1+i\,\theta_2^{(0)}\right)\right]}{(\Gamma-\theta_1)}. \qquad (5.5.7)$$

Then

$$\tilde{U}_{n+1}\,(q,\,q';\,\theta_1,\,\Gamma;\,\lambda^{(0)};\,\theta_2^{(0)}) = \int\limits_{\vartheta=\vartheta_1}^{\vartheta_2}\int\limits_{\Lambda_1^{(1)}(\vartheta)}^{\Lambda_1^{(2)}(\vartheta)} + \int\limits_{\vartheta_1}^{\vartheta_3}\int\limits_{\Lambda_1^{(3)}(\vartheta)}^{\Lambda_1^{(1)}(\vartheta)} + \int\limits_{\vartheta_3}^{\vartheta_4}\int\limits_{\Lambda_1^{(4)}(\vartheta)}^{\Lambda_1^{(1)}(\vartheta)} + \int\limits_{\vartheta_2}^{\vartheta_4}\int\limits_{\Lambda_1^{(1)}(\vartheta)}^{\Lambda_1^{(5)}(\vartheta)} \cdot$$

$$\frac{\Lambda_1{}''\,(q_1)\,q_1{}^2}{1-\dfrac{q_1{}^2}{a_1{}^2}-[\Lambda_1{}'\,(q_1)]^2\,q_1{}^2}\,\frac{\partial U_n}{\partial q_1}\,dq_1\,d\vartheta_1, \qquad (5.5.8)$$

where

$$\vartheta_1 = \theta_1 + i\,\theta_2^{(0)}, \qquad\qquad \Lambda_1^{(1)}\,(\vartheta) = T - i\,\frac{\Lambda-\Delta}{\Gamma-\theta_1}\,\vartheta,$$

$$\vartheta_2 = \frac{\theta_1+\Gamma}{2} + i\,\theta_2^{(0)} + i\left(\frac{\Lambda-\Delta}{2}\right), \quad \Lambda_1^{(2)}\,(\vartheta) = \theta_1 + i\,\theta_2^{(0)} + i\,(\Lambda - i\,\lambda^{(0)}) - \vartheta,$$

$$\vartheta_3 = \frac{\theta_1+\Gamma}{2} + i\,\theta_2^{(\theta)} - i\left(\frac{\Lambda-\Delta}{2}\right), \quad \Lambda_1^{(3)}\,(\vartheta) = \vartheta - \theta_1 - i\,\theta_2^{(0)} + i\,(\Lambda - i\,\lambda^{(0)}),$$

$$\vartheta_4 = \Gamma + i\,\theta_2^{(0)}, \qquad\qquad \Lambda_1^{(4)}\,(\vartheta) = \Gamma + i\,\theta_2^{(0)} + i\,(\Delta - i\,\lambda^{(0)}) - \vartheta,$$

$$\Lambda_1^{(5)}\,(\vartheta) = \vartheta - \Gamma - i\,\theta_2^{(0)} + i\,(\Delta - i\,\lambda^{(0)}).$$

Further,

$$\tilde{V}_0 = \frac{i\,(\Lambda-\Delta)\left\{\dfrac{\Lambda_1{}''\,(q)\,q^2\,U - (q\,\Lambda_1{}'\,(q))^3\,U}{1-\dfrac{q^2}{a^2}-[\Lambda_1{}'\,(q)]^2\,q^2} - \dfrac{U_q}{\Lambda_1{}'\,(q)}\right\} + (\Gamma-\theta_1)\,U_\theta}{[(\Gamma-\theta_1)^2 + (\Delta-\Lambda)^2]}, \qquad (5.5.9)$$

and $\tilde{V}_{n+1}$ follows the law of (5.5.8).

We then get

$$\tilde{W}\,(q,\,q';\,\theta_1,\,\Gamma;\,\lambda^{(0)},\,\theta_2^{(0)}) =$$

$$= \tilde{U}\,(q,\,q';\,\theta_1,\,\Gamma;\,\lambda^{(0)},\,\theta_2^{(0)})\,\log\,[(\Gamma-\theta_1)^2 + (\Delta-\Lambda)^2] + \tilde{V}\,(q,\,q';\,\theta_1,\,\Gamma;\,\lambda^{(0)},\,\theta_2^{(0)}).$$

$$(5.5.10)$$

In order to obtain the solution of the non-homogeneous equation (5.3.1), where $N_0 = N_0\,(\Phi^{(0)})$, in a neighborhood N_μ which lies completely in a sub- or a supersonic region, we apply the method of fundamental solutions, and according to (5.4.16), we write

$$\Phi^{(1)} = \frac{1}{2}\int\int\limits_{\Gamma^2 + [\Lambda_1{}'(q')]^2 < 1} W\,(q,\,q';\,\theta_1,\,\Gamma;\,\lambda^{(0)},\,\theta_2^{(0)}) \cdot$$

$$\cdot G\,(q')\,\Lambda_1{}'\,(q')\,[\Lambda_1{}''\,(q') - q'\,(\Lambda_1{}'\,(q'))^3]^{-1}\,dq'\,d\Gamma + \Phi_1^{(0)}, \qquad (5.5.11)$$

where $N_0 (\Phi^{(0)})$, is considered as a function of q' and Γ rather than q and θ. Here, $\Phi_1^{(0)}$ is again an arbitrary solution.

By the procedure described in sections V. 3 and V. 4, we can, starting from a solution $\Phi^{(0)}$ of an axially symmetric flow, determine in every neighborhood N_μ a solution, say $\Phi_\mu^{(1)}$, of equation (5.1.5) for $n = 1$. As we stressed before, in doing this we can use tables of functions (5.5.5), (5.5.6), (5.5.7), (5.5.8), etc., and expressions (5.3.9), (5.3.15), (5.3.18), which can be computed once and for all. We remind the reader that in every N_μ the function $N (\Phi^{(0)})$ is replaced by $N_0 (\Phi^{(0)})$, see (5.1.7), and that naturally $c_\mu = \dfrac{1}{2} (\Phi_{v\,\mathrm{max}}^{(0)} + \Phi_{v\,\mathrm{min}}^{(0)})$ will change from region to region. On the other hand, the tables mentioned above can be prepared for $c = 1$, and the values obtained from the tables have to be divided by c_μ.

Each of the solutions $\Phi_\mu^{(1)}$ involves an arbitrary function $\Phi_\mu^{(0)} = \sum\limits_{v=-\infty}^{\infty} X_v^{(\mu)}\, q^{|v|}\, e^{i\,v\,\theta}\, F_{|v|}$ [see (5.2.9)]. We note that in (5.2.9), the symbol B_v appears instead of $X_v^{(\mu)}$, and we wish to indicate how to determine the constants $X_v^{(\mu)}$ in order that the solution obtained in this manner will represent but one solution in the whole domain $N = \sum\limits_{\mu} N_\mu$ in which $\Phi^{(0)}$ has originally been defined.

We assume here that we have only one domain N_μ, say N_1, which lies in the transonic region, and the remaining N_μ, say $N_2, N_3, N_4, \ldots$, belonging to the supersonic region. Since we wish to determine a solution of the axially symmetric flow which differs as little as possible from the plane solution $\Phi^{(0)}$, we determine the $X_v^{(1)}$ in such a way that

$$\int\limits_{N_1} \int [\Phi_1^{(0)} - \Phi_1^{(1)}]^2 \, dq\, d\theta = \mathrm{min}. \qquad (5.5.12)$$

This condition permits us to determine the unknown $X_v^{(1)}$. Let us denote (5.3.10) into which $X_v^{(1)}$ have been substituted by $\Phi_1^{(1)}$. Let us denote the function obtained by $\Phi_1^{(0)} + \Phi_1^{(1)} = \varphi_1^{(1)}$. The function $\varphi_1^{(1)}$ and its derivative assume certain values on the line $A\,B$ of intersection N_1 and N_2, and now we determine the constants $X_v^{(2)}$ in the solution $\Phi_2^{(1)}$ so that on the segment $A\,B$ the obtained value of $\varphi_2^{(1)} = \Phi_2^{(0)} + \Phi_2^{(1)}$ and its derivative differ as little from $(\Phi^{(1)})$ and its derivatives as possible. This procedure can be continued through all N_μ so that we get a solution $\varphi^{(1)}$ in the whole domain. We should like to note that if the domain involves neighborhoods N belonging to the subsonic region, we can determine the constants $X_v^{(\mu)}$ by using the method of analytic continuation, which we shall not discuss in detail at this time.

The second approximation $\varphi^{(2)}$ can be obtained, in principle, in the same manner as $\varphi^{(1)}$. We note, however, that presumably in most instances it will be sufficient to determine $\varphi^{(1)}$.

6. General Remarks

We described above a method of starting from a given solution of (5.1.3) to obtain functions which can be considered as approximations of (5.1.2). Every $\Phi_\mu^{(n)}$ can be inserted into (5.1.2); we obtain then a non-homogeneous equation

$$L (\Phi^{(n)}) - N (\Phi^{(n)}) = M (u, v), \qquad (5.6.1)$$

and from the magnitude of $M (u, v)$, one can make some conclusions about the accuracy of the approximation in every N_μ.

On the other hand, it is of interest to give at least some sufficient criteria when the $\Phi^{(n)}$ converge, as $n \to \infty$, to the solution of (5.1.2). A discussion of this subject is beyond the scope of the present work, and the reader interested in this item may refer to Bergman's original paper[5].

Part VI

Singularities

1. General Remarks

As mentioned above, singularities play an important role in hodograph transformations. In a brief résumé of Bergman's method given in the present work, it is impossible to discuss all the details and items on the subject of singularities. Hence, below only general and more important features will be briefly outlined. A reader, advanced in mathematics, may find more details in the original literature, particularly in [6].

2. Properties of Bergman's Operator in the Subsonic Range

It is obvious that investigations in fluid dynamics require, in addition to the study of regular solutions, the investigation of singularities, for example, sources, sinks, vortices, doublets, and so on. Since, further, the image in the hodograph (or an allied) plane of a flow pattern is not necessarily uni-valent (uni-valent functions are sometimes called schlicht functions) various aerodynamical problems lead to the study of branch points as well as to the study of the corresponding Riemann surfaces.

As mentioned in previous chapters we have several procedures for generating the functions Φ and Ψ in compressible fluid flow. Chaplygin, who introduced the hodograph method into the theory of compressible fluid, applied the method of separation of variables. If, in the power series development for the stream function of an incompressible fluid, the powers of the speed, q, are replaced by certain hypergeometric functions of q, then, as Chaplygin has shown, the series obtained in this manner is a solution of the compressibility equation. We designate this procedure as the Chaplygin operator, P_C, sometimes denoted as P_1. In [7] and [8] Bergman introduced a new operator which generates solutions Φ and Ψ of the compressibility equations. This is defined in general as the operator P_B or P_2. P_B can obviously be applied to finite sums and in some cases to infinite ones, producing potentials and stream functions of a compressible fluid flow. As mentioned above, in a joint investigation, Bers and Gelbart[9], independently of Bergman, found the same operator, which they applied in [9]. They term the functions obtained Σ-monogenic.

The operator P_C and P_B can be applied only to a power series development of an analytic function of a complex variable. They transform these series into potential

<hr>

[5] S. Bergman: Determination of axially symmetric flow patterns of a compressible fluid. J. Math. Physics **29**, 133—145 (1950).

[6] G. S. S. Ludford: The behavior at infinity of the potential function of the flow of a perfect gas in two dimensions. Report 23a, Nav. Res. Contr. N5ori 76/XVI NR 043 046, Harvard Univ. (1950). — Two notes on the singularities of a compressible flow in the pseudo-logarithmic plane. Report 23b, Nav. Res. Contr. N5ori 76/XVI NR 043 046, Harvard Univ. (1950). — The behavior at infinity of the potential function of a two-dimensional subsonic compressible flow. Report 23, Nav. Res. Contr. N5ori 76/XVI NR 043 046, Harvard Univ., J. Math. Physics **30**, (3), 117—130 (1951).

[7] S. Bergman: The hodograph method in the theory of compressible fluids. Supplement to "Fluid Dynamics" by R. v. Mises and K. O. Friedrichs, Brown University (1941/42).

[8] S. Bergman: A formula for the stream function of certain flows. Proc. Nat. Acad. Sci., U. S. A. **29**, 276—281 (1943).

[9] L. Bers and A. Gelbart: On a class of differential equations in mechanics of continua. Quart. Appl. Math. **1**, 168—188 (1943). — On a class of functions defined by partial differential equations. Trans. Amer. Math. Soc. **56**, No. 1, 67—93 (1944).

and stream functions of both subsonic and supersonic flows. The potential and stream function of a supersonic flow each satisfy an equation of hyperbolic type. Thus P_C and P_B transform solutions of an equation of elliptic type (Laplace's equation) into solutions of a hyperbolic one. It is well known that the solutions of both equations have a quite different character, and therefore these operators cannot preserve various properties of the functions upon which they act.

Operators P_C and P_B can be applied only to functions which in the hodograph plane are defined in a circle with the center at the origin. Often we are interested in considering stream functions in domains different from circles. The methods for analytic continuation of solutions obtained in this manner are studied comparatively little. As is known, a power series is defined only for a certain value of the independent variable, and hence is convergent inside the circle with the center at the origin of radius equal to that value.

Moreover, complex potentials of an incompressible fluid flow have branch points, poles and logarithmic singularities, around points other than the origin, in the hodograph plane.

Operators, described so thoroughly in the present work, to be denoted henceforth as P_k, or $P_{3,k}$, $k = 1, 2$ (operators of the first and second kind), when specialized to the case of functions Φ and Ψ which satisfy the compressibility equations, have the advantage that they can be applied directly to functions (and not only to their power series developments), and they generate solutions which are defined in an arbitrary simply-connected domain (and not necessarily in a circle with the center at the origin, which is a domain of convergence for a power series).

These operators generate solutions which are defined in any simply-connected domain which includes the origin. They can easily be extended so as to act on functions possessing the singularities of various types around points other than the origin, in the hodograph plane.

The operators $P_{3,k}$ (or P_k) transform analytic functions of a complex variable into stream functions or potential functions of subsonic flows, and differentiable functions of one real variable into stream functions of supersonic flows, in both cases preserving many properties of the functions to which the operator has been applied, thus serving as a useful tool in the investigation of flow patterns of a compressible fluid. In contrast to operators of type $P_{3,k}$, which have been introduced in order to "translate" properties of solutions of simpler equations into properties of more complicated ones, the operators P_C and P_B (denoted sometimes as P_1 and P_2) can be used only in a few cases as a tool for investigation of the properties of the functions which they generate.

It should be indicated that by using the Chaplygin operator P_C it is possible to obtain various important types of singularities of supersonic flows and various singularities at stagnation points. On the other hand, certain singularities, which can be obtained easily by using Bergman's operator cannot be directly obtained by Chaplygin's method. If we wish to represent these functions in a larger domain using Chaplygin's hypergeometric functions, we need several series developments, each of which represents the function in another part of the neighborhood of the singular point. It should be stressed that in considering the potentials, Φ, and stream functions, Ψ, there exists a basic difference between the incompressible and compressible fluid case: In the first case an analytic function $f = \Phi + i\Psi$ of a complex variable can be interpreted as a flow of an incompressible fluid, the flow being in general defined in the whole physical (x, y)-plane or the logarithmic (λ, θ)-plane, $\lambda = \log q$, that is, for all values of λ and θ. In the second case, the functions defined on Riemann surfaces arise in a quite natural manner, representing flow in the physical plane which has physical significance. But if we limit ourselves to the subsonic case, and attempt to develop in the (λ, θ)-plane a theory of compressible flows analogous to the theory of analytic functions of a complex variable, and in particular to develop an analogue to the theory of algebraic functions and their integrals, then Φ and Ψ are defined only for $\lambda < 0$. To the value $\lambda = 0$, there corresponds $M = 1$. For $M > 1$ the flow becomes supersonic, the equations for Φ and Ψ become hyperbolic, and the functions have basically different properties. As mentioned and used in previous chapters, if one wishes to apply operators to the flows which are partially subsonic and partially supersonic, it is useful to introduce as the class of functions to which the operator is to be applied, functions of $Z = \lambda + i\,\theta$, or $\overline{Z} = \lambda - i\,0$, defined in the (λ, θ)-plane or (M, θ)-plane. For $M < 1$, Z is complex, and for $M > 1$ it becomes real.

As has been shown in [10], the operators $P_{3,k}$, $k = 1, 2$, preserve the location of the singular points as well as the order of infinity. They transform branch points of analytic functions of a complex variable into branch points of the same order of functions satisfying the linear differential equations, so that the operator $P_{3,k}$ can be used successfully for generating and investigating these singularities. On the other hand, the operator $P_{3,k}$ when applied to analytic functions $g(Z)$ of a complex variable which posses a pole, yields solutions of differential equations which are infinite of the same order as $g(Z)$ and at the same point, but which are no longer single-valued. If we apply the operator $P_{3,k}$ to functions with logarithmic singularity whose real part is single-valued, we obtain functions whose real as well as imaginary parts are multi-valued. On the other hand, in connection with the transition to the physical plane the equation of single-valuedness of at least one of the quantities Φ or Ψ is of great importance, and therefore the question arises of defining procedures which generate real solutions of the compressibility equations which are logarithmic or infinite of the n-th order and single-valued. Briefly, it is convenient to use operators $P_{3,k}$ only in the case of branch points, defining the correspondence in the case of poles and logarithmic singularities in a different way.

Each of the operators, mentioned above, can be interpreted as a correspondence rule which associates with the complex potential of an incompressible fluid flow, defined in a domain B, a stream function of a compressible fluid flow which is defined in a domain B'. All three operators act primarily on functions which are regular in the domain in which they are considered, and they produce solutions which are regular in the domain in which they are defined.

It is natural to require that, insofar as possible, the stream functions and the potential functions of compressible fluid flows behave in the pseudo-logarithmic plane at singular points in the same manner as the corresponding functions for incompressible fluids in the logarithmic plane; that is, they become infinite of the same order, the geometrical structure of the stream lines and potential lines in a sufficiently small neighborhood of the singularity is essentially the same, and so on.

Let us now explain why single-valuedness of at least one of the components (that is, of the potential or stream function) is of importance for our purposes. Poles and logarithmic singularities in the logarithmic plane are, as a rule, images of doublets, vortices, sinks or sources at infinity in the physical plane. It is natural to require for compressible fluids that at infinity the speed is constant and that the motion has the same character as in the case of an incompressible fluid. If the potential or the stream function is known in the pseudo-logarithmic plane, we can come back

---

[10] S. Bergman: Certain classes of analytic functions of two real variables and their properties. Trans. Amer. Math. Soc. **57**, 299—331 (1945).

to the physical plane by using known expressions for x and y, cited in previous sections. It follows from those equations, that if the derivatives of at least one of the functions, Φ or Ψ, are single-valued, then, if we move around the singular point, the functions $x = x(u, v)$, $y = y(u, v)$ can each increase at most by a constant, say X, Y. In most cases we have at infinity a linear combination with constant coefficients of these singularities, and the constants have to be so chosen that the quantities X and Y vanish, and therefore $x(u, v)$ and $y(u, v)$ are single-valued. This is exactly the case for an incompressible fluid. If this property were not preserved, the flow would completely change its character. Indeed, suppose that $x(u, v)$ and $y(u, v)$ are multi-valued: then the functions $u(x, y)$, and $v(x, y)$ are periodic, that is, in the physical plane every value u, v, is assumed infinitely often. The question whether other types of singularities—in particular those for which both components are multi-valued—are of interest in hydrodynamical applications, remains open. The first step in answering this question would consist in studying the behavior of flow patterns in the logarithmic and physical planes in the neighborhood of these singularities.

3. The Behavior of Certain Types of Flows in the Case of an Incompressible Fluid

Following the line of our approach, that is, applying the principle of correspondence between incompressible and compressible fluid planes described in previous sections, we first have to describe the behavior in the logarithmic plane of various types of flows of an incompressible fluid, and in particular those possessing singularities such as sinks, sources, vortices, and so on.

Let us consider the flow of an incompressible fluid around a closed curve, a profile, say, and let us assume that neither the velocity nor the circulation vanishes at infinity. The complex potential in the physical (z-plane) plane may be expressed for $|z|$ sufficiently large, in the form

$$w(z) = \vec{q}_0 z + m \log z + \sum_{1}^{\infty} a_n z^{-n}, \qquad (6.3.1)$$

$$\vec{q}_0 \neq 0, \qquad m \neq 0,$$

where $\vec{q}_0$ is the velocity vector at infinity and the real constant m/i (m is an imaginary quantity) is the strength of the vortex at infinity. For the velocity vector $\vec{q}$ we obtain by differentiation with respect to z

$$\vec{q}(z) = \vec{q}_0 + m/z - \sum_{n=1}^{\infty} n\, a_n z^{n-1}. \qquad (6.3.2)$$

The expression for the complex potential in the logarithmic plane is [11]

$$W_1(\dot{\tilde{Z}}) = m\left[-(\bar{\tilde{Z}} - \bar{\tilde{Z}}_0) - \log i\,(\bar{\tilde{Z}} - \bar{\tilde{Z}}_0)\right] \qquad (6.3.3)$$

$$+ \text{Power series in } i\,(\bar{\tilde{Z}} - \bar{\tilde{Z}}_0),$$

where $\dot{\tilde{Z}} = \tilde{\lambda} + i\,\theta$, $\tilde{\lambda} = \log q$, $\tilde{Z}_0 = \dot{\lambda}_0 + i\,\theta_0$, $\tilde{\lambda}_0 = \log q_0$. The independent complex variable $\tilde{Z}$ may also be assumed to be equal to $\bar{Z} = \lambda - i\,\theta$; theoretically this change has no effect upon the considerations presented below.

In the most general case we can derive the stream function in a form:

$$\Psi^+ = \mu_0^{(1)} \Psi^{(1,\,1)} + \mu_0^{(2)} \Psi^{(1,\,2)} + \mu_0^{(3)} \Psi^{(L,\,1)} + \mu_0^{(4)} \Psi^{(L,\,2)} + \Psi_1.$$

If there is only circulation in the physical plane and no source, then $\mu_0^{(2)}$ and $\mu_0^{(3)}$ are determined solely by the circulation. Similarly $\mu_0^{(1)}$ and $\mu_0^{(4)}$ are determined solely by the source in the physical plane. This means that Ψ^+ splits into two pairs of singularities $(\Psi^{(1,1)}, \Psi^{(L,2)})$ and $(\Psi^{(1,2)}, \Psi^{(L,1)})$. The first pair is absent if there is no source in the physical plane, and the second if there is no circulation. In case there are both source and circulation, all the terms in (6.3.4) must be preserved. Thus we have:

(1) (q_0, θ_0) determines location of singularities in pseudo-logarithmic plane,

(2) source strength alone in physical plane determines $\mu_0^{(1)}$ and $\mu_0^{(4)}$,

(3) circulation alone in physical plane determines $\mu_0^{(2)}$ and $\mu_0^{(3)}$.

Assuming the most general case, we can derive the stream function in a sufficiently small neighborhood of $i\widetilde{\overline{Z}} = i\widetilde{\overline{Z}}_0$ in the form:

$$\Psi^+ = \sum_1^4 G_i + \ldots, \tag{6.3.4}$$

$$G_1 = \mu_0^{(1)} (\widetilde{\lambda} - \widetilde{\lambda}_0) \left[(\widetilde{\lambda} - \widetilde{\lambda}_0)^2 + (\theta - \theta_0)^2\right]^{-1}, \tag{6.3.4a}$$

etc., where the dots indicate terms which go to zero as $\widetilde{\overline{Z}} \to \widetilde{\overline{Z}}_0$ and therefore can be neglected in considering the behavoir of Ψ^+ in a sufficiently small neighborhood of $i\widetilde{\overline{Z}} = i\widetilde{\overline{Z}}_0$. This point corresponds to the point at infinity in the z-plane.

Thus we see that $W_1(\widetilde{Z})$ and Ψ^+ have singular points for $i\widetilde{\overline{Z}} = i\widetilde{\overline{Z}}_0$. One may verify that $z = z(\theta, \widetilde{\lambda}) = x + iy$, is single-valued at the singular point $\theta_0, \widetilde{\lambda}_0$. Following our principle of introducing (in the pseudo-logarithmic plane) for the compressible fluid, functions which have a behavior similar to that of the corresponding functions in the logarithmic plane for the incompressible fluid case, we see that we should consider flow patterns whose stream functions Ψ possess, as singularity at $Z = Z_0$, a linear combination of a logarithmic singularity and two independent singularities of the first order. Suppose a flow, which symbolically we shall denote as $\overline{\gamma}(\Phi, \Psi)$ (of a compressible or incompressible fluid) has a singularity at the point (λ_0, θ_0). If at least one of the functions, Φ or Ψ, is single-valued at this point, the singularity of $\overline{\gamma}$ at (λ_0, θ_0) will be said to be of type S. For example, the flow $\widetilde{\Phi} = \log\left[(\theta - \theta_0)^2 + (\widetilde{\lambda} - \widetilde{\lambda}_0)^2\right]/2$, $\Psi = \arctan\left[(\theta - \theta_0)(\lambda - \lambda_0)^{-1}\right]$, represents a flow possessing a singularity of type S. As we indicated above, this property plays an essential role for our purposes. In one of the following sections we shall describe a method of generating required singularities of type S in a compressible fluid flow.

4. The behavior of a subsonic compressible flow at infinity

To explain the behavior of a subsonic compressible flow at infinity, let us discuss some examples. At the point $a = \alpha + i\beta$, α, β, real, of the hodograph plane which corresponds to the point $z = (x + iy) \to \infty$ of the physical plane, the stream function has a singularity. This fact leads to the study of the singularities of functions satisfying equations in the pseudo-logarithmic plane. Let us put down the equation of a compressible fluid flow in forms

$$\Psi_{\lambda\lambda}{}^* + \Psi_{\theta\theta}{}^* + 4F_m(2\lambda)\,\Psi^* = 0, \tag{6.4.1}$$

or

$$\Psi_{Z\overline{Z}}{}^* + F_m\,\Psi^* = 0. \tag{6.4.2}$$

A solution of the equation (6.4.2) may be written in the form

$$\Psi^*(\lambda, \theta) = Im\left[g(\bar{Z}) + \sum_{n=1}^{\infty} \frac{(2n)!}{2^{2n} n!} Q^{(n)}(2\lambda) \cdot \int_0^{\bar{Z}} \cdots \int_0^{\bar{Z}_{n-1}} g(\bar{Z}_n) d\bar{Z}_n \cdots d\bar{Z}_1\right]. \quad (6.4.3)$$

Since $g(\bar{Z})$ is an arbitrary function we may choose it so as to fulfill certain conditions. If we know that the point corresponding to $z \to \infty$ should be a branch point, then we choose $g = (a - \bar{Z})^{-1/2}$ and we obtain a singularity which possesses the desired features. Indeed, let us insert g into (6.4.3) and the function

$$\Psi(\lambda, \theta) = H\,Im\left\{(a - \bar{Z})^{-1/2} - Q^{(1)}[(a - \bar{Z})^{1/2} - a^{1/2}] + \right.$$

$$\left. + \frac{3}{2} Q^{(2)}[(a - \bar{Z})^{3/2} - a^{1/2}\bar{Z} - \frac{2}{3} a^{3/2}] + \ldots\right\}, \quad (6.4.4)$$

with $\bar{Z} = \lambda - i\theta$, is a stream function which is two-valued at point a and becomes infinite as $(a - \bar{Z})^{-1/2}$. If, however, the point a is not a branch point, that is, if (6.4.3) is applied to the function $g = (a - \bar{Z})^{-1}$, (a pole), say, then

$$\Psi(\lambda, \theta) = H\,Im\{(a - \bar{Z})^{-1} + (1/2) Q^{(1)}[\log(a - \bar{Z}) - \log a] + \ldots\}, \quad (6.4.5)$$

is obtained, which is not a single-valued function because $\log(a - \bar{Z})$ is a many valued function. Similarly the function $g = \log(a - \bar{Z})$ will originate a many-valued function. The function (6.4.5) can, however, be made one-valued by replacing the many-valued term $(\theta = \arg \bar{Z})$ by its mean value (in the sheet under consideration). Clearly, this new function will no longer be an exact solution of the equation in the pseudo-logarithmic plane, but in many instances it will not differ very much from an exact solution. Naturally, this procedure may be refined.

5. Fundamental Solutions in the Subsonic Range

It is, however, of interest from a theoretical point of view to determine exact solutions of (6.4.2) which are single-valued and have a logarithmic singularity at point a. Clearly, it is sufficient to find functions for equation (6.4.2) which possess a logarithmic singularity. A function

$$W^*(\lambda, \theta; \lambda_0, \theta_0) = W(Z, \bar{Z}; Z_0, \bar{Z}_0) =$$

$$= A(Z, \bar{Z}; Z_0, \bar{Z}_0)\log|\bar{Z} - \bar{Z}_0| + B(Z, \bar{Z}; Z_0, \bar{Z}_0), \quad (6.5.1)$$

which considered as a function of Z, $\bar{Z}$, satisfies equation (6.4.2) in the whole plane, except at the point $\bar{Z} = \bar{Z}_0$, is termed a "fundamental solution" of equation (6.4.2) with the affix at $\bar{Z} = \bar{Z}_0$. Then, clearly, $\Psi = W(Z, \bar{Z}; a, \bar{a})$ represents a desired stream function with a singularity at $\bar{Z}_0 = a$. The corresponding flow can be said to have a pseudo-vortex (we may also call it a flow with a "vortex-like" singularity at λ_0, θ_0). The names "vortex-like", "source-like", etc., are used although there is no direct relation between a source-like singularity and a source in the physical plane. In the case of an incompressible fluid (and in the physical plane) the stream function of a vortex flow is given by the expression $m\log(z - z_0)$, and in the case of a source the potential function is given by $m\log(z - z_0)$, m being a constant. We note that in general A and B are not analytic functions of one complex variable Z (or Z_0). In order to emphasize this, they are often denoted as $A(Z, \bar{Z}; Z_0, \bar{Z}_0)$, $B(Z, \bar{Z}; Z_0, \bar{Z}_0)$, respectively.

In order to find the functions A and B a new variable will be introduced, namely

$$\bar{Z}_1 = \bar{Z} - \bar{Z}_0. \tag{6.5.2}$$

Equation (6.4.2) then becomes

$$\Psi^*_{Z_1 \bar{Z}_1} + F_m \, \Psi^* \, (Z_1 + Z_0, \bar{Z}_1 + \bar{Z}_0) = 0. \tag{6.5.3.}$$

A fundamental solution of (6.5.3) with the affix at $\bar{Z}_1 = 0$, will be a fundamental solution of (6.4.2) with the affix at $\bar{Z} = \bar{Z}_0$. The functions A and B are given by the equations:

$$A = 1 - \int_0^{Z_1} \int_0^{\bar{Z}_1} F_m \, dZ_2 \, d\bar{Z}_2 + \int_0^{Z_1} \int_0^{\bar{Z}_1} F_m \left[\int_0^{Z_2} \int_0^{\bar{Z}_2} F_m \, dZ_3 \, d\bar{Z}_3 \right] dZ_2 \, d\bar{Z}_2 + \dots, \tag{6.5.4}$$

$$B = \int_0^{Z_1} \int_0^{\bar{Z}_1} G \, dZ_2 \, d\bar{Z}_2 - \int_0^{Z_1} \int_0^{\bar{Z}_1} F_m \left[\int_0^{Z_2} \int_0^{\bar{Z}_2} G \, dZ_3 \, d\bar{Z}_3 \right] dZ_2 \, d\bar{Z}_2 + \dots, \tag{6.5.5}$$

$$G = (A_{\bar{Z}_1}/Z_1) + A_{Z_1}/\bar{Z}_1). \tag{6.5.6}$$

Not only the function W (6.5.1) satisfies equation (6.4.2); the expression

$$\frac{\partial W^*}{\partial \theta} = A_1 \, (\theta - \theta_0) \, [(\lambda - \lambda_0)^2 + (\theta - \theta_0)^2]^{-1/2} +$$

$$+ A_2 \log \left([\lambda - \lambda_0)^2 + (\theta - \theta_0)^2 \right] + A_3, \tag{6.5.7}$$

where A_i, $i = 1, 2, 3$, are entire functions, is also a solution of that equation.

In the case where the pressure-density relation is of the (simplified) form $p = A + B/\varrho$, the functions W^* and $\partial W^*/\partial \theta$ are

$$(1/2) \log \left[(\lambda - \lambda_0)^2 + (\theta - \theta_0)^2 \right], \tag{6.5.8}$$

and

$$(\theta - \theta_0) \, [(\lambda - \lambda_0)^2 + (\theta - \theta_0)^2]^{-1}, \tag{6.5.9}$$

respectively. It is noted that in this particular case new singularities are obtained by differentiating W^* with respect to λ.

Let us differentiate $\Psi = \Psi^{(L)} = W \, (Z, \bar{Z}; Z_0, \bar{Z}_0)$ with respect to the real and imaginary parts of the parameter Z. We obtain single-valued functions $\Psi^{(1, 1)}$ and $\Psi^{(1, 2)}$ with an infinity of the first order. Flow patterns generated by $\Psi^{,(L)} \, \Psi^{(1, 1)}$, $\Psi^{(1, 2)}$ have a simple physical meaning: $\Psi^{(L)}$ yields a vortex at infinity (in the physical plane), $\Psi^{(1, 1)}$, $\Psi^{(1, 2)}$ yield components of a doublet at infinity.

There are instances where it is more convenient to operate with the potential function Φ, rather than with the stream function Ψ. Functions Φ and Ψ satisfy the system of equations

$$\Phi_\theta = \Psi_H, \qquad l \, (H) \, \Psi_\theta = - \, \Phi_H. \tag{6.5.10}$$

Introducing the variable λ, we obtain

$$\Phi_\theta = \varrho^{-1} \, (1 - M^2)^{1/2} \, \Psi_\lambda, \qquad \varrho^{-1} \, (1 - M^2)^{1/2} \, \Psi_\theta = - \, \Phi_\lambda. \tag{6.5.11}$$

Differentiating the first equation (6.5.11) with respect to θ and the second with respect to λ, and applying a few transformations yields the equation

$$\Phi_{\lambda\lambda} + \Phi_{\theta\theta} - 4 \, N \, \Phi_\lambda \equiv 4 \, [\Phi_{Z\bar{Z}} - N \, \Phi_Z] = 0, \tag{6.5.12}$$

$$Z = \lambda + i \, \theta, \qquad \bar{Z} = \lambda - i \, \theta, \tag{6.5.13}$$

$$N = (- \, 1/8) \, (k + 1) \, M^4 \, (1 - M^2)^{-3/2}. \tag{6.5.14}$$

Equation (6.5.12) possesses a fundamental solution of the form (6.5.1); in this case a flow with a "source-like" singularity is obtained. In the physical plane it yields a source (or sink) at infinity.

6. Modified Integral Operator in the Subsonic Range

As mentioned above, in connection with the transition to the physical plane, it is of importance to have singularities for which at least one of the functions Ψ (stream function) or Φ (potential function) is single-valued at the singular point. Sometimes it is necessary to modify the operator described in previous chapters. If we have a function g (for example, a complex potential) which possesses poles and logarithmic singularities, we decompose it writing

$$g = g^* + \sum_{k=1}^{m} A_k (Z - Z_k)^{-\nu_k} + \sum_{k=m}^{n} A_k \log (Z - Z_k), \qquad (6.6.1)$$

where A_k are constants and the ν_k are positive integers. The function g^* possesses as its only singularities algebraic branch points.

Consider the equation (6.4.2). The integral operator may be chosen in the form of equation (6.4.3). Applying the integral operator to the first part of (6.6.1), we obtain a solution of (6.4.2), say $\Psi_1 = I\,m\,[p\,(g^*)]$. By introducing the fundamental solutions $\Psi^{(L)} = W$ in form of (6.5.1), we obtain solutions of (6.4.2) which have the same behavior as $\log (\overline{Z} - \overline{Z}_0)$. Adding to Ψ_1 the functions with singularities described above we obtain a solution of the compressibility equation, possessing the same characteristic features as g. We shall refer to this procedure for determining stream functions of a compressible fluid as the "modified integral operator". In the case of non-symmetric flow patterns with the velocity vector $\vec{q}_0 \exp (i\,\theta_0)$ at infinity $(z \to \infty)$, we obtain as a rule the stream function (in the λ, θ-plane) in the form

$$\Psi = A_0 \Psi^{(L)} + A_1 \Psi^{(1,\,1)} + A_2 \Psi^{(1,\,2)} + \Psi_1, \qquad (6.6.2)$$

where $\Psi_1 = \Psi_1(\overline{Z})$ is a function which has branch points of positive order as its only singularities, and which can be obtained by the operator from a suitably chosen function $g^*(\overline{Z})$ of one complex variable. In the case of symmetric flow patterns the representation (6.6.2) simplifies. In the latter case we can directly apply the operator to the function

$$g(\overline{Z}) = \sum_{\nu=-1}^{\infty} A\,(\overline{Z} - \overline{Z}_0)^{\nu/2}. \qquad (6.6.3)$$

The introduction of the modified integral operator enables us to develop a theory of compressible fluid flow which in the subsonic case strongly resembles that of an incompressible fluid (when we use the hodograph method). In the following chapters a generalization of two results of the latter theory will be discussed, namely: (a) the condition that a stream function defines a flow pattern around a closed curve; and (b) a generalization of the Blasius formulas.

7. Subsonic Compressible Fluid Flows with Singularities of Type S

Classical methods in the theory of partial differential equations yield almost immediately two independent flows with logarithmic singularities of type S. Assuming the equations of the stream function and the velocity potential of a compressible fluid flow in Z-plane or $i\,\overline{Z} = \theta + i\,\lambda$-plane, one can find the two pairs of fundamental solutions of those equations in the pseudo-logarithmic plane:

$$\Psi^{(L,\,1)}, \qquad \Psi^{(L,\,2)}, \qquad \Phi^{(L,\,1)}, \qquad \Phi^{(L,\,2)}.$$

The superscript L symbolizes that either the stream-function or the potential function has a logarithmic singularity. By means of a simple procedure, one may show that the flows $\bar{\gamma}^{(L,\,k)}$ $(\Phi^{(L,\,k)},\ \Psi^{(L,\,k)})$, $k = 1, 2$, are flows with singularities of type S and that both components $\Phi^{(L,\,k)}$, $\Psi^{(L,\,k)}$ of the $\bar{\gamma}^{(L,\,k)},$ $s, k = 1, 2$, are single-valued. Infinitely many new flows may be obtained from the first two linearly independent flows by a differentiation, namely

$$\bar{\gamma}^{(n,\,k)}\left(\frac{\partial^n \Phi^{(L,\,k)}}{\partial \theta^n}, \ \frac{\partial^n \Psi^{(L,\,k)}}{\partial \theta^n} \right), \quad n = 1, 2, \ldots \tag{6.7.1}$$

The components of those flows are derivatives of n-th order of the first potential and stream-functions, and they are infinite of n-th order at the point (θ_0, λ_0). The question may arise as to whether for some index n (and hence for all higher indices) the two flows $\bar{\gamma}^{(n,\,1)}$ and $\bar{\gamma}^{(n,\,2)}$ are not essentially the same: that is, whether functions $\Psi^{(n,\,1)}$ and $\Psi^{(n,\,2)}$ (and likewise $\Phi^{(n,\,1)}$ and $\Phi^{(n,\,2)}$) are not linearly dependent. The proof that this is not the case, as well as all the other proofs of statements cited in the present section, are given in [11].

A problem of considerable interest is that of determining a flow of a compressible fluid around a given profile, or at least around a profile the shape of which approximates the given profile. Since, in many instances, by reasoning from the incompressible case, the approximate image in the hodograph (λ, θ)-plane is known (the image in the logarithmic plane of an incompressible fluid flow around a profile is supposed to be used as a first approximation of the image in the (λ, θ)-plane of the flow of a compressible fluid around a similar profile), it is possible to consider, instead of the above problem, the question of determining a flow for a given hodograph. The behavior at the point of the hodograph corresponding to $z = \infty$ is prescribed. Clearly, instead of the image in the hodograph plane, the image in the (λ, θ)-plane may be used. By means of formulas given in previous sections, it is possible to determine a function $\Psi_1(\lambda, \theta)$ satisfying the equation of a compressible fluid flow which possesses the required behavior at $z = \infty$. Naturally, $\Psi_1(\lambda, \theta)$ for the point $(\lambda_\infty, \theta_\infty)$ corresponding to the point $z = \infty$ of the physical plane, must have a singularity which satisfies the conditions indicated in section VI. 8 (below), in order that the flow in the physical plane will be a flow around a closed curve. Function $\Psi_1(\lambda, \theta)$ is not as yet the required stream-function, since it does not assume constant values on the boundary of the domain. But it is possible to determine this function in an approximate way by means of a procedure given in [12], p. 45.

8. The Transition from the Pseudo-logarithmic to the Physical Plane in the Subsonic Range

We shall discuss below the conditions that $\Psi(\lambda, \theta)$ defines in the physical plane a flow around a closed curve. As has been indicated above, the introduction of singularities of type S enables us to define in the pseudo-logarithmic plane, flow patterns which have in this plane a behavior similar (in the sense indicated above) to that of the flow pattern of an incompressible fluid in the logarithmic plane. The problem arises of determining the conditions that the flow defined in this manner in the pseudo-logarithmic plane will represent a flow around a closed curve in the physical plane. In the following discussion we shall present necessary and sufficient conditions to assure this.

If a flow has a singularity of type S at (λ_0, θ_0) then either Φ or Ψ is a single-valued function, and, according to the statements and equations cited in previous sections,

[11] S. Bergman: Two-dimensional subsonic flows of a compressible fluid and their singularities. Trans. Amer. Math. Soc. **61**, 452—498 (1947).

[12] S. Bergman: Methods for determination and computation of flow patterns of a compressible fluid. N. A. C. A., T. N. No. 1018 (1946).

x and y increase by constants if we go around the singular point (λ_0, θ_0). The constants X, Y, by which the functions $x(\lambda, \theta)$ and $y(\lambda, \theta)$ increase (if we move once counterclockwise around a singularity of type S of a flow $\bar{\gamma}$), will be denoted by T-periods. (The letter T is suggested by the expression "transition to the physical plane.") These constants X, Y, can be determined for the flows $\bar{\gamma}^{(L, 1)}$, $\bar{\gamma}^{(1, k)}$, (for details see [11]). Following the principle of correspondence, explained above, we introduce stream-functions which in the pseudo-logarithmic plane have a behavior similar to that of stream-functions of a incompressible fluid in the logarithmic plane. The images of a flow of an incompressible fluid in the logarithmic plane and that of a compressible fluid in the pseudo-logarithmic plane extend in general to infinity, since at a stagnation point $q = 0$ and therefore $\widetilde{\lambda} = \log q$ tends to $-\infty$. Below we shall consider only flows satisfying the following conditions:

(a) The stagnation point (or points), if any, lies on the boundary (but not in the interior) of the flow;

(b) the flow (that is, the stream and potential functions) possesses only a finite number of singularities (including branch points), so that for $|Z| \gg R$ the stream-function is regular at every point of the domain (call it P);

(c) the stream-function Ψ as well as Ψ_Z and $\Psi_{\bar{Z}}$ exist and are continuous on the boundary except perhaps at a finite number of points.

Assume that the stream-function Ψ of $\bar{\gamma}$ is defined in a (not necessarily schlicht) domain P of the pseudo-logarithmic plane and can be represented there by

$$\Psi = \mu_0^{(1)} \Psi^{(1, 1)} + \mu_0^{(2)} \Psi^{(1, 2)} + \mu_0^{(3)} \Psi^{(L, 1)} + \mu_0^{(4)} \Psi^{(L, 2)} + \Psi_1. \qquad (6.8.1)$$

Assume that Ψ_1 satisfies hypotheses (a), (b), (c), and that the functions $\Psi^{(1, 1)}$, $\Psi^{(1, 2)}$, etc., are stream-functions possessing at the point λ_0, θ_0 singularities of the type described in the section VI. 7. The point λ_0, θ_0 corresponds to $z \to \infty$. Under the hypotheses (a), (b), (c) and some additional assumptions, the following theorem can be proved ([11] and [6]):

T h e o r e m. The necessary and sufficient condition in order that the stream-function (6.8.1) defined in a simply-connected domain P represents a flow in the physical plane around a closed curve is that the following equations are fulfilled (called "the consistency relations"):

$$\mu_0^{(1)} X^{(1, 1)} + \mu_0^{(2)} X^{(1, 2)} + \mu_0^{(3)} X^{(L, 1)} + \mu_0^{(4)} X^{(L, 2)} = 0, \qquad (6.8.2)$$

$$\mu_0^{(1)} Y^{(1, 1)} + \mu_0^{(2)} Y^{(1, 2)} + \mu_0^{(3)} Y^{(L, 1)} + \mu_0^{(4)} Y^{(L, 2)} = 0, \qquad (6.8.3)$$

$X^{(1, k)}$, $Y^{(1, k)}$ being T-periods of $\bar{\gamma}^{(1, k)}$, $k = 1, 2$, $X^{(L, k)}$, $Y^{(L, k)}$ those of $\bar{\gamma}^{(L, k)}$.

We may derive another form of the conditions in order that the obstacle in the physical plane (which is formed by the image of the boundary curve h in the simply-connected domain H in the hodograph plane) be a closed curve. Assume that the stream-function Ψ defined in this domain H has a singularity at the point a (the image of $z = \infty$). It is assumed that the point $q = 0$ is not an interior point of domain H. Clearly, the necessary and sufficient condition in order that the image of H in the physical plane be single-valued is that

$$\int_l \varrho_0 \exp(i\,\theta)\, \varrho^{-1} \{[-(1 - M^2)\, q^{-2}\, \Psi_\theta + i\,\Psi_q\, q^{-1}]\, dq + [\Psi_q + i\,\Psi_\theta\, q^{-1}]\, d\theta\} = 0, \qquad (6.8.4)$$

where l is any simple closed curve lying entirely in $H + h$. The symbol $(H + h)$ means that at most the curve l can coincide with h. Since the integrand of (6.8.4) is a complete differential, the value of the integral does not change if l is continuously deformed without leaving $H + h$ and without passing through the singular point a. Thus, in particular, if the boundary curve h is chosen for l, it follows that $d\Psi = \Psi_q \, dq + \Psi_\theta \, d\theta = 0$, that is $d\theta = -\left(\Psi_q/\Psi_\theta\right) dq$ along h, since h, being the boundary curve of the body, must be a streamline. If this expression is inserted into (6.8.4), the condition is obtained that the image of h in the physical plane be a closed curve. Notice that H must be simply-connected.

On the other hand the relation (6.8.4) can also be written in a different form which is often more suitable for applications. Let us assume that the stream-function Ψ can be represented in the neighborhood of a (the image of $z = \infty$) in the form

$$\Psi = \sum_{n=0}^{k} A_n \, W^{(n)} + \Psi_1, \qquad (6.8.5)$$

where A_n are constants, Ψ_1 is a function which is regular at the point a and $W^{(n)}$ are fundamental solutions [see (6.5.1)]:

$$W^{(0)}(q, \theta; q_0, \theta_0) = W^*(\lambda, \theta; \lambda_0, \theta_0), \qquad (6.8.6)$$

$$W^{(n)}(q, \theta; q_0, \theta_0) = \frac{\partial W^{*(n)}(\lambda, \theta; \lambda, \theta_0)}{\partial \theta^n}. \qquad (6.8.7)$$

With every singularity $W^{(n)}$, $n = 0, 1, 2, \ldots$ there may be associated a complex number

$$R(W^{(n)}) = X_n + i \, Y_n, \qquad (6.8.8)$$

which is equal to the left-hand side of (6.8.4) with $\Psi_\theta = W_\theta^{(n)}$ and $\Psi_q = W_q^{(n)}$, and where l is an arbitrary simple closed curve around a, the direction of integration being such that a always lies to the left. The alternative form of conditions (6.8.2) and (6.8.3) is that

$$\sum_{n=0}^{k} A_n \, R(W^{(n)}) = 0. \qquad (6.8.9)$$

The evaluation of the quantities $R(W^{(n)})$ in the special case, when $k = 1$, is given in [13].

9. Generalization of the Blasius Formulae in the Subsonic Range

As is known, the behavior of the flow at infinity determines the force and the moment acting upon a profile immersed in a fluid. As one may show, the results obtained in previous chapters enable one to obtain for the force F and the moment M_{z0} about a given point $z_0 = (x_0, y_0)$ the formulas analogous to those holding in the incompressible case. For an incompressible fluid these quantities are given by the formulas of Blasius:

$$F = \frac{1}{2} i \, \varrho_0 \int_C \left(\frac{dw}{dz}\right)^2 dz, \qquad (6.9.1)$$

$$M_{z0} = -\operatorname{Re}\left[\frac{1}{2} \varrho_0 \int_C (z - z_0)\left(\frac{dw}{dz}\right)^2 dz\right]. \qquad (6.9.2)$$

Here ϱ_0 is the density of the fluid, w is the complex potential of the flow and C a control contour. Formulae (6.9.1) and (6.9.2) can be expressed in terms of the coordinates $\widetilde{\lambda}$, θ of the logarithmic plane. Bergman and Ludford generalized those formulas by taking into account the compressibility of the fluid.

---

[13] S. **Bergman**: On two-dimensional flows of compressible fluids. N. A. C. A., T. N. No. 972 (1945).

Moreover, Ludford proved the validity of the generalization of the Kutta-Joukowski formulae to the subsonic compressible case:

$$F_x = 0, \qquad F_y = \varrho_0 \, U \, \Gamma. \tag{6.9.3}$$

The moment M about the origin is:

$$M = \varrho_0 \, U^2 \, D_2, \tag{6.9.4}$$

with D_2 given in [6].

10. An Integral Formula Representing a Subsonic Flow Inside a Domain in Terms of Its Values on the Boundary

The Cauchy integral formula, cited in one of the previous sections, is one of the most powerful tools in the study of analytic functions of a complex variable. In the subsonic case there exists an analogue of this formula, which enables us to express the values of Φ and Ψ inside a domain B of the (λ, θ)-plane in terms of the values of Φ and Ψ on the boundary of B. The following theorem will be given without the proof:

Theorem: Let $\bar{\gamma}\,(\Phi, \Psi)$ be a flow which is regular in the bounded domain B (lying entirely within the subsonic region) whose boundary C is a simple closed curve. Then at every point (λ_0, θ_0) of B

$$2\,\pi\,\Phi\,(\lambda_0, \theta_0) = [H\,(2\,\lambda_0)]^{-1} \int_C \left[-\,H^2\,\Phi\,\frac{\partial \Phi^{(L,\,2)}}{\partial n} + \Psi\,\frac{\partial\,(H^2\,l^{1/2}\,\Phi^{(L,\,2)})}{\partial s} \right] ds, \tag{6.10.1}$$

$$2\,\pi\,\Psi\,(\lambda_0, \theta_0) = H\,(2\,\lambda_0) \int_C \left[-\,H^{-2}\,\Psi\,\frac{\partial \Psi^{(L,\,1)}}{\partial n} + \Phi\,\frac{\partial\,(H^{-2}\,l^{-1/2}\,\Psi^{(L,\,1)})}{\partial s} \right] ds, \tag{6.10.2}$$

where ∂/∂_n indicates differentiation in the direction of the inward normal, and $\Phi^{(L,\,2)}$ and $\Psi^{(L,\,1)}$ are the fundamental solutions.

11. A Proof of an Auxiliary Lemma in the Subsonic Range

In the following, the proof will be outlined that in the subsonic range the function $F\,(2\,\lambda)$ can be approximated by the polynomial

$$F_m\,(2\,\lambda) = \sum_{s=0}^{m} \alpha_s^{(m)} \exp\,(2\,s\,\lambda), \qquad \alpha_0^{(m)} = 0, \tag{6.11.1}$$

in every interval $(-\infty, \lambda_0)$, $\lambda_0 < 0$, and the method of determining the $F_m\,(2\,\lambda)$ will be indicated.

If $2\,\lambda$ is replaced by $\log X$, then $F\,[2\,\lambda\,(X)]$ is a continous function of X in the interval $(0, X_0)$, $X_0 < 1$, and by classical results, it is obvious that it can be approximated by a polynomial in X. It will be seen that it will not suffice merely to approximate F, but in addition to this it will be possible to require that any given number of higher derivatives of F be approximated in the interval $(-\infty, \lambda_0)$ by the corresponding derivatives of F_m. It is, however, of interest to give a more explicit form of the approximating polynomials. This will make it possible to determine the corrections of $Q^{(n)}$ given in previous sections on subsonic flow which have to be made in order to obtain functions $Q^{(n)}$ corresponding to a given F_m.

If M increases steadily from 0 to 1, λ increases from $-\infty$ to 0. Since the relation $M \leftrightarrow \lambda$ is a one-to-one correspondence, to every λ_0, $\lambda_0 < 0$ there corresponds an $M_0 = M\,(\lambda_0) < 1$. Therefore, if $\lambda \,\epsilon\,(-\infty, \lambda_0)$ then $M\,(\lambda)\,\epsilon\,(0, M_0)$. For $M \leq M_0$, the function

$$F = [64\,(1 - M^2)^3]^{-1}\,(k + 1)\,M^4\,[-(3\,k - 1)\,M^4 - 4\,(3 - 2\,k)\,M^2 + 16], \qquad (6.11.2)$$

may be approximated by a polynomial F_n of the $(2\,n + 8)$-th degree in M^2,

$$F_n\,(M) = (64)^{-1}\,(k + 1)\,M^4\,[-(3\,k - 1)\,M^4 - 4\,(3 - 2\,k)\,M^2 + 16] \cdot$$

$$\cdot \left[\sum_{\nu = 1}^{n} (\nu^{-3})\,(-1)^\nu\,M^{2\nu} \right]. \qquad (6.11.3)$$

Only a finite number of powers of M^2 appears in (6.11.3). It will now be shown that M^2 can be developed in the uniformly convergent series

$$M^2 = \sum_{\nu = 1}^{\infty} \beta_\nu\,X^\nu, \qquad (6.11.4)$$

$$X = 2\left[\frac{(k + 1)^{1/2} - (k - 1)^{1/2}}{(k + 1)^{1/2} + (k - 1)^{1/2}} \right] \left(\frac{k + 1}{k - 1} \right)^{1/2} \exp\,(2\,\lambda), \qquad (6.11.5)$$

with $\lambda < 0$. Instead of considering M^2, it is well to introduce

$$s = 1 - (1 - M^2)^{1/2}. \qquad (6.11.6)$$

Since $M^2 = 1 - (1 - s)^2$, it will suffice to determine the series for s. As one may easily verify, by simple calculations one obtains

$$X = \left(\frac{s}{2 - s} \right) \left\{ 2\, \frac{(h^{-1} + 1 - s)}{(h^{-1} - 1 + s)} \left(\frac{h^{-1} - 1}{h^{-1} + 1} \right)^{1/h} \right\}, \qquad (6.11.7)$$

$$h = \left(\frac{k - 1}{k + 1} \right)^{1/2}, \quad k > 1. \qquad (6.11.8)$$

It is possible to prove that the image of the area of the circle $|s| < 1$ is a schlicht domain in the X-plane. That is, one can show ([13], p. 58 and 59) that the boundary curve of the image of $|s| < 1$ is a curve which does not intersect itself. By classical theorems of the theory of functions, the domain bounded by this curve is schlicht. Clearly, it includes in its interior the domain $|X| < 1$. Since the image of $|s| < 1$ is schlicht and includes $|X| < 1$, the inverse function $s = s\,(X)$ is regular in $|X| < 1$ and can be expanded in $|X| < 1$ in the form of an infinite series

$$s\,(X) = \sum_{\nu = 1}^{\infty} \beta_\nu\,X^\nu. \qquad (6.11.9)$$

With some additional proofs it is possible to show that (6.11.9) yields the required approximation.

12. Singularities in Transonic Flow

In one of the previous sections we indicated a procedure for constructing a mixed flow (partially supersonic) around an obstacle the boundary of which is a closed curve. If the boundary is prescribed, then probably, in many instances, no solution of the problem exists. Unfortunately, no definite results in this direction are known. However, investigations of problems of similar type seem to indicate that the above is the case. Therefore the problem arises of finding necessary and sufficient conditions in order that a mixed flow pattern may exist around an obstacle. In considering this problem, two cases have to be distinguished[4]:

1. The behavior of the flow at infinity is completely prescribed (for example, a uniform flow with no vorticity), or,

2. the flow at infinity possesses certain properties, but any solution will be considered as admissible if its behavior at infinity is of a certain type.

The second problem is to determine when the solution is unique and, further, under what conditions it is "stable". This situation suggests considering a problem which is much simpler mathematically (but still quite difficult), and, in certain instances, can give the type of answer desired by a research engineer, namely, the question whether to a given hodograph there exists a flow possessing the prescribed behavior at infinity. Often it will suffice to know not whether a flow pattern exists around some prescribed (in the physical plane) profile, but only whether such a flow exists with prescribed (or approximately prescribed) velocity distribution. The whole investigation may be shifted to the hodograph plane, and the problem considered as formulated above. As an example, a flow will be considered in order to determine whether to a given hodograph there corresponds some mixed flow. While, as is known in the case of a subsonic flow, it is more convenient to operate in the (λ, θ)-plane (pseudo-logarithmic plane), in the case of a mixed flow it is convenient to return to the (H, θ)-plane, since $\lambda(M)$ becomes imaginary for $M > 1$. As λ and H are connected by the relation $d\lambda = [l(H)]^{-1/2} dH$, $l(H) = (1 - M^2)\varrho^{-2}$, it does not present any theoretical difficulty to replace the variable λ by the variable H, in functions $S_k(\lambda, \theta; \lambda_0, \theta_0)$, $k = 1, 2, 3$, which are assumed to represent a pseudo-vortex and two pseudo-doublets, respectively (see section VI. 5). Thus, functions are obtained which shall be denoted by

$$T_k(H, \theta; H_0, \theta_0) = S_k(\lambda, \theta; \lambda_0, \theta_0). \qquad (6.12.1)$$

Let us now assume that there exists a function which at the point (H_0, θ_0) (which is the image of the point at infinity in the physical plane) behaves like

$$A_1 T_1(H, 0; H_0, \theta_0) + A_2 T_2(H, \theta; H_0, \theta_0) + A_3 T_3(H, \theta; H_0, \theta_0) + R(H, \theta), \quad (6.12.2)$$

where $R(H, \theta)$ is a regular function of H, θ. Naturally, A_i, $i = 1, 2, 3$, must satisfy the conditions discussed above, in order that in the physical plane a flow around a closed curve is obtained (6.8.9). Now, the question arises whether or not a function $R(H, \theta)$ can be found such that the expression (6.12.2) vanishes on the boundary line of the hodograph. Without going into too many details of a more advanced mathematical nature, let us note that one can find a necessary condition for the solution of this problem (for more details see [4], pp. 40—44, or [14]).

Obviously, if the function $\Psi(x, y)$ is to yield a flow pattern around a closed profile in the physical plane, it is necessary that

(a) the streamline $\Psi(x, y) = 0$ (or a part of this streamline) yield a closed curve in the physical plane;

(b) that every point of the physical plane be covered at most once (that is that the mapping onto the (x, y)-plane be schlicht). It is remarked, however, that if the whole image in the (x, y)-plane is not schlicht, the flow pattern obtained may nevertheless be of physical interest since in some instance it is required to determine flow patterns covering only part of the (x, y)-plane or to use only a part of the flow pattern.

The necessary and sufficient conditions in the subsonic case that the boundary in the hodograph plane when mapped onto the physical plane yield a closed curve in the latter plane have been established in previous chapters for flow patterns that are not necessarily symmetric. In the subsonic case, the determinant of the mapping from the hodograph to the physical plane $\partial(x, y)/\partial(u, v)$ is always positive. It is therefore sufficient and necessary, in order that the mapping be schlicht, that the boundary curve in the physical plane be schlicht, and that the mapping satisfy, at infinity, certain regularity conditions which will be stated below. In some cases the schlichtness

---

[14] S. Bergman: On tables for the determination of transonic flow patterns. Hans Reissner Anniversary Volume: Contributions to Applied Mechanics. J. W. Edwards (1949).

of the boundary curve follows from the symmetry of the image in the hodograph plane and a few other conditions[14]. The mapping must satisfy the following conditions at infnity: As a doubly covered circle about the point A (which represents the image of infinity in the hodograph plane) shrinks to a point, its image in the inverted physical plane $x' + i\,y' = (x + i\,y)^{-1}$ must be schlicht and also shrink to a point.

In generalizing these results to transonic flow patterns, one must be on guard against the following difficulties:

(a) the functional determinant $\partial(u, v)/\partial(x, y)$ ist not necessarily positive,

(b) the mixed character of the equation of flow may give rise to hyperbolic singularities which do not appear in the subsonic case.

Part VII

Review of Other Methods

Below a brief review of some other methods in the theory of compressible fluid flow is given in order that the reader may have a clear picture of some possibilities at the present time in that field. The author particularly emphasizes the fact that not all available methods are included in the present brief review but only some of them for illustrative purposes. All shock-phenomena and -theories, the methods of characteristics, limiting lines, a thorough discussion of the conical field theory, unsteady flow, special two- and three-dimensional airfoil theories, etc., are excluded from this review.

1. Methods of Expansion in Powers of the Mach Number

Janzen[1] (1913) starting with the well-known solution for the incompressible fluid flow as the initial approximation obtained the first approximation to the compressible fluid flow. He employed the method of successive approximations in Mach number. The method may be briefly described. Through insertion of the initial, or first, approximation into the right-hand side of compressible fluid flow equation, written in the form $\Phi_{xx} + \Phi_{yy} = F\,(M\,(x, y))$, the first (or second) approximation is computed on the base of the previously given boundary conditions. The differential equation, cited above, is linear and therefore simpler than the rigorous non-linear equation. In principle the method may be continued in an analogous way to any desired further approximations, where the question of convergence from case to case must be studied. The amount of calculation is considerable and increases with each further approximation. Janzen and Rayleigh[2] (1916) have applied the method to the compressible flow about a circular cylinder. Rayleigh applied it also to the flow past a sphere. Both Janzen and Rayleigh calculated only first approximations in each case. It turns out that the iteration process converges only as long as the critical velocity is not reached anywhere; the method therefore provides only the pure subsonic flow around a body. The failure of the method when applied to a flow with supersonic velocity has no physical, but only a purely formal mathematical base; since, of course, the equations of the iteration process throughout are of the elliptic type, they cannot lead to solution of the hyperbolic potential equations of supersonic flow.

---

[1] O. Janzen: Beitrag zu einer Theorie der stationären Strömung kompressibler Flüssig-keiten. Physik. Z. **14**, 639—643 (1913).

[2] Lord Rayleigh: On the flow of compressible fluid past an obstacle. Phil. Mag. (6) **32**, 1—6 (1916), or Papers, 6, 402.

Lamb[3] (1928) applying Janzen-Rayleigh method of successive approximations calculated the first approximation to the velocity potential of the circulatory flow past a circular cylinder.

Poggi[4] (1932, 1934) applying basically the method of expansion in a series replaced the compressible fluid flow by an incompressible fluid flow due to a suitable distribution of sources throughout the region of flow. He calculated the second approximation to the flow around a circular cylinder but did not take into account all the important and necessary items.

Pistolesi[5] (1935) calculated the second approximation to the flow around a circular cylinder but also in an incomplete form.

Prandtl[6] (Volta-Congress 1935) proposed a change in the Janzen-Rayleigh iteration process. The original Janzen-Rayleigh treatment starts with an incompressible flow as the initial approximation. According to Prandtl, the linearized flow should be used as a first approximation of an iteration method, whence with Mach numbers considerably different from zero. By this method there results from the very beginning a better approximation to the desired rigorous solution.

Hooker[7] (1936) employed the Janzen-Rayleigh method for the determination of the effects of compressibility upon the air flow past elliptic cylinders and a symmetrical Joukovski's section at Mach number 0.5. Only the case of the fluid streaming past the cylinders in the direction of their major axes was considered. The fineness ratio of the ellipse to which the method can be applied is limited. The method is an approximate one and is restricted to lower Mach numbers. The local velocity of sound cannot ever be exceeded at any point. In general the method may be applied to two-dimensional flow solutions around slender cylindrical bodies, the effect of compressibility being merely to distort slightly the streamlines from the positions they would have occupied if the fluid had been incompressible.

Imai[8] (1938) applying Janzen-Rayleigh method of successive approximations first found a solution of the flow of a compressible fluid past a circular cylinder correct to the second approximation. This same problem has been also rediscussed by Tamada and Saito[9] (1939) by means of Poggi's method and Imai's result has been confirmed.

Kaplan[10] (1938 and 1939) applied Poggi's method for the determination of the effects of compressibility upon the flow past symmetrical Joukovsky profiles and past circular and elliptic cylinders at zero angle of attack and with no circulation.

[3] H. Lamb: On the flow of a compressible fluid past an obstacle. R. and M. No. 1156, British A. R. C. (1928).

[4] L. Poggi: Campo di velocita in una corrente piana di fluido compressibile. L'Aerotecnica 12, 1579—1593 (1932). Caso dei profili ottenuti con representazione conforme dal cerchio ed in particolare dei profili Joukowsky. L'Aerotecnica 14, 532—549 (1934).

[5] E. Pistolesi: Atti dei convegni fondazione Allessandro Volta (Roma) 5, 283—328 (1935).

[6] L. Prandtl: Atti del 5. Convegno Volta, Rome 169—197 (1935).

[7] S. G. Hooker: The two-dimensional flow of compressible fluids at subsonic speeds past elliptic cylinders. R. and M., British A. R. C. No. 1684, 100—115 (1936).

[8] I. Imai: On the flow of a compressible fluid past a circular cylinder. Proc. Physico-Math. Soc. Japan 20, 636—645 (1938).

[9] K. Tamada and Y. Saito: Note on the flow of a compressible fluid past a circular cylinder. Proc. Physico-Math. Soc. Japan 21, 403—409 (1939).

[10] C. Kaplan: Compressible flow about symmetrical Joukowski profiles. N. A. C. A., Tech. Rep. No. 621 (1938). Two-dimensional subsonic compressible flow past elliptic cylinders. N. A. C. A., Tech. Rep. No. 624 (1938). A theoretical study of the moment on a body in compressible fluid. N. A. C. A., Rep. No. 671, 581—598 (1939).

Comparison with Hooker's treatment shows that for thick elliptic cylinders the two methods agree very well. The method of Poggi has the advantage in comparison with the original Janzen-Rayleigh method that it not only permits an unrestricted thickness ratio of the ellipse but also reduces the labor of computation. Kaplan's solution of the flow around a circular cylinder contains a few incorrect terms.

Pistolesi[11] (1939) gave the correct values of the second approximation to the flow around a circular cylinder in a full agreement with the results of Imai, and Tamada and Saito.

Tamada[12] (1939 and 1940) and Kaplan[13] (1940) attacked the problem of the flow of a compressible fluid past a sphere by means of the method of expansion in powers of the Mach number. Tamada applied Poggi's method and calculated the second approximation. In connection with this type of flow there are interesting experimental results obtained by Pasqualini[14] (1927 and 1931).

Lamla[15] (1940) attacked the problem of the flow past a circular cylinder by this method.

Imai and Aihara[16] (1940) applied an improved Janzen-Rayleigh method to the subsonic compressible fluid flow past an elliptic cylinder.

Tomotika and Tamada[17] (1940) applied M^2-expansion (Janzen-Rayleigh) to a flow around an elliptic cylinder. They derived a general method for any angle of attack.

Tomotika and Umemato[18] (1941) used Poggi's method to calculate the flow past a symmetrical Joukovsky airfoil by means of a transformation from a circle.

Imai[19] (1941) calculated the flow of a compressible fluid past a circular cylinder using Janzen-Rayleigh method exactly to M^6. Similarly in 1944[20] he applied it to an elliptic cylinder.

Eser[21] (1942) in his dissertation calculated the third approximation (M^6) to the flow around a circular cylinder.

Kaplan[22] (1942) improved the Janzen-Rayleigh method and discussed the use of residue theory for treating the subsonic flow of a compressible fluid.

---

[11] E. Pistolesi: Aerotechnica 19, 1064—1068 (1939).

[12] K. Tamada: On the flow of a compressible fluid past a sphere. Proc. Physico-Math. Soc. Japan 21, 403—409 (1939). Ibid., 743. Further studies on the flow of a compressible fluid past a sphere. Ibid. 22, 519—525 (1940).

[13] C. Kaplan: The flow of a compressible fluid past a sphere. N. A. C. A., T. N. No. 762 (1940).

[14] C. Pasqualini: Atti 4⁰ Congresso Internazionale di Navigazione Aerea (Roma) 4 (1927). Rivista Aeronautica 7, 443—457 (1931).

[15] E. Lamla: Die symmetrische Potentialströmung eines kompressiblen Gases um einen Kreiszylinder im Kanal im unterkritischen Gebiet. Luftfahrtforschung 17, 329—331 (1940).

[16] I. Imai and T. Aihara: On the subsonic flow of a compressible fluid past an elliptic cylinder. Rep. Aeron. Res. Inst. Tokyo Imp. Univ. No. 194, 15 (1940).

[17] S. Tomotika and K. Tamada: Studies on the subsonic flow of a compressible fluid past an elliptic cylinder. Rep. Aeron. Res. Inst. Tokyo Imp. Univ. No. 201, 15, 481—551 (1940).

[18] S. Tomotika and H. Umemato: On the subsonic flow of a compressible fluid past a symmetrical Joukowski aerofoil. Aeron. Res. Inst. Tokyo Imp. Univ. 16, 35—125 (1941).

[19] I. Imai: On the flow of a compressible fluid past a circular cylinder, II. Proc. Physico-Math. Soc. Japan 23, 180 (1941).

[20] I. Imai: Note on the velocity distribution round an elliptic cylinder at high speeds. Proc. Physico-Math. Soc. Japan 26, 71 (1944).

[21] F. Eser: Dissertation, München (1942).

[22] C. Kaplan: On the use of residue theory for treating the subsonic flow of a compressible fluid. N. A. C. A., Rep. No. 728 (1942).

Heaslet[23] (1944) using Janzen-Rayleigh method obtained the potential function for flow, with circulation, of a compressible fluid about a circular cylinder in series form including terms of the order M^4.

Goldstein and Lighthill[24] (1944) applied Janzen-Rayleigh method for finding the first-order effect of compressibility on the potential flow past a solid body when the irrotational incompressible flow is known (by use of the complex variable).

Imai[25] (1947) applied the Janzen-Rayleigh method to the calculation of the flow past a flat plate exactly to M^8.

Wendt[26] (1948) treated this method by use of complex variables. Ehlers[27] (1949) discussed various applications of Janzen-Rayleigh method.

2. Method of Expansion in Powers of a Thickness Parameter (Small Perturbations)

Ackeret[28] (1925) linearized the equation of the velocity potential in the supersonic case (hyperbolic type) applying the small perturbation method and derived the equation for lift and drag.

Ackeret[28] (1928) applied the iteration process consisting in developing the velocity potential function or the stream-function in powers of a geometrical characteristic parameter of the body (like thickness or camber) to the flow past elongated bodies of small thickness. In that way he obtained an approximation that applies to the entire subsonic range of velocity. He applied also this same method to flow along a wave-shaped wall. The method of expansion in powers of the amplitude of the wave was used. Different names are attached to this method and its variations: Ackeret iteration process, small perturbations method or Prandtl-Busemann method[29] (1929). A brief discussion on the subject of two methods given above may be interesting. Janzen-Rayleigh development of the velocity potential or the stream function in a power series of the stream Mach number, and the Ackeret iteration method are complementary in nature. The complementary nature of these two methods lies in the fact that the Janzen-Rayleigh procedure yields accurate results in the case of thick bodies, for which the critical stream Mach numbers are low; whereas the Ackeret iteration process yields accurate results in the case of slender bodies, for which the critical Mach numbers are in the neighborhood of unity.

Prandtl-Glauert formula[30] (1928–30) is based on the small perturbations method. With the assumption that the disturbances introduced by a solid body to the parallel

<hr>

[23] M. A. Heaslet: Compressible potential flow with circulation about a circular cylinder. N. A. C. A., A. R. R. No. 4 A 08 (1944), or N. A. C. A., Rep. No. 780 (1944).

[24] S. Goldstein and M. J. Lighthill: Two-dimensional compressible flow past a solid body in unlimited fluid or symmetrically placed in a channel. Phil. Mag. **35**, 549—568 (1944).

[25] I. Imai: On the hodograph method in the theory of compressible flow (in Japanese). Ooyôsûgaku-Rikigaku (The Applied Mathematics and Mechanics) **1**, 147 (1947).

[26] H. Wendt: The Janzen-Rayleigh approximation for the calculation of subsonic flows. Heidelberger Akad. Wiss., math.-nat. Kl., Nr. 7 (1948).

[27] F. B. Ehlers: Methods of linearization in compressible flow. Part. I. Janzen-Rayleigh Method. Air Material Command Technical Report No. F-TR-1180A-ND (1948), Dayton, Ohio.

[28] J. Ackeret: Luftkräfte auf Flügel, die mit größerer als Schallgeschwindigkeit bewegt werden. ZFM. 16, 72—74 (1925). Über Luftkräfte bei sehr großen Geschwindigkeiten, insbesondere bei ebenen Strömungen. Helv. Physica Acta 1, 301—322 (1928).

[29] L. Prandtl und A. Busemann: Näherungsverfahren zur zeichnerischen Ermittlung von ebenen Strömungen mit Überschallgeschwindigkeiten. Stodola-Festschrift, Zürich (1929).

[30] L. Prandtl: Remarks on paper by A. Busemann entitled, „Profilmessungen bei Geschwindigkeiten nahe der Schallgeschwindigkeit (im Hinblick auf Luftschrauben)". Jb. W. G. L. 95—99 (1928). Über Strömungen, deren Geschwindigkeiten mit der Schallgeschwindigkeit vergleichbar sind. J. Aeron. Res. Inst. Univ. Tokyo No. 65, 14 (1930). — H. Glauert: The effect

flow are small, the velocity potential equation can be linearized and solved in certain cases by means of simple methods. An example of this procedure is just the well-known Prandtl-Glauert formula. From the nature of this formula it is obvious that it is applicable to flows over bodies of small thickness ratio. Basically, the Prandtl-Glauert formula may be considered as the first term in an expansion in powers of a thickness parameter.

Taylor[31] (1932) discussed the application of Ackeret's supersonic theory to certain problems in aeronautics.

Busemann and Walchner[32] (1933) and Busemann[33] (1935) extended the Ackeret's procedure by including the second order approximation when it was found that certain of the first-order results were not in complete accord with experiments.

Riabouchinsky[34] (1938) applied the method of small perturbations to three-dimensional problems.

Göthert[35] (1940) applied also the method of small perturbations to two- and three-dimensional problems.

Görtler[36] (1940) applied the method of expansion in powers of a thickness parameter to the flow past profiles of a small camber in the subsonic range.

Hantsche and Wendt[37] (1942) applying this method used the results of the second approximation. They discussed also the regions where the transformation is not allowed (stagnation points).

Goldstein and Young[38] (1943) applied the development in powers of a thickness parameter to wind-tunnel interference. They also listed more important results of the linearized theory correctly except for some points concerning axisymmetrical flow which Sears has cleared up (1946 and 1947).

Hantsche[39] (1943) treated the case of an elliptic cylinder at zero incidence in a uniform stream according to Ackeret's process.

Kaplan[40] (1943) attacked the problem of the flow of a compressible fluid past

of compressibility on the lift of an airfoil. Proc. Roy. Soc. London, Ser. A 118, 113—119 (1928); also, R. and M., British A. R. C. No. 1135 (1928).

[31] G. I. Taylor: The flow of air at high speeds past curved surfaces. British A. R. C., R. and M. No. 1381 (1930).

[32] A. Busemann und O. Walchner: Profileigenschaften bei Überschallgeschwindigkeit. Forsch. Gebiete Ingenieurwes. 4, 87—92 (1933). Also: R. T. P. Translation No. 1786, British Ministry of Aircraft Production.

[33] A. Busemann: Aerodynamischer Auftrieb bei Überschallgeschwindigkeit. Luftfahrtforschung 12, 210—220 (1935).

[34] D. Riabouchinsky: Équations approchées des mouvements troisdimensionnels d'un fluide parfait compressible. C. r. acad. sci., Paris 206, 472—475 (1938).

[35] B. Göthert: Ebene und räumliche Strömung bei hohen Unterschallgeschwindigkeiten. Lilienthal-Gesellschaft für Luftfahrtforschung, Bericht 127, 97—101 (1940). Also: Plane and three-dimensional flow at high subsonic speeds. N. A. C. A., T. M. No. 1105 (1946).

[36] H. Görtler: Gasströmungen mit Übergang von Unterschall- zu Überschallgeschwindigkeiten. Z. angew. Math. Mechan. 20, 254—262 (1940).

[37] W. Hantsche und H. Wendt: Der Kompressibilitätseinfluß für dünne, wenig gekrümmte Profile bei Unterschallgeschwindigkeiten. Z. angew. Math. Mechan. 22, 72—86 (1942). Also: R. T. P. Translation No. 2198, British Ministry of Aircraft Production; Roy. Aero. Soc. J. 1943, 406—407.

[38] S. Goldstein and A. D. Young: The linear perturbation theory of compressible flow with application to wind tunnel interference. British A. R. C. 6865, Aero. 2262, FM. 601 (1943), or British A. R. C., R. and M. 1909 (1943).

[39] M. Hantsche: Die Prandtl-Glauertsche Näherung als Grundlage für ein Iterationsverfahren zur Berechnung kompressibler Unterschallströmungen. Z. f. angew. Math. Mechan., 23, 185—199 (1943).

[40] C. Kaplan: The flow of a compressible fluid past a curved surface. N. A. C. A., ARR. No. 3K02 (L 320) (1943) or N. A. C. A., Rep. No. 768 (1943).

a curved surface by means of an iteration process based on that of Ackeret. The first approximation which leads to the Prandtl-Glauert rule, is based on the assumption that the flow differs but little from a pure translation. The iteration process then consists in improving this first approximation in order that it will apply to a flow differing from pure translatory motion to a greater degree. This same method was applied to flow past a circular cylinder (1944)[41].

Imai[20, 42] (1944) developed the so-called thin-wing-expansion method which is essentially an extension of the linear theory of Prandtl-Glauert and applied it to a flow around an elliptic cylinder.

Lock[43] (1944) applied Busemann's second order approximation to evaluate the forces on supersonic aerofoils.

Ward[44] (1945) using the linearized equation attacked the problem of a compressible flow in a tube of slightly varying cross-section.

Tsien and Lees[45] (1945) presented the application of the Prandtl-Glauert method to the three-dimensional flow and to the calculation of wall interference in wind-tunnel testing at high speeds. Their approach also involves the derivation of the Biot-Savart law for the compressible flow by the Prandtl-Glauert formula and the induced velocity field of the helical vortex, which represents approximately the trailing vortex shed from a propeller blade.

Lees[46] (1946), Laitone[47] (1946, 1947), Sears[48] (1946, 1947), Young[49] (1949) obtained in various ways the correct compressibility correction factors in applying the Prandtl-Glauert rule to three-dimensional flow and wind-tunnel wall effects.

Kaplan[50] (1946 and 1947) applied an extended form of the Ackeret iteration method to calculate the compressible flow at high subsonic velocities past an elliptic cylinder.

Tsien[51] (1946) derived similarity rules for hypersonic flows.

Kármán[52] (1947) derived the similarity laws of transonic flows.

---

[41] C. Kaplan: The flow of a compressible fluid past a circular arc profile. N. A. C. A., ARR. No. L4G15 (L 216) (1944), or N. A. C. A., Rep. No. 794 (1944).

[42] I. Imai: Two-dimensional aerofoil theory for compressible fluids (in Japanese). Rep. Aeron. Res. Inst. Tokyo Imp. Univ. No. 294 (1944).

[43] C. N. H. Lock: Examples of the application of Busemann's formula to evaluate the aerodynamic force coefficients on supersonic aerofoils. R. and M. No. 2101, British A. R. C. (1944).

[44] G. N. Ward: A note on compressible flow in a tube of slightly varying cross-section. Aeron. Res. Council, R. and M. No. 2183 (1945).

[45] H. S. Tsien and L. Lees: The Glauert-Prandtl approximation for subsonic flows of a compressible fluid. J. Aeron. Sci. 12, 173—187 (1945).

[46] L. Lees: A discussion of the application of the Prandtl-Glauert method to subsonic compressible flow over a slender body of revolution. N. A. C. A., T. N. No. 1127 (1946).

[47] E. V. Laitone: The subsonic and supersonic flow fields of slender bodies. Sixth International Congress for Applied Mechanics (1946). (Paris.) The subsonic and supersonic two-dimensional flow fields about slender bodies. Cornell Aeron. Lab. Publication No. 6 (1946). The subsonic axial flow about a body of revolution. Quart. Appl. Math. 5, 227—331 (1947).

[48] W. R. Sears: On compressible flow about bodies of revolution. Quart. Appl. Math. 4, 191—193 (1946). A second note on compressible flow about bodies of revolution. Quart. Appl. Math. 5, 89—91 (1947).

[49] A. D. Young: On the compressibility correction factor for axially symmetric bodies. Readers' Forum, J. Aeron. Sci. 16, 703 (1949).

[50] C. Kaplan: Effect of compressibility at high subsonic velocities on the lifting force acting on an elliptic cylinder. N. A. C. A., T. N. No. 1118 (1946), or N. A. C. A., Rep. No. 834 (1946). Effect of compressibility at high subsonic velocities on the moment acting on an elliptic cylinder. N. A. C. A., T. N. No. 1218 (1947).

[51] H. S. Tsien: Similarity laws of hypersonic flows. J. Math. Physics 25, No. 3, 247—251 (1946).

[52] Th. von Kármán: Supersonic aerodynamics—principles and applications. J. Aeron. Sci. 14, 182—190 (1947).

Kaplan[53] (1948) used Tsien's method (to derive similarity rules for hypersonic flows) to derive von Kármán's similarity rules for transonic flows. A slight generalization is introduced by the inclusion of k, the ratio of specific heats, as a parameter. It is further shown that the obtained formula can also be obtained by means of the Prandtl-Glauert small-perturbation method.

Ward[54] (1948) by using the Heaviside operational method applied to the linear perturbation theory calculated the supersonic external and internal flow past a quasi-cylindrical tube.

Ward[55] (1949) solved the linearized potential problem of supersonic flow for nearly plane wings of arbitrary shape of planform and section.

Carriére[56] (1949) considered a motion slightly varied from a given potential supersonic flow.

Ehlers[57] (1949) discussed the general application of the Prandtl-Glauert type of linearization in subsonic and supersonic flows.

Ward[58] (1950) discussed the bases and scope of the linearized theory of compressible fluid flow with special reference to steady supersonic flow.

3. Variational Methods

Bateman[59] (1929) was probably the first to apply the variational principle to certain problems in fluid dynamics.

Braun[60] (1932) investigated more deeply those possibilities.

Pinl[61] (1941) and Behrbohm and Pinl[62] (1941) discussed some phases of this problem and pointed out a relationship between the hydrodynamical problems and the theory of minimal surfaces.

Truesdell[63] (1946) proved that Behrbohm and Pinl's result is in fact not a new linearization of a non-linear equation but rather a transformation of an old one obtained by means of the Legendre transformation.

---

[53] C. Kaplan: On similarity rules for transonic flows. N. A. C. A., T. N. No. 1527 (1948).

[54] G. N. Ward: The approximate external and internal flow past a quasi-cylindrical tube moving at supersonic speeds. Quart. Mechan. a. Appl. Math. 1, 225—245 (1948).

[55] G. N. Ward: Supersonic flow past thin wings. I. General Theory. Quart. J. Mechan. a. Appl. Math. 2, 136—152 (1949). Supersonic flow past thin wings. II. Flow-reversal theorems. Quart. J. Mechan. a. Appl. Math. 2, 374—384 (1949).

[56] P. Carriére: Sligthly varied supersonic flows. C. r. acad. sci., Paris 228, 1632—1634 (1949).

[57] F. B. Ehlers: Methods of linearization in compressible flow. Part III. Prandtl-Glauert method. Headqtrs. Air Mater. Comm. Dayton Transl. Tech. Rep. No. F-TR-11800-ND (1949).

[58] G. N. Ward: The bases and scope of the linearized theory of compressible fluid flow with special reference to steady supersonic flow. Symposium on theoretical compressible flow (June 1949), Naval Ordnance Laboratory, White Oak, NOLR 1132 (Issued July 1950), 35—41.

[59] H. Bateman: Notes on a differential equation which occurs in the two-dimensional motion of a compressible fluid and the associated variational problem. Proc. Roy. Soc. London, Ser. A 16 (1930). The irrotational motion of a compressible fluid. Proc. Nat. Acad. Sci., U. S. A. 16 (1930).

[60] G. Braun: Die ebene kompressible Potentialströmung als Variations- und Eigenwertproblem. Dissertation Göttingen (1932). Also: Ann. Physik (5) 15, 645—676 (1932).

[61] M. Pinl: Zur geometrischen Deutung und Transformation der Grundgleichung der ebenen kompressiblen Potentialströmung. Z. angew. Math. Mechan. 21, 80—85 (1941).

[62] H. Behrbohm und M. Pinl: Zur Theorie der kompressiblen Potentialströmungen I. Neue Linearisierung der Grundgleichung der ebenen adiabatisch kompressiblen Potentialströmung. Z. angew. Math. Mechan. 21, 193—203 (1941). Also R. T. P. Translation No. 1625.

[63] C. Truesdell: On Behrbohm and Pinl's linearization of the equation of two-dimensional steady polytropic flow of a compressible fluid. Proc. Nat. Acad. Sci., USA 32, 289—293 (1946).

Wang[64] (1948), Wang and Rao[65] (1950), Wang and de los Santos[66] (1949) and de los Santos[67] (1949) applied the variational method (following the Rayleigh-Ritz procedure) to several problems like the calculation of two-dimensional subsonic compressible flows past arbitrary bodies, past bodies of revolution, transonic flows, etc.

Manwell[68] (1949) discussed a method of variation as applied to both in- and compressible fluid flow in the subsonic region.

Wang and Chou[69] (1950) discussed the application of Biezeno-Koch method to compressible fluid flow problems. As Courant pointed out this is essentially a special case of the variational method.

4. Hodograph Method

Molenbroeck[70] (1890) linearized the equations of motion by means of transformation, in which potential and stream functions themselves are used as functions of the velocity coordinates.

Chaplygin[71] (1904) was the first to construct a set of particular solutions of the linearized equation in the hodograph plane in the subsonic range. In this method the stream function is represented by an infinite series of particular solutions, and such a series converges only within a part of the domain in which the flow is defined. A hodograph of a flow around a profile is (in general) a multiply covered domain, the branch-points of which are not necessarily located either at the origin or at infinity. On the other hand, the Chaplygin solutions yield flows which in the hodograph plane either are single-valued or multi-valued with a branch-point as the origin or at infinity. This comes out from the nature of the hypergeometric series which represents solutions. In order to represent such flow patterns, several series developments, each of which represents the stream function ψ under consideration in a certain part of the domain in which ψ is defined, are needed. Another suggestion of Chaplygin was to use the condition that the quantity $s^2 = (1 - M^2)\,\varrho^{-2}$ (see section I. 2) is constant. Under this assumption the function $f(\lambda)$ (1.2.5) vanishes identically since $s' = 0$. Equation (1.2.4) reduces to a Laplacien which has a standard solution. One may easily notice that this great simplification of the equation in the hodograph plane results in $k = -1$. This is obvious from (1.3.11).

Meyer[72] (1908) presented the well-known expansion around the corner (called Prandtl-Meyer expansion).

[64] C. T. Wang: Variational method in the theory of compressible fluid. J. Aeron. Sci. **15**, 675—685 (1948).

[65] C. T. Wang and G. V. R. Rao: A study of the nonlinear characteristics of compressible flow equations by means of variational methods. J. Aeron. Sci. **17**, 343—348 (1950).

[66] C. T. Wang and S. de los Santos: Approximate solutions of compressible flows past bodies of revolution by variational method. (To be published.)

[67] S. de los Santos: Subsonic compressible flow past bodies of revolution by the variational method. Thesis (Eng. Sc. D.), New York Univ. (1949).

[68] A. R. Manwell: A method of variation for flow problems. I. Quart. J. Math. Oxford Ser. **20**, 166—189 (1949).

[69] C. T. Wang and P. C. Chou: Application of Biezeno-Koch method to compressible fluid flow problems. J. Aeron. Sci., Readers' Forum **17**, No. 9, 599—600 (1950).

[70] P. Molenbroek: Über einige Bewegungen eines Gases mit Annahme eines Geschwindigkeitspotentials. Arch. Math. Physik (2) **9**, 157—195 (1890).

[71] S. A. Chaplygin: On gas jets. Scientific Memoirs, Moscow University, Phys.-Math. Section **21**, 1—127 (1904), or N. A. C. A., T. M. No. 1063 (1944).

[72] Th. Meyer: Über zweidimensionale Bewegungsvorgänge in einem Gas, das mit Überschallgeschwindigkeit strömt. Dissertation Göttingen (1908), or Mitt. Forsch.-Arb. Ing.-Wes. **62** (1908).

Demtchenko[73] (1932) and Busemann[74] (1937) clarified the meaning of the specific value $k = -1$, proposed by Chaplygin. They found that this really means that the pressure is a linear function of $1/\varrho$, i. e., $p = \sigma \varrho^{-1}$, where σ is a constant. Thus the curve $p = p(1/\varrho)$ is approximated in Chaplygin's approach by a tangent. However, both Demtchenko and Busemann limited themselves to the use of the tangent at the state of the gas corresponding to the stagnation point (conditions at rest), i. e., $M = 0$, $s^2 = 1/\varrho_0^2$, shere ϱ_0 corresponds to the conditions at rest. As a result their approach can only be applied to a flow with velocities up to about 0.4 to 0.5, the velocity of sound. Using this method Demtchenko and Busemann attacked some problems in the subsonic range.

Kármán[75] (1941) and Tsien[76] (1939) improved the approach, described above, in order to enable one to apply the hodograph method to larger velocities. They suggested that the theory can be generalized by using the tangent at the state of the gas corresponding to undisturbed parallel flow. Thus the range of usefulness of the theory can be greatly extended and we can obtain a better agreement with tests at speeds above 0.4 velocity of sound.

Using this assumption, Tsien[76] (1939) proposed his method of solving equation of flow in the subsonic range by the use of hodograph transformation. In the final step of his method an arbitrary analytic function must be chosen to which the function $g(\bar{Z})$ in (1.9.2) may be considered to be in a certain analogy. Using this procedure, it is not possible to predict whether the chosen function will give the desired shape of the solid boundary and flow pattern. In other words, this procedure, in common with all hodograph methods, still suffers the difficulty of boundary conditions. However, the resulting shape of the body can be ascertained approximately by starting with the function so chosen as to give the flow of an incompressible fluid over the desired body shape in a plane which may be considered to be a transformation of the physical plane. Using this method Tsien solved as an example the flow around an elliptic cylinder at zero angle of attack. Let us compare Tsien's method with some others. The difference between the various theories lies in the assumptions which are made to simplify the calculations. As explained above, Prandtl and Glauert treated the flow over a body of small thickness ratio. Kaplan and Hooker (Hooker's results are very close to those of Kaplan) assumed that the Mach number of the undisturbed flow is small, so that terms containing the third and higher powers of the Mach number can be neglected. Tsien's theory is essentially an improvement of the Prandtl-Glauert theory, so that the effect of large disturbances to the parallel flow is approximately taken into account. Therefore, for flow over thin sections at high Mach numbers, the result of Tsien's theory should agree well with the Prandtl-Glauert theory, especially at points not too close to the stagnation point. The results of Kaplan and Hooker should show smaller effect of the compressibility due to their second order approximation. For flow over thick sections at lower Mach numbers, the situation is reversed. In this case results of Tsien's theory should give better agreement with the results obtained by Kaplan and Hooker than with those obtained from the Prandtl-Glauert theory. The above reasoning is substantiated by diagrams

---

[73] B. Demtchenko: Sur les mouvements lents des fluides compressibles. C. r. acad. sci., Paris **194**, 1218—1220 (1932).

[74] A. Busemann: Hodographenmethode der Gasdynamik. Z. angew. Math. Mechan. 17, 73—79 (1937).

[75] Th. von Kármán: Compressibility effects in aerodynamics. J. Aeron. Sci. **8**, 337—356 (1941).

[76] H. S. Tsien: Two-dimensional subsonic flow of compressible fuids. J. Aero. Sci. **6**, 399—407 (1939).

presented in Tsien's paper. The amount of labor involved in Tsien's method is probably much less than in the successive approximations devised by Janzen-Rayleigh, Poggi and Walther, especially at higher Mach numbers. The main difficulty of this method lies in its application to flow involving circulation. The problem is more difficult and requires further study.

Ringleb[77] (1940) using hodograph transformation calculated the flow around a sharp edge.

Tamada[12] (1940) based his calculations on Kármán-Tsien idea, i. e., approximated the (p, ϱ^{-1})-curve by a tangent at the point p_1, ϱ_1^{-1}, and applied the Tsien's method. Starting from the well-known solution for the incompressible fluid flow past a circular cylinder, the compressible fluid flow past the corresponding body shape in the physical plane was calculated. The obtained shape is nearly circular. The deformed profile was calculated for $M = 0.4$. It was thus found that the deformed profile is very similar to the original circular profile, but the radius of the former is somewhat smaller than that of the latter. The calculated results are compared with those calculated by using the second approximation terms and also with the experimental results of Taylor, showing a fair agreement at those low Mach numbers.

Christianovitch[78] (1941) derived a method for solution a Cauchy problem for a partial differential equation of mixed elliptico-hyperbolic type.

Bergman[79] (1942) proposed an extension of Kármán-Tsien method. Kármán-Tsien method is inconvenient in that it can be applied if $s^2 = (1 - M^2)\, \varrho^{-2}$ is assumed to be constant, i. e., under the hypothesis of a linear pressure volume relation. However the operator introduced in the present work enables us to get rid of this restriction. The relation between the pressure and the volume can always be approximated by $p = P\,(\varrho^{-1})$, where P is a polynomial in ϱ^{-1}. For the sake of simplicity we take

$$p = (c_0 + b_0\, \varepsilon) - (c_1 + b_1\, \varepsilon)\, \varrho^{-1} + \varepsilon\, \varrho^{-3}. \qquad (7.4.1)$$

For example, the curve of the form (7.4.1) which possesses the common tangent and the same curvature as the isentropic curve at the point $(p_1, 1/\varrho_1)$ is given by the formula

$$p = \frac{1}{3}\,(k + 1)\,(k + 3)\,p_1 - \frac{1}{2}\,k\,(k + 3)\,p_1\,\varrho_1\,\varrho^{-1} + \frac{1}{6}\,k\,(k + 1)\,p_1\,\varrho_1{}^3\,\varrho^{-3}. \qquad (7.4.2)$$

Equating the formulas (7.4.1) and (7.4.2) results in obtaining the proper value of the quantity $\varepsilon = \varepsilon_0$, say. Consider equation

$$\psi_{HH} + l\,(H, \varepsilon)\,\psi_{\theta\theta} = 0, \qquad (4.5.2)$$

in which it is assumed that l depends also upon the parameter ε. As explained above, a solution of this equation is given by the imaginary or the real part of (4.5.16). Consider now all the functions as depending upon ε also. Thus the function E_1 (4.5.17) now is

$$E_1\,(Z', \bar{Z}', t, \varepsilon) = R\,(Z', \bar{Z}', \varepsilon)\,E_1{}^*\,(Z', \bar{Z}', t, \varepsilon). \qquad (7.4.3)$$

For sufficiently small $Z', \bar{Z}', \varepsilon$ and $|t| \leqq 1$, E_1 is an analytic function of $Z', \bar{Z}', \varepsilon, t$. We can therefore develop E_1 in a series

$$E_1 = E^{(0)}\,(Z', \bar{Z}', t) + \varepsilon\,E^{(1)}\,(Z', \bar{Z}', t) + \ldots \qquad (7.4.4)$$

[77] F. Ringleb: Über die Differentialgleichungen einer adiabatischen Gasströmung und den Strömungsstoß. Dtsch. Math. 5, 377—384 (1940).

[78] Christianovitch: On supersonic gas flows. CAHI Rep. No. 543 (1941) (in Russian).

[79] S. Bergman: The hodograph method in the theory of compressible fluids. Supplement to "Fluid Dynamics" by R. v. Mises and K. O. Friedrichs. Brown University. 1941/42.

Since ε will vary only in a small range $[0, \varepsilon^{(0)}]$, say, we neglect the coefficients of order higher than 1, and write

$$\psi(Z', \bar{Z}', \varepsilon) = \psi_0(Z', \bar{Z}') + \varepsilon\,\psi_1(Z', \bar{Z}') + \ldots, \qquad (7.4.5)$$

where

$$\psi_0(Z', \bar{Z}') = Re\left\{\int_{-1}^{+1} E^{(0)}(Z', \bar{Z}', t)\, f\left[\frac{1}{2}Z'(1 - t^2)\right](1 - t^2)^{-1/2}\, dt\right\}, \qquad (7.4.6\,a)$$

$$\psi_1(Z', \bar{Z}') = Re\left\{\int_{-1}^{+1} E^{(1)}(Z', \bar{Z}', t)\, f\left[\frac{1}{2}Z'(1 - t^2)\right](1 - t^2)^{-1/2}\, dt\right\}. \qquad (7.4.6\,b)$$

Suppose now that $(4.5.2)$ is the equation of the stream function $\psi(Z', \bar{Z}', \varepsilon)$ of motion of a gas with the pressure-volume relation given by $(7.4.1)$. For $\varepsilon = 0$, which corresponds to $p = -c_1\,\varrho^{-1}$ or $k = -1$, the coefficient l in $(4.5.2)$ becomes constant. If a domain B is given in the physical plane, we can determine (using the operator's method) a domain B^* and the stream function $\psi(Z', \bar{Z}')$ of motion of a gas with the pressure-volume relation $p = -c_1\,\varrho^{-1}$. Since for $\varepsilon = 0$, $\psi = \psi_0$ $(7.4.5)$, we have the first approximation to ψ. Next we find the function f associate of ψ (for details see [79]), which introduced into $(7.4.6\,b)$ gives the value of ψ_1 and next the value of ψ $(7.4.5)$. Inserting the proper numerical value ε_0 for ε we obtain the stream function $\psi(Z', \bar{Z}', \varepsilon_0)$ which corresponds to the pressure-volume relation $(7.4.1)$ or $(7.4.2)$. The function $\psi(Z', \bar{Z}', \varepsilon_0)$ will no longer assume a constant value on the boundary line b^* of B^*. However, since ε_0 is small, there will exist in most cases lines $\psi(Z', \bar{Z}', \varepsilon_0) = $ constant, which only slightly differ from b^*. Next we determine the corresponding domain B^{**} in the physical plane. In many instances it will only slightly differ from B. The method can be improved and for details in that respect see [79].

Temple and Yarwood[80] (1942) attempting to set up a correspondence between incompressible and compressible flows of the nature of correction factors, derived a velocity correction formula similar to those of Kármán-Tsien or Garrick-Kaplan.

Gelbart[81] (1943), Bers and Gelbart[82] (1943, 1944), and Bers[83] (1945) applied the theory of sigma-monogenic functions to study a class of complex functions the role of which in gas-dynamics is comparable to that of analytic functions in the theory of incompressible flows. Their method in the years 1943–45 was restricted to subsonic flows. The method permits the transformation of a two-dimensional incompressible flow around a closed profile into a subsonic compressible flow around another closed profile. The profile distortion is small for small values of the stream Mach number. In the case of the actual pressure-density relation an approximate method of solving the problem is indicated. If Kármán-Tsien linearized equation is assumed the transformation is carried out completely and in a closed form not only for flow without

[80] G. Temple and J. Yarwood: The approximate solution of the hodograph equations for compressible flow. Rep. No. 8. M. E. 3201, British R. A. E. (June 1942). Compressible flow in a convergent-divergent nozzle. R. and M. No. 2077, British A. R. C. (1942).

[81] A. Gelbart: On a function-theory method for obtaining potential-flow patterns of a compressible fluid. N. A. C. A., A. R. R. No. 3027 (1943).

[82] L. Bers and A. Gelbart: On a class of differential equations in mechanics of continua. Quart. Appl. Math. 1, 168—188 (1943). On a class of functions defined by partial differential equations. Trans. Amer. Math. Soc. 56, No. 1, 67—93 (1944).

[83] L. Bers: On a method of constructing two-dimensional subsonic compressible flows around closed profiles. N. A. C. A., T. N. No. 969 (1945). On the circulatory subsonic flow of a compressible fluid past a circular cylinder. N. A. C. A., T. N. No. 970 (1945).

circulation but also for flows with circulation. Using the linearized pressure-volume relation Bers computed the circulatory subsonic flow around an infinite circular cylinder.

Tsien and Fejer[84] (1944) examined the general applicability of the Kármán-Tsien formula from purely geometric considerations.

Garrick and Kaplan[85] (1944) developed the elementary basic solutions of the equations of motion of a compressible fluid in the hodograph plane in order to provide a basis for comparison in the form of velocity correction formulas, of corresponding compressible and incompressible flows. The known approximate results of Chaplygin, Kármán and Tsien, Temple and Yarwood, and Prandtl and Glauert are unified by means of a simple analysis.

Garrick and Kaplan[86] (1944) utilized the differential equation of Chaplygin's jet problem to give a systematic development of particular solutions of the hodograph flow equations, which extends the treatment of Chaplygin into the supersonic range and completes the set of particular solutions. The graphical flow patterns associated with certain particular solutions (of index $k = \pm 1/2$, $\pm 3/2$, and ± 2) of the Chaplygin's equation are contained in the paper by Kraft and Dibble[87] (1944).

Frankl[88] (1944) derived a method for solution of a Cauchy problem for a partial differential equation of mixed elliptico-hyperbolic type. He showed that the continuation of the subsonic flow into the supersonic region is not uniquely determined.

Coburn[89] (1945) discussed the applicability of the Kármán-Tsien idea in the supersonic range. He also had shown that when the Kármán-Tsien relation can be used, the characteristics form a Tschebyscheff net.

Poritsky[90] (1946 and 1948) extended the Kármán-Tsien method by approximating the equation of state by means of not one but several straight-line segments in the $(p, 1/\varrho)$-plane. This was applied by him to the flow from a point-source.

Maccoll[91] (1946) presented some investigations on compressible flow at sonic speeds.

Bers[92] (1946), Lin[93] (1946), Gelbart[94] (1947), Bartnoff and Gelbart[95] (1947) followed the idea of applying the simplified pressure-density relation so that the approximation theories can be developed which make use of the Cauchy-Riemann equations.

[84] H. S. Tsien and A. Fejer: A method for predicting the transonic flow over airfoils and similar bodies from data obtained at small Mach numbers. Army Air Forces (Dec. 1944).

[85] I. E. Garrick and C. Kaplan: On the flow of a compressible fluid by the hodograph method. I. Unification and extension of present day results. N. A. C. A., A. C. R. L 4 C 24 (1944), or N. A. C. A. Rep. No. 789 (1944).

[86] I. E. Garrick and C. Kaplan: On the flow of a compressible fluid by the hodograph method. II. Fundamental set of particular flow solutions of the Chaplygin differential equation. N. A. C. A., A. R. R. No. L 4 I 29, War Time Rep. No. L 147 (1944), or N. A. C. A., Rep. No. 790 (1944).

[87] H. Kraft and C. G. Dibble: Some two-dimensional adiabatic compressible flow patterns. J. Aeron. Sci. 11, 283—298 (1944).

[88] F. I. Frankl: On the Cauchy problem for equations of mixed elliptico-hyperbolic type with initial data on the parabolic line. Bull. Acad. Sci. URSS 8, 195—245 (1944) (in Russian).

[89] N. Coburn: The Kármán-Tsien pressure-volume relation in the two-dimensional supersonic flow of compressible fluids. Quart. Appl. Math. 3, 106—116 (1945).

[90] H. Poritsky: An approximate method of integrating the equations of compressible fluid flow in the hodograph plane. Sixth International Congress of Applied Mechanics, Paris (1946). Polygonal approximation method in the hodograph plane. J. Appl. Mechan., Trans. Amer. Soc. Mechan. Eng. 16, 123—133 (1949).

[91] J. W. Maccoll: Investigations of compressible flow at sonic speeds. Sixth International Congress for Applied Mechanics (1946) (Paris.).

[92] L. Bers: Velocity distribution on wing sections of arbitrary shape in compressible potential flow. I. Symmetric flows obeying the simplified density-speed relation. N. A. C. A., T. N. No. 1006 (1946).

Tsien's and Kuo's[96] (1946) approach consists of several parts. In the first part the general properties of hypergeometric functions of large order are investigated in preparation for the proof of the convergence given in the second part. The general solution constructed by the Chaplygin's method is really an existence theorem. The extremely slow convergence of the series makes numerical calculation very difficult if not impossible. This, in fact, constitutes the main difficulty of the method. In the third part of their work Tsien and Kuo overcame this difficulty by using the asymptotic properties of the hypergeometric functions. The result is the separation of the solution in the hodograph plane into two parts. One part is of closed form and is the product of a universal function of the velocity and the same solution as for incompressible flow but with a velocity distortion or velocity correction. The other part is an infinite series which converges rapidly everywhere except in a small region on both sides of a critical circle with a radius equal to the velocity of sound in the hodograph plane. In practice, by using only a few terms of the infinite series, one can obtain a satisfactory result. As an example, the motion of air past a cylindrical body was considered. The flow patterns for two freestream Mach numbers (0.6 and 0.7) have been calculated. It should be noticed that there is considerable distortion in the shape of the body in the compressible flow from that in the incompressible flow. This paper of Tsien and Kuo had been criticized by Lighthill[97] (1947) and Cherry[98] (1947). In his answer Kuo[99] (1949) showed that the criticism was not justified.

Reutter[100] (1947) proposed asymptotic solutions of Chaplygin's equation.

Cherry[98] (1947) found a family of exact solutions for the two-dimensional flow of a compressible perfect fluid about a cylinder. The work is restricted to the case where the circulation is zero and the speed at large distances from the cylinder is subsonic; but there is no restriction that the speed near the cylinder be subsonic. The family of solutions involves the hypergeometric functions and an infinite set of constants, upon the values of which depends the shape of the cylinder; but the question of so disposing these constants as to suit a prescribed shape is not here entered upon. The flow-field was divided into a few parts and the analytic continuation is largely applied. The key to the matter lies in a theorem whereby a hypergeometric

---

[93] C. C. Lin: On an extension of the von Kármán-Tsien method to two-dimensional subsonic flows with circulation around closed profiles. Quart. Appl. Math. 4, No. 3 (1946).

[94] A. Gelbart: On subsonic compressible flows by a method of correspondence. I. Methods for obtaining subsonic circulatory compressible flows about two-dimensional bodies. N. A. C. A., T. N. No. 1170 (1947).

[95] S. Bartnoff and A. Gelbart: On subsonic compressible flows by a method of correspondence. II. Application of methods to studies of flow with circulation about a circular cylinder. N. A. C. A., T. N. No. 1171 (1947).

[96] H. S. Tsien and Y. H. Kuo: Two-dimensional irrotational mixed subsonic and supersonic flow of a compressible fluid and the upper critical Mach number. N. A. C. A., T. N. No. 995 (1946).

[97] M. I. Lighthill: The hodograph transformation in transonic flow. I. Symmetrical channels. Proc. Roy. Soc. London, Ser. A 191, 323—341 (1947). The hodograph transformation in transonic flow. II. Auxiliary theorems on the hypergeometric functions $\Psi_n(\tau)$. Detto, 341—351. The hodograph transformation in transonic flow. III. Flow around a body. Detto, 352—369.

[98] T. M. Cherry: Flow of a compressible fluid about a cylinder. Proc. Roy. Soc. London, Ser. A 192, 45—79 (1947). Relation between Bergman's and Chaplygin's methods of solving the hodograph equation. Quart. Appl. Math. 9, 1, 92—94 (1951). Asymptotic expansions for the hypergeometric functions occurring in gas-flow theory. Proc. Roy. Soc. London, Ser. A 202, 1071, 505—522 (1950).

[99] Y. H. Kuo: On the stability of transonic flows. Proc. symposia appl. math. 1, 72—73, Amer. Math. Soc. (1949).

[100] F. Reutter: Über eine angenäherte quasilineare Potentialgleichung der ebenen kompressiblen Strömung und ihre mittels der Legendre-Transformation zu gewinnenden Lösungen. Z. angew. Math. Mechan. 25/27, 156—157 (1947).

function whose parameters are large is expressed by a convergent series whose leading term represents the function assymptotically.

Lighthill[97] (1947) in a very extensive study applies hypergeometric functions to the solution of hodograph equation in the case of a flow in a symmetrical sub, -trans, -super, -sonic channel (Part I). The construction and application of the singular solutions is thoroughly discussed and presented. Because of an extensive application of hypergeometric functions the second part is devoted entirely to auxiliary theorems on the hypergeometric functions. In the third part Lighthill making wide use of hypergeometric function presents the solution of the flow past a body. The solution is given in terms of integrals in the physical plane for the incompressible flow and can therefore be used when only data of the most numerical kind are available concerning this flow (to which the solution reduces when the Mach number tends to zero). It is shown how, when an analytic series (of a very general type) is available in the incompressible flow, the solution can be continued into the supersonic region. The solution contains an arbitrary function; so the different possible determinations of this function lead to an infinity of solutions of the compressible flow problem, all tending to the given incompressible flow as the Mach number tends to zero. It is shown that when circulation is absent all these solutions give a possible physical picture. When circulation is present, however, all the solutions but one give a physical flow which does not close up behind the body. The single solution which gives a physically sensible result in this case is determined by Lighthill. In the fourth part (Ferguson and Lighthill[101], 1947–48) the tables of various kinds are given.

Kuo[102] (1948) slightly modified the method developed earlier by Kuo and Tsien and extended it to include flows with circulation. The essential feature of the modified method is that in analytic continuation of the solution the alteration of the singularities of the incompressible solution due to the presence of the hypergeometric functions has been taken into account. It was found that for finite Mach number the only case in which the nature of the singularities of the incompressible solution can remain unchanged is for a ratio of specific heats equal to -1. The method is applied to the flow about an elliptic cylinder of thickness ratio 0.60 and for Mach number 0.60. Bergman, in his review of the paper[103], pointed out that a few results are not formulated in a sufficiently exact manner.

Oyama[104] (1948) investigated the compressible flow with circulation past a profile derived from an elliptic cylinder of thickness ratio 1/10, at various angles of incidence on the basis of the approximate method of Kármán-Tsien.

Goldstein, Lighthill and Craggs[105] (1948) proposed an alternative method of solution to the problem of finding, by use of the hodograph transformation steady, plane isentropic flows around contours. For illustrative purposes it is set out for the case when the incompressible flow is the symmetrical one about a circle. The flow with circulation was attacked by Lighthill[106] (1948).

[101] D. F. Ferguson and M. J. Lighthill: The hodograph transformation in transonic flow. IV. tables. Proc. Roy. Soc. London, Ser. A 192, 135—142 (1947/48).

[102] Y. H. Kuo: Two-dimensional irrotational transonic flows of a compressible fluid. N. A. C. A., T. N. No. 1445 (1948).

[103] S. Bergman: Review of the paper: Two-dimensional irrotational transonic flows of a compressible fluid. By Y. H. Kuo, N. A. C. A., T. N. No. 1445 (1948): in Math. Reviews 11, No. 3, 223 (1950).

[104] S. Oyama: Application of the hodograph method to the compressible flow with circulation past an elliptic cylinder (in Japanese). Rep. Inst. Sci. Tech., Univ. Tokyo 2, 63 (1948).

[105] S. Goldstein, M. J. Lighthill and J. M. Craggs: On the hodograph transformation for high-speed flow. I. A flow without circulation. Quart. J. Mech. a. Appl. Math. 1, 344—357 (1948).

Ehlers[107] (1949) discussed some hodograph methods: Chaplygin's, Lighthill's, Kuo's and Tsien's, etc.

Timman[108] (1949) and Frankl[109] (1949) proposed asymptotic solutions of Chaplygin's equation.

Tollmien[110] (1949) established relations between the local coordinates and the velocity components without using a velocity potential or a stream-function at all. This is the so-called "direct" hodograph method.

Imai[111] (1949) expressed the fundamental equations in the hodograph plane in terms of new variables suggested by the W. K. B. method well-known in quantum mechanics. These new variables are just the same as those explored previously by Bergman. In the case of subsonic flow past an obstacle approximate treatment similar to Kármán-Tsien's method can be developed which is believed to be very accurate even near the critical Mach number. In the supersonic case the equations of motion in terms of new variables have the advantage of giving immediately the equations for the characteristics. Similarly like in Bergman's method an auxiliary function must be used to represent the flow around the given body in an incompressible fluid. Some numerical examples are given in the paper by Imai and Hasimoto[112] (1950).

Mathur[113] (1950) in his thesis (under the direction of the author) assumes the pressure-volume relation in the form

$$p = A - \sum_1^n B_i/\varrho^i. \tag{7.4.7}$$

Using the variable $\tau = M^2 (2\beta + M^2)^{-1}$, $\beta = (k-1)^{-1}$, the Chaplygin's equation can be transformed into the form:

$$4\tau^2 \psi_{\tau\tau} + [4 + 2f(\tau)]\tau\psi_\tau + [1 - f(\tau)]\psi_{\theta\theta} = 0, \tag{7.4.8}$$

where $f(\tau) = q^2/a^2$. By means of the substitution

$$\psi = \tau^{k/2} Y_k(\tau) \cos k\theta, \quad k = 1, 2, \ldots n, \ldots \tag{7.4.9}$$

equation (7.4.8) transforms into

$$\tau^2 Y_k'' + (A_0 + A_1\tau + A_2\tau^2)\tau Y_k' + C_0 Y_k = 0, \tag{7.4.10}$$

where A_0, A_1, A_2, C_0, are constants. By means of another transformation, namely

$$Y_k = \tau^a e^{bt} y(\tau), \tag{7.4.11}$$

[106] M. J. Lighthill: On the hodograph transformation for high-speed flow. II. A flow with circulation. Quart. J. Mech. a. Appl. Math. 1, 442—450 (1948).

[107] F. B. Ehlers: Methods of linearization in compressible flow. Part II. Hodograph method. Headqtrs. Air Mat. Comm. Dayton, Transl. Tech. Rep. No. F-TR-1180 B-ND (1948).

[108] R. Timman: Linearization of the equations of two-dimensional subsonic compressible flow by means of complex characteristics. Proc. Seventh Int. Congr. Appl. Mech. 4, 28—42 (1948). Asymptotic formulae for special solutions of the hodograph equation in compressible flow (in English). Nal. Lucht. Lab. Amsterdam Rap. No. R. 46 (Apr. 1949).

[109] F. I. Frankl: Asymptotic resolution of Chaplygin's functions. Air Mat. Com. Dayton, Transl. No. F-TS-1212-1 A (1949).

[110] W. Tollmien: The direct hodograph method in the theory of the flow of compressible fluids. H. Reissner Anniversary Volume, 89—110, (1949), Edwards Bros., Ann Arbor.

[111] I. Imai: Application of the W. K. B. method to the flow of a compressible fluid, I. J. Math. Physics 28, 173 (1949).

[112] I. Imai and H. Hasimoto: Application of the W. K. B. method to the flow of a compressible fluid, II. J. Math. Physics 28, 205 (1950).

[113] P. N. Mathur: On compressible fluid flow by hodograph method utilizing a polynomial. Master's Degree Thesis, Dept. Aer. Eng., Univ of Illinois (1950). On the solution of Chaplygin's equation by means of Kummer's formula. Proceedings of the Midwestern Conferences on Fluid Dynamics. I. Univ. of Illinois (1950). Edwards Brothers, Ann Arbor, Mich., 99—108.

with $a = \text{const.}$, $b = \text{const.}$, equation (7.4.10) can be given in the form

$$\bar{\tau}\, y'' - (\lambda - \bar{\tau})\, y' - \alpha\, y = 0, \qquad\qquad (7.4.12)$$

where $\alpha = \text{const.}$, $\lambda = \text{const.}$, $\bar{\tau} = -\, b\,\tau$. Equation (7.4.12) has a standard solution by means of confluent hypergeometric series (Kummer's formula). From the solved particular examples it seems that Mathur's method gives a better approximation than Poritsky's approach.

Loewner[114] (1950) developed a transformation theory of systems of partial differential equations which allows the construction of classes of pressure-density relations depending on arbitrarily many parameters for which the equations governing the flow can be transformed into an essentially simpler form, namely, into the Cauchy-Riemann equations in the subsonic region, into the system corresponding to the wave equation in the supersonic case, and finally into that form corresponding to the Tricomi equation in the transonic region. The transition from one system of differential equations to the other is always such that only the solving of ordinary differential equations is required in order to find solutions of the more complicated system from corresponding solutions of the simpler one.

Bers[115] (1950) discussed simple, sufficient conditions under which a two-dimensional steady, compressible flow can be continued across the line of Mach number 1 as a continuous supersonic flow. Methods for the actual computation of the flow are described.

Weinstein[116] (1950) derived a new approach for the transonic flow based on the generalized axially symmetric potential theory which enables one to consider, instead of isolated examples, some flows satisfying prescribed boundary conditions. The paper deals with solutions in the hodograph plane.

Munk[117] (1950) discussed the Rankine flow in the hodograph plane.

Ludford[118] (1950) discussed the singularities in compressible flow.

Clauser[119] (1950) presented solutions of two-dimensional compressible fluid flows having arbitrarily specified pressure distributions for gases with gamma equal to minus one.

v. Krzywoblocki[120] (1951) transformed Chaplygin's equation into Fuch-Frobenius normal form by means of simple substitutions, preserving the original isentropic pressure-density relation.

---

[114] C. Loewner: A transformation theory of the partial differential equations of gas dynamics. N. A. C. A., T. N. No. 2065 (1950).

[115] L. Bers: On the continuation of a potential gas flow across the sonic line. N. A. C. A., T. N. No. 2058 (1950).

[116] A. Weinstein: Transonic flow and generalized axially symmetric potential theory. Symposium on theoretical compressible flow (June 1949), Naval Ordnance Laboratory, White Oak, NOLR 1132 (Issued July 1950), 73—82.

[117] M. M. Munk: The Rankine gas flow in the hodograph plane. Symposium on theoretical compressible flow (June 1949), Naval Ordnance Laboratory, White Oak, NOLR 1132 (Issued July 1950), 43—51. Or: Quart. Appl. Math. 8, 387—392 (1951).

[118] G. S. S. Ludford: The behavior at infinity of the potential function of the flow of a perfect gas in two dimensions. Report 23a, Nav. Res. Contr. N 5 ori 76/XVI NR 043 046, Harvard Univ. (1950). Two notes on the singularities of a compressible flow in the pseudo-logarithmic plane. Report 23b, Nav. Res. Contr. N 5 ori 76/XVI NR 043 046, Harvard Univ. (1950). The behavior at infinity of the potential function of a two-dimensional subsonic compressible flow. Report 23, Nav. Res. Contr. N 5 ori 76/XVI NR 043 046, Harvard Univ., J. Math. Physics (3) 30, 117—130 (1951).

[119] F. H. Clauser: Two-dimensional compressible flows having arbitrarily specified pressure distributions for gases with gamma equal to minus one. Symposium on theoretical compressible flow, June, 1949, U. S. Naval Ordnance Laboratory, White Oak, Nd., NOLR 1132 (1950).

[120] M. Z. v. Krzywoblocki: On the transformation of Chaplygin's equation into Fuchs-Frobenius normal form. J. Phys. Soc. Japan 6, No. 6, 452—453 (1951).

Cherry[98] (1951) had shown how Chaplygin's form of solution can be converted into Bergman's.

5. Slender Bodies and Bodies of Revolution

v. Kármán and Moore[121] (1932) calculated the flow past a slender body of revolution by source and sink method.

Ferrari[122] (1937) attacked the problem of the flow past the body of revolution.

Tsien[123] (1938) extended Kármán-Moore method to bodies inclined to the direction of the undisturbed stream.

Sauer[124] (1942) attacked the problem of the flow past an inclined body of revolution.

Busemann[125] (1942) pushed forward the idea of the conical flow.

Lighthill[126] (1945) attacked the problem of the flow past the body of revolution.

Sears[48] (1946, 1947) discussed the problem of the flow past a body of revolution by means of the linearized theory.

Hayes[127] (1946), Bartels and Laporte[128] (1946) attacked the problem of the flow past conical bodies by means of the linearized theory.

Jones and Margolis[129] (1946) and Laitone[47] (1947) attacked the same problem.

Young and Kirkby[130] (1947) applied Prandtl-Glauert method to three-dimensional bodies.

Lighthill[131] (1948) calculated the lift and moment on a slender pointed body of revolution whose axis is inclined at a small angle α to the uniform supersonic stream by means of the expansion in powers of the angle of incidence α, including the terms of order α^3.

Broderick[132] (1949) derived a method for solving approximately the complete equation for isentropic axi-symmetrical flow of a gas past bodies of revolution by means of expansion the velocity potential in powers of the thickness ratio of the body. The method was applied to the flow past semi-infinite cone (1949)[133].

---

[121] Th. von Kármán and N. B. Moore: The resistance of slender bodies moving with supersonic velocities with special reference to projectiles. Trans. Amer. Soc. Mechan. Eng. **54**, 303—310 (1932).

[122] C. Ferrari: Campi di corrente ipersonora attorna a solidi di rivoluzione. L'Aerotecnica **17**, 507—518 (1937).

[123] H. S. Tsien: Supersonic flow over an inclined body of revolution. J. Aeron. Sci. **5**, 480—483 (1938).

[124] R. Sauer: Überschallströmung um beliebig geformte Geschoßspitze unter kleinem Anstellwinkel. Luftfahrtforschung **19**, 148—152 (1942).

[125] A. Buseman: Conical supersonic flow with axial symmetry. Luftfahrtforschung **19**, 137—144 (1942). Also: R. T. P. Translation No. 1598 or Roy. Aeron. Soc. J. (1942).

[126] M. J. Lighthill: Supersonic flow past bodies of revolution. R. and M. No. 2003, British A. R. C. (1945).

[127] W. D. Hayes: Linearized Supersonic flows with axial symmetry. Quart. Appl. Math. **4**, 255—261 (1946).

[128] R. C. F. Bartels and O. Laporte: Investigation of the supersonic flow over conical bodies with angles of attack or yaw by means of the linearized theory. Sixth International Congress for Applied Mechanics (1946) (Paris.).

[129] R. T. Jones and K. Margolis: Flow over a slender body of revolution at supersonic velocities. N. A. C. A., T. N. No. 1081 (1946).

[130] A. D. Young and S. Kirkby: Application of the linear perturbation theory to compressible flow about bodies of revolution. Coll. Aeron. Cranfield Rep. No. 11 (1947).

[131] M. J. Lighthill: Supersonic flow past slender pointed bodies of revolution at yaw. Quart. J. Mechan. a. Appl. Math. **1**, 76—89 (1948).

[132] J. B. Broderick: Supersonic flow round pointed bodies of revolution. Quart. J. Mechan. a. Appl. Math. **2**, 98—120 (1949).

[133] J. B. Broderick: Supersonic flow past a semi-infinite cone. Quart. J. Mechan. a. Appl. Math. **2**, 121—128 (1949).

Ward[134] (1949) attacked the supersonic flow past slender pointed bodies but not necessarily of revolution.

Whitham[135] (1950) discussed the behavior of supersonic flow past a body of revolution, far from the axis.

6. Southwell's and other Methods

Southwell's method was applied to compressible fluid problems. Without going into too many details let us only mention the works of Southwell himself[136] (1940) and his followers like Christopherson[137] (1938), Emmons[138] (1943, 1944, 1946, 1948), Fox[139] (1944), Green[140] (1944), etc.

Here and there one may find other methods than those of classical types mentioned above.

Crocco[141] (1949) developed a new method for adiabatic but non-isentropic gas.

Guderley[142] (1942, 1948) and Guderley and Yoshihara[143] (1949, 1951) applied special transformations to particular solution of the transonic flow equation (with the approximations usually made in the derivation of the transonic similarity law) to linearize the equation and to obtain the flow pattern and pressure coefficient. A discussion of transonic flow singularities was also included.

7. General Remarks

From the brief review of various methods, outlined above, it seems that the hodograph method probably offers the greatest possibilities. From the various methods of solutions of the hodograph equation, there are some which by themselves, from

[134] G. N. Ward: Supersonic flow past slender pointed bodies. Quart. J. Mechan. a. Appl. Math. 2, 75—97 (1949).

[135] G. B. Whitham: The behavior of supersonic flow past a body of revolution, far from the axis. Proc. Roy. Soc. London, Ser. A 201, 89—109 (1950).

[136] R. V. Southwell: Relaxation methods in engineering science. Oxford: Clarendon Press. 1940.

[137] Christopherson and R. V. Southwell: Relaxation methods applied to engineering problems. III. Problems involving two independent variables. Proc. Roy. Soc. London, 168, 317—350 (1938).

[138] H. W. Emmons: The numerical solution of heat conduction problems. Trans. Amer. Soc. Mechan. Eng. 65, 607—615 (1943). The numerical solution of compressible fluid flow problems. N. A. C. A., T. N. No. 932 (1944). The theoretical flow of a frictionless, adiabatic, perfect gas inside of a two-dimensional hyperbolic nozzle. N. A. C. A., T. N. No. 1003 (1946). Flow of a compressible fluid past a symmetrical airfoil in a wind tunnel and in free air. N. A. C. A., T. N. No. 1746 (1948).

[139] L. Fox and R. V. Southwell: On the flow of a gas through a nozzle with velocities exceeding the speed of sound. Proc. Roy. Soc. London, Ser. A 183, 38—54 (1944).

[140] J. R. Green and R. V. Southwell: Relaxation methods applied to engineering problems. IX. High-speed flow of compressible fluid through a two-dimensional nozzle. Philos. Trans. Roy. Soc., London, Ser. A 239, 367—386 (1944).

[141] L. Crocco: Una nuova funzione potenziale per lo studio del moto bidimensionale non-isentropico dei gas. Aerotecnica 29, 347—355 (1949). Transformations of the hodograph flow equation and the introduction of two generalized potential functions. N. A. C. A., T. N. No. 2432, 81, Aug. (1951).

[142] G. Guderley: Rückkehrkanten in ebener kompressibler Potentialströmung. Z. angew. Math. Mechan. 22, 121—126 (1942). Considerations on the structure of mixed subsonic-supersonic flow patterns. U. S. Air Force Report, F-TR-2168-ND, Oct. (1947). Singularities at the sonic velocity. Wright Field Report F-TR-1171-ND, June (1948).

[143] G. Guderley and H. Yoshihara: The flow over a wedge profile at Mach number one. U. S. A. F. Technical Report No. 5783, July (1949). On axial-symmetric transonic flow patterns. U. S. A. F. Technical Report No. 5797, Sept. (1949). The flow over a wedge profile at Mach number 1. J. Aeron. Sci. 17, 723—735 (1950). An axial-symmetric transonic flow pattern. Quart. Appl. Math. 8, (4), 333—339 (1951).

the nature of their construction seem to offer some greater possibilities both from the mathematical as well as from an applicable standpoint than do the others. Lighthill[144] (1949) mentioned that only the methods of Bergman, Cherry and himself are correct. In his review of one Bergman's papers[145] (1949) he wrote that the analysis of Tricomi's equation has not been previously extended so far as in the paper of Bergman. Bers[146] (1950) wrote: "The so-called correspondence method was first used by Chaplygin for jet problems (where, however, it is identical with the solution of the so-called direct problem). The correspondence method may be expressed in various analytical forms; however, except for the very original method of Bergman the basic procedure is always the same, as is clearly pointed out by Gelbart[81] (1943) (but those remarks of Gelbart could not refer to methods of Cherry and Lighthill published later). It consists of associating with a solution of the Cauchy-Riemann equations (representing an incompressible flow) a solution of Chaplygin's hodograph equation. In some cases the resulting compressible flow is of the same character as the initial incompressible flow."

From these opinions expressed independently by other authors it is quite obvious that Bergman's method is today considered to be one of the strongest. The question of two types of functions appearing in this method: those which can be calculated once and for all; and those which must be recalculated in each particular case, may be solved with some effort. This statement may be confirmed by the calculation of the long tables for the conical flow[147] (1947). Its weak side lies in its complexity.

Part VIII
Review of Tables and Particular Formulas

1. Subsonic Flow

The expression for $[1 - M^2 (2 \lambda)]^{1/2}$ as a function (exponential expansion) of λ, (which expression occurs in section I.3 and in subsequent parts of the present work) is given in [148]. The quantity λ is given by (1.3.13).

The evaluation of the several first coefficients $Q^{(n)}$ (1.9.3) which appear in the solution of equation (1.7.7) is given in [148]. The values of the derivatives ψ_q and ψ_θ are given in [148].

The values of $H, Q^{(n)}$, for $k = -0.5$ and $k = 1.4$ are given numerically in tables IIa and IIb, [148]. Also their diagrams are given in that report.

The proof that f (1.3.11) can be approximated by a polynomial f_m (1.7.6) is given in [148].

The representation of f_m by means of M^2, and of M^2, or $S = 1 - T$, by means of a uniformly convergent series in exp (2λ) is given in [148].

The representations $T = T (\exp (2 n \lambda))$ and $T^{-1} = T^{-1} (\exp (2 n \lambda))$ are given in [149] jointly with the tables of the coefficients.

[144] M. J. Lighthill: Methods for predicting phenomena in the high-speed flow of gases. J. Aeron. Sci. 16, 69—83 (1949).

[145] M. J. Lighthill: Review of Bergman's paper: Two-dimensional transonic flow patterns. Amer. J. Math. 70 (1948). Math. Reviews 10, 752—753 (1949).

[146] L. Bers: Velocity distribution on wing sections of arbitrary shape in compressible potential flow. III. Circulatory flow obeying the simplified density-speed relation. N. A. C. A., T. N. No. 2056 (1950).

[147] Z. Kopal: Tables of supersonic flow around cones. M. I. T., Cambridge, Mass. (1947).

[148] S. Bergman: On two-dimensional flows of compressible fluids. N. A. C. A., T. N. No. 972 (1945).

[149] S. Bergman: Methods for determination and computation of flow patterns of a compressible fluid. N. A. C. A., T. N. No. 1018 (1946).

An example showing the application of the formula (1.9.9) for $k = -0.5$ and $k = 1.4$ is given in [150]. The function $g(\bar{Z})$ is assumed to be equal to $\frac{1}{2}[(1 - 2\exp\bar{Z})^{-1/2} + (1 - 2\exp\bar{Z})^{1/2}]$. The example shows clearly three steps of calculations:

I. Computation of the functions $g^{(n)}(\bar{Z})$.

II. Construction of the flow in the pseudo-logarithmic plane for $\alpha = 0.1$.

III. Transition from the pseudo-logarithmic plane to the physical plane by means of formula (1.8.1). In this same report [150] there are tabulated the functions $Q^{(n)}$ (which can be calculated once and for all). Also their diagrams are given.

An application of punch-card machines to the solution of a problem is given in [149]. Equation (1.9.16) is used in its approximate and exact forms. In the Appendix I of that report the method of obtaining the polynomials in $\exp(2\lambda)$ of the approximation function $f_m(2\lambda)$ is given.

A method for the numerical determination of the coefficients of the integral operator (1.9.2) by the use of punch card machines is given in [151]. The procedure of the numerical determination of the system of functions $Q^{(n)}$ is thoroughly explained in that report.

Determination of those coefficients of the integral operator by interpolatory means is thoroughly presented in [152].

An example of how to determine a basic solution is given in [149]. A thorough and detailed discussion and determination of a flow pattern past an obstacle of oval shape is given in that report. The values of the following functions are included into that report: $T, M, \lambda, H, dH/d\lambda, Q^{(n)}, Q^{(n)\prime}, n = 1, 2, 3, 4, 5, 6$, for $k = 1.405$. These functions are calculated once and for all and can always be used. In this report also the functions $g^{[n]}$, $n = 1, 2, 3, 4$, are calculated for a flow around an ellipse in an incompressible fluid. A method for the determination of the stream function in the (λ, θ)-plane and in the physical plane is explained.

Reference [153] contains the values of the functions $A^{(n)}$, $n = 1, 2, 3, \ldots 8$, and their derivatives, as well as the values of the function $F(\lambda)$ and its derivative. Tables 1 to 4 in that report give the values of $T = T(\exp(2\lambda))$, $T^{-1} = T^{-1}(\exp(2\lambda))$, $F = F(\exp(2n\lambda))$, $M, B, q/a_0$ and λ. The first four functions $Q^{(n)}$ can be expressed in closed forms as functions of T, while in the expressions for the last four functions $Q^{(n)}$, $n = 5$ to 8, the integrals appear. However, the integrands in these cases can be written in closed forms, so that at least numerical computation of the integrals will not prove too difficult. Auxiliary functions for the computation of $Q^{(n)}$ are on p. 78 and Table 5 of that report. The values of the coefficients in equations (1.9.23a, b) are given on pp. 37 and 75 of [153]. The fundamental solutions for $k = -1$ are given in [148] (see 6.5.8).

2. Supersonic Flow

The numerical values of $\Lambda = \Lambda(M)$, equation (3.2.8) are given in [149]. Attention should be called to the fact that in this report $\Lambda(M) = \beta(M)$.

[150] S. Bergman: Graphical and analytical methods for the determination of a flow of a compressible fluid around an obstacle. N. A. C. A., T. N. No. 973 (1945).

[151] S. Bergman and L. Greenstone: Numerical determination by use of special computational devices of an integral operator in the theory of compressible fluids. I. Determination of the coefficients of the integral operator by use of punch card machines. J. Math. Physics **26**, 1—9 (1947).

[152] R. Isaacs: Numerical determination by use of special computational devices of an integral operator in the theory of compressible fluids. II. Determination of the coefficients of the integral operator by interpolatory means. J. Math. Physics **26**, 165—181 (1947).

[153] S. Bergman: On supersonic and partially supersonic flows. N. A. C. A., T. N. No. 1096 (1946).

An example of numerical calculation of the functions $V_1{}^*$ and $V_2{}^*$, representing the solutions of (3.9.2) is given in [153]. In Table 9 of that report the values of the functions, M_i, $i = 1, 2, 3, 4$, (3.10.1) and (3.10.2) are given.

The tables of quantities B, $2\,\Lambda$, F_1, $\Gamma^{(1)}$, $\Gamma^{(2)}$, (3.5.1), (3.5.2) are given in [154].

3. Transonic Flow

Equations for $dS_\nu{}^{(k)}/d\tau$, $dF_\nu/d(1 - \tau)$, and $dF^*_\nu/d(1 - \tau)$, referring to (4.4.17) are given in [148].

The expressions for $Q^{(n)}$ $(2\,\lambda)$, $n = 1, 2, \ldots$ are given in [153].

In [155] are given the tables of the type I of the functions

$$\psi^{(\nu)}\,(\eta, \theta;\, \eta_0)\ \text{for}\ \eta_0 = -\,0.02, \qquad (4.9.20)$$

$$\varphi^{(\nu)}\,(\eta, \theta;\, -\,0.02), \qquad (4.9.30)$$

and a conversion table from the variable η to M and vice versa. In that paper there are also given the tables referring to a numerical calculation of the flow pattern around an oval-shaped obstacle.

4. General Remarks

Since in his papers Bergman does not use a consistent notation, for the sake of enabling one to compare various formulas in his and von Mises' and Schiffer's papers, below we shall cite some formulas and symbols.

In [156] the following notations are used:

$$z = \theta + i\,\lambda = i\,\bar{Z}, \quad \check{z} = \theta - i\,\lambda = -\,iZ, \qquad (8.4.1)$$

$$\frac{\partial}{\partial z} = \frac{1}{2}\left(\frac{\partial}{\partial \theta} - i\,\frac{\partial}{\partial \lambda}\right) = -\,\frac{1}{2}\,i\,\frac{\partial}{\partial \bar{Z}}, \qquad (8.4.2\mathrm{a})$$

$$\frac{\partial}{\partial \check{z}} = \frac{1}{2}\left(\frac{\partial}{\partial \theta} + i\,\frac{\partial}{\partial \lambda}\right) = \frac{1}{2}\,i\,\frac{\partial}{\partial Z}. \qquad (8.4.2\mathrm{b})$$

The fundamental formula of Bergman is (1.9.2) which may be represented as

$$\psi^* = Re\left[g\,(Z) + \sum_{n+1}^{\infty} 2^{-2n} Q^{(n)}\,B\,(n, n+1)^{-1} \cdot (-1)^{n-1} \int_{\mathfrak{z}=0}^{Z} (Z - \mathfrak{z})^{n-1} g\,(\mathfrak{z})\,d\mathfrak{z}\right], \qquad (8.4.3)$$

where

$$B\,(n, n+1) = \frac{\Gamma\,(n)\,\Gamma\,(n+1)}{\Gamma\,(2\,n+1)} = \frac{n!\,(n-1)!}{2\,n!}, \qquad (8.4.3\mathrm{a})$$

and the formula in v. Mises' and Schiffer's notations is:

$$\psi^* = Re\left[g\,(Z) + \sum_{n=1}^{\infty} G_n\,\frac{(-1)^n}{(n-1)!\,2\,n} \cdot \int_0^Z g\,(t)\,(Z - t)^{n-1}\,dt\right]. \qquad (8.4.4)$$

The table, given below, enables one to compare the notation of Bergman and v. Mises and Schiffer.

	(2.15)		(2.9)	(2.6)	(2.2)
Bergman[157], p. 861	M^2	$4\,F$	$2\,N$	λ	l
v. Mises[158], p. 254	m	f	n	λ	z^2
	2.(3′)		2.(8)	4.(12)	1.(3)

[154] S. Bergman: On two-dimensional supersonic flows. N. A. C. A., T. N. No. 1875 (1949).

[155] S. Bergman: On tables for the determination of transonic flow patterns. Hans Reissner Anniversary Volume; Contributions to Applied Mechanics. J. W. Edwards. 1949.

[156] S. Bergman: Two-dimensional subsonic flows of a compressible fluid and their singularities. Trans. Amer. Math. Soc. **61**, 452—498 (1947).

[157] S. Bergman: Two-dimensional transonic flow patterns. Amer. J. Math. **70**, 856—891 (1948).

[158] R. von Mises and M. Schiffer: On Bergman's integration method in two-dimensional compressible fluid flow. Advances in Applied Mechanics, Vol. I. New York: Academic Press Inc. 1948.

Part IX

General Remarks

In previous sections we presented Bergman's integral operator method in some of its variations (sub-, trans-, supersonic, axially symmetric flow, etc.). But for a mathematically advanced reader, it is obvious that the method presents some further possibilities which may be realized in the future. In the present section we shall briefly present a few of them.

1. Application of the Hodograph Method in the Three-dimensional Case and to Flows with Shocks

Assume a three-dimensional irrotational flow of a compressible fluid and introduce the notion of the velocity potential, $\Phi_x = u$, $\Phi_y = v$, $\Phi_z = w$. The continuity equation has the form

$$(\varrho\, \Phi_x)_x + (\varrho\, \Phi_y)_y + (\varrho\, \Phi_z)_z = 0. \tag{9.1.1}$$

Introduce a new unknown function:

$$\lambda = x\,u + y\,v + z\,w - \Phi. \tag{9.1.2}$$

Inserting into (9.1.1) the quantities $\Phi_x = u$, etc., and combining (9.1.1) and (9.1.2) the following equation in λ results:

$$(\varrho_u\, u + \varrho)\, D_1 - \varrho_v\, u\, D_2 + \varrho_w\, u\, D_3 - \varrho_u\, v\, D_4 + (\varrho_v\, v + \varrho)\, D_5 -$$
$$- \varrho_w\, v\, D_6 + \varrho_u\, w\, D_7 - \varrho_v\, w\, D_8 + (\varrho_w\, w + \varrho)\, D_9 = 0, \tag{9.1.3}$$

where the symbols D_i, $i = 1$ to 9, denote determinants of the form

$$D_1 = \begin{vmatrix} \lambda_{v\,v} & \lambda_{v\,w} \\ \lambda_{v\,w} & \lambda_{w\,w} \end{vmatrix}, \text{ etc.} \tag{9.1.4}$$

There now arises the problem of determining particular solutions of (9.1.3). Clearly, this can be done by using the series development

$$\sum_{m,\,n,\,p} A_{mnp}\, u^m\, v^n\, w^p, \tag{9.1.5}$$

which satisfy equation (9.1.3). Such a series development which represents (in the hodograph space) the potential function Φ of a possible flow pattern of a compressible fluid converges only in the neighborhood of the origin. However, there exist methods of determining Φ in the whole region of the real (u, v, w)-space where Φ is regular. Such a representation, for instance, is given in many cases by

$$\Phi\,(x,\,y,\,z) = \lim_{k\,\to\,0} \sum_{m,n,\,p} A_{mnp}\, u^m\, v^n\, w^p\, \{\Gamma\,[1 + k\,(m + n + p)]\}^{-1}. \tag{9.1.6}$$

Another way to solve a three-dimensional problem, namely the case of an axially symmetric flow, was demonstrated in section V. But, in general, the three-dimensional case up to now cannot be solved in such an elegant manner as a two-dimensional flow. (For more details see[148].)

There are indications that it may be possible to extend Bergman's method to flows with shocks (discontinuities).

2. The Method of Orthogonal Functions and the Theory of Integral Operators

As we indicated in previous sections, many problems in the theory of compressible fluids may be solved by the hodograph method which enables us to consider a linear equation instead of a non-linear one. In particular, this method enables us to handle the mixed case. On the other hand, there are many problems in the two-dimensional case where it seems to be preferable to work directly in the physical plane. Further, in many instances, for examples, in the case of a three-dimensional motion, we are unable to reduce the equation to a linear one. Hence, it is of importance to develop a method of working directly in the physical plane, in which case we have to solve a non-linear partial differential equation.

Bergman and Schiffer[159] developed certain considerations which enable us to use methods of the theory of complex orthogonal functions for the solution of linear and non-linear partial differential equations. Naturally, it is impossible in the case of a non-linear equation, to introduce a linear operator which transforms in a bi-uniform manner analytic functions of one complex variable into solutions of this equation, and vice versa. But it is possible to derive certain properties of solutions of non-linear partial differential equations, in particular, those which appear in the theory of compressible fluids.

Suppose at first that the density ϱ is known at every point of a flow past a profile P. Then the potential function Φ can be defined as the solution of the differential equation

$$\Delta\Phi + \varrho^{-1}(\varrho_x\,\Phi_x + \varrho_y\,\Phi_y) = 0, \qquad \Delta\Phi = \Phi_{xx} + \Phi_{yy}, \tag{9.2.1}$$

which is defined in B (the exterior of P), satisfies on the boundary C of B the condition

$$\Phi_n = 0, \tag{9.2.1a}$$

(since along the contour there is only the tangential velocity component), and at infinity possesses a singularity representing a doublet. Instead of Φ we can introduce the function $\Phi^* = \Phi/h$, $h = \varrho^{-1/2}$. Then Φ^* satisfies in B the equation

$$L^*(\Phi^*) = \Delta\Phi^* + 4\,P_1\,\Phi^* = 0, \tag{9.2.2}$$

$$P_1 = -\,4\,\varrho^{-1/2}\,\Delta(\varrho^{1/2}), \tag{9.2.2a}$$

satisfies the boundary condition

$$B_1(\Phi^*) = \Phi_n{}^* - q_1\,\Phi^* = 0, \tag{9.2.3}$$

$$q_1 = -\,h^{-1}\,(\partial h/\partial n), \tag{9.2.3a}$$

and has a singularity at infinity. Let us assume that $P_1 < 0$. The determination of Φ can be reduced to the determination of a function Φ^* which satisfies (9.2.2) in B and satisfies the condition $B_1(\Phi^*) = p$, say, on the boundary C. This can be achieved by the use of the theory of orthogonal functions. Let $a(x, y)$ and $b(x, y)$ be two functions twice differentiable in $B + C$ and such that

$$\iint\limits_B (a_x + b_y)\,dx\,dy = \int\limits_C g(s)\,ds, \tag{9.2.4}$$

and

$$a\cos(n, x) + b\cos(n, y) = g(s). \tag{9.2.5}$$

We introduce now a complete system $\{\Phi_\nu{}^*\}$, $\nu = 1, 2$, of solutions of (9.2.2) which are orthogonalized so that

$$R(\Phi_\nu{}^*, \Phi_\mu{}^*) = \iint\limits_B \left[\frac{\partial\Phi_\nu{}^*}{\partial x}\frac{\partial\Phi_\mu{}^*}{\partial x} + \frac{\partial\Phi_\nu{}^*}{\partial y}\frac{\partial\Phi_\mu{}^*}{\partial y} - 4\,P_1\,\Phi_\nu{}^*\,\Phi_\mu{}^* - (a\,\Phi_\nu{}^*\,\Phi_\mu{}^*)_x - \right.$$

$$\left. - (b\,\Phi_\nu{}^*\,\Phi_\mu{}^*)_y\right]dx\,dy = \delta_{\nu\mu}, \tag{9.2.6}$$

[159] S. Bergman and M. Schiffer: A representation of Green's and Neumann's functions in the theory of partial differential equations of second order. Duke Math. J. 14, 609—638 (1947). On Green's and Neumann's functions in the theory of partial differential equations. Bull. Amer. Math. Soc. 53, 1141—1151 (1947). Kernel functions in the theory of partial differential equations of elliptic type. Duke Math. J. 15, 535—566 (1948).

where

$$\delta_{\nu\mu} = 1, \quad \text{for } \nu = \mu, \qquad (9.2.7)$$

$$\delta_{\nu\mu} = 0, \quad \text{for } \nu \neq \mu.$$

Then every solution Φ^* of (9.2.2) in B can be written in the form

$$\Phi^* = \sum_{\nu=1}^{\infty} a_\nu \Phi_\nu^*, \qquad (9.2.8)$$

$$a_\nu = \int_C \{\Phi_\nu^* [(\Phi^*)_n - q\,\Phi_\nu^*]\}\, ds. \qquad (9.2.8a)$$

This result enables us to determine the desired potential function Φ.

In order to solve the non-linear equation, we apply the method of successive approximations in the following manner. At first we determine as a first approximation, the potential function $\Phi^{(1)}$ for the incompressible fluid case. By using then the formula

$$\varrho^{(2)} = [1 - 1/2\,(k - 1/2)\,a_0^{-1}\,(\Phi_x^{(1)2} + \Phi_y^{(1)2})], \qquad (9.2.9)$$

we determine the second approximation for the density. Using $\varrho^{(2)}\,(x, y)$ we determine the second approximation for Φ, say $\Phi^{(2)}$. Substituting it in the right-hand side of (9.2.9) we obtain the third approximation $\varrho^{(3)}$, and so on. As it can be shown, the procedure converges in certain cases, in which it yields the desired solution of the non-linear equation.

3. General Remarks

On the base of a brief review (Part VII) of other methods available in the theory of compressible fluids we have shown that Bergman's method is one of the most general. Generally, Bergman's solutions are convergent in a much larger domain than all the expansions which involve hypergeometric functions. The presented method may serve in some cases as the starting point for methods of solution in more complicated cases, such as three-dimensional flows, etc. There are indications that it will be possible to find a solution valid in the entire plane of the exact compressibility equation (in the sub-, trans-, and supersonic regions). It may also be possible to extend the method to the case of flows with shocks. Since in all those cases the fundamental method will remain the same, the same tables if calculated precisely once and for all, can be used. This is undoubtedly a great advantage of this method.

Part X

List of Tables

Some tables will be given below. No attempt will be made to include long tables of various functions and coefficients which may be found in special papers by Bergman, as listed in Part VIII.

1. Subsonic Flow

(a) $T\,(2\,\lambda) = 1 - M^2\,(2\,\lambda)^{1/2} =$

$$= 1 - X - (1/2)\,(2\,k + 1)\,X^2 - (1/4)\,(4\,k^2 + 6\,k + 3)\,X^3$$
$$- (1/24)\,(24\,k^3 + 68\,k^2 + 76\,k + 29)\,X^4$$
$$- (1/48)\,(48\,k^4 + 212\,k^3 + 392\,k^2 + 328\,k + 103)\,X^5$$
$$- (1/480)\,(480\,k^5 + 2976\,k^4 + 7968\,k^3 + 10788\,k^2 + 7266\,k + 1935)\,X^6$$
$$- (1/2880)\,(2880\,k^6 + 23472\,k^5 + 84232\,k^4$$
$$+ 162124\,k^3 + 173940\,k^2 + 98086\,k + 22675)\,X^7 - \ldots$$
$$X = 2 \left[\frac{(k+1)^{1/2} - (k-1)^{1/2}}{(k+1)^{1/2} + (k-1)^{1/2}}\right] \left(\frac{k+1}{k-1}\right)^{1/2} \exp\,(2\,\lambda).$$

(b)　$$Q^{(1)} = -4 \int_{-\infty}^{\lambda} F \, d\lambda = -2 \int_{i}^{T(2\lambda)} F \, \frac{d(2\lambda)}{dT} \, dT = -(1/8) \int_{i}^{T} \left[\frac{5(1+k)}{T^6} - \frac{12\,k}{T^4} + \right.$$

$$\left. + \frac{2(3k-7)}{T^2} + 4(k+2) - (3k-1)\,T^2 \right] \frac{T^2}{(T^2-1)(1-h^2\,T^2)} \, dT =$$

$$= \frac{(k+1)}{8} \left[-\frac{5}{3\,T^3} + \frac{2\,k}{(k+1)\,T} + \frac{(1-3k)}{(k-1)}\,T + \right.$$

$$\left. + \frac{4}{h\,(k+1)^2\,(k-1)} \log \frac{1+(k-1)^{1/2}(k+1)^{-1/2}\,T}{1-(k-1)^{1/2}(k+1)^{-1/2}\,T} \right]_{T=1}^{T},$$

$$Q^{(2)} = \frac{4}{3}\,F + \frac{1}{6}\,Q^{(1)\,2},$$

(c)　$$\psi_\theta = H\,[Re\,g_z + (1/2)\,Q^{(1)}\,Re\,g + (3/4)\,Q^{(2)}\,Re\,g^{(1)} + \ldots],$$

$$\psi_q = R^{(0)}\,Im\,g_z + R^{(1)}\,Im\,g + (1/2)\,R^{(2)}\,Im\,g^{(1)}$$

$$+ (3/4)\,R^{(3)}\,Im\,g^{(2)} + \ldots \frac{\Gamma(2n-1)}{2^{2n-2}\,\Gamma(n)}\,R^{(n)}\,Im\,g^{(n-1)} + \ldots,$$

$$R^{(0)} = H\,\frac{d\lambda}{dq}, \qquad R^{(1)} = H_q + (1/2)\,HQ^{(1)}\,\frac{d\lambda}{dq},$$

$$R^{(n)} = \left[(HQ^{(n-1)})_\lambda + \frac{2\,n\,(2\,n-1)}{4\,n}\,HQ^{(n)} \right] \frac{d\lambda}{dq},$$

$$n = 2, 3, \ldots$$

$$Re = \text{real part}, \quad Im = \text{imaginary part}.$$

(d)　$$F_m\,(M) = \frac{(k+1)}{64}\,M^4\,[-(3k-1)\,M^4 - 4(3-2k)\,M^2 + 16] \cdot$$

$$\cdot \left[\sum_{\nu=1}^{n} (\nu^{-3})\,(-1)^\nu\,M^{2\nu} \right],$$

$$M^2 = \sum_{k=1}^{\infty} \beta_k\,X^k,$$

$$s = 1 - (1-M^2)^{1/2},$$

$$X = \frac{s}{(2-s)} \left\{ 2\,\frac{(h^{-1}+1-s)}{(h^{-1}-1+s)} \left[\frac{(h^{-1}-1)}{(h^{-1}+1)} \right]^{1/h} \right\},$$

$$h = (k-1)^{1/2}(k+1)^{-1/2}, \quad k > 1, \quad s = \exp(i\,\Phi),$$

$$X = (5 - 4\cos\Phi)^{-1/2}\,M^{1/(2h)},$$

$$M = \frac{(h^{-1}+1)^2 + 1 - 2(h^{-1}+1)\cos\Phi}{(h^{-1}-1)^2 + 1 + 2(h^{-1}-1)\cos\Phi},$$

$$X = s - (1/2)(2k+1)\,s^2 + (1/4)(4k^2+2k-1)\,s^3$$

$$- (1/24)(24k^3 + 8k^2 - 14k - 1)\,s^4 +$$

$$+ (1/48)(48k^4 + 4k^3 - 44k^2 + 2k + 5)\,s^5$$

$$- (1/480)(480k^5 - 104k^4 - 572k^3 + 148k^2 + 126k - 25)\,s^6$$

$$+ (1/2880)(2880k^6 - 1584k^5 - 3944k^4 + 2212k^3 + 1140k^2 -$$

$$- 602k - 5)\,s^7 + \ldots,$$

$$s = X + (1/2)(2k+1)\,X^2 + (1/4)(4k^2 + 6k + 3)\,X^3 +$$

$$+ (1/24)(24k^3 + 68k^2 + 76k + 29)\,X^4 + (1/48)(48k^4 + \ldots)\,X^5$$

$$+ \ldots \text{ similar to (a) with opposite signs.}$$

(e)　$$T = \sum_{n=0}^{\infty} A_n\,X^n = \sum_{n=0}^{\infty} a_n\,X_1^n,$$

$$T^{-1} = \sum_{n=0}^{\infty} B_n X^n = \sum_{n=0}^{\infty} b_n X_1^n.$$

X is given by formula (a); $X_1 = \exp(2\,\lambda)$.

For $k = 1.4$:

$X = 0.239 \exp(2\,\lambda)$.

n	$-A_n$	$-a_n$	B_n	b_n
0	-1	-1	1	1
1	1	.2392	1	.2392
2	1.9	.1087	2.9	.1659
3	4.81	.0658	9.61	.1315
4	13.939	.0456	33.869	.1108
5	43.68	.0342	123.696	.0968
6	144.02	.0270	462.39	.0865
7	492.11	.0220		.0786
8		.0185		.0724
9		.0158		.0672
10		.0138		.0629

(f) $F = \sum_{n=0}^{\infty} C_n \exp(2\,n\,\lambda)$,

C_0	C_1	C_2	C_3	C_4	C_5	C_6	C_7	C_8	C_9
0.0000	0.0000	0.1373	0.2858	0.4333	0.6073	0.7241	0.8678	1.011	1.153

(g) Fundamental solutions for $k = -1$ (see 6.5.8):

$$W^{(0)} = (1/2) \log\left[(\lambda - \lambda_0)^2 + (\theta - \theta_0)^2\right],$$

$$W^{(1)} = (\theta - \theta_0)\left[(\lambda - \lambda_0)^2 + (\theta - \theta_0)^2\right]^{-1},$$

$$q_0 + i\,\theta_0 = a_0, \qquad \lambda = (1/2) \log\left\{\frac{[1 + (q/a_0)^2]^{1/2} - 1}{[1 + (q/a_0)^2]^{1/2} + 1}\right\},$$

$$\lambda_0 = (1/2) \log\left\{\frac{[1 + (q_0/a_0)^2]^{1/2} - 1}{[1 + (q_0/a_0)^2]^{1/2} + 1}\right\},$$

$$W^{(01)} = W_\lambda^{(0)} = (\lambda - \lambda_0)\left[(\lambda - \lambda_0)^2 + (\theta - \theta_0)^2\right]^{-1}.$$

2. Transonic Flow

Conversion of M to $\eta = H$ (4.5.1):

M	η	M	η
.60	$-.1977$	.92	$-.0336$
.62	$-.1857$	.94	$-.0245$
.64	$-.1750$	.96	$-.0165$
.66	$-.1616$	.98	$-.0083$
.68	$-.1541$	1.00	0.0000
.70	$-.1407$	1.01	.0044
.72	$-.1307$	1.02	.0079
.74	$-.1195$	1.03	.0120
.76	$-.1087$	1.04	.0160
.78	$-.0978$	1.05	.0197
.80	$-.0887$	1.06	.0234
.82	$-.0787$	1.07	.0274
.84	$-.0692$	1.08	.0312
.86	$-.0603$	1.09	.0350
.88	$-.0513$	1.10	.0387
.90	$-.0421$		

Part XI

Examples

To illustrate the theory presented above, we shall outline briefly a few examples.

1. Subsonic Flow

We shall outline the construction of the compressible fluid flow generated by the function[150]:

$$g(\bar{Z}) = 1/2\,[(1 - 2\exp\bar{Z})^{+\,1/2} + (1 - 2\exp\bar{Z})^{-\,1/2}]. \tag{11.1.1}$$

In the case of an incompressible fluid, this function would lead to the circulation-free flow past a circular cylinder. In the case of a compressible fluid, a flow past a cylinder of somewhat distorted section is obtained. The actual computation consists of the following steps:

    I. Computation of the function $g^{(n)}(\bar{Z})$.

    II. Construction of the flow in the pseudo-logarithmic plane.

    III. Transition from the pseudo-logarithmic plane to the physical plane.

Step I

The functions $g^{(n)}(\bar{Z})$ are defined by (1.9.2a). In general, it will be impossible to evaluate the integrals in closed form:

$$g^{(n+1)}(\bar{Z}) = \int_0^{\bar{Z}} g^{(n)}(\bar{Z})\,d\bar{Z} = \int_0^{\lambda-i\theta} [\Phi^{(n)}(\lambda,\theta) + i\,\psi^{(n)}(\lambda,\theta)]\cdot(d\lambda - i\,d\theta) = \int_0^{\lambda,\theta} (\Phi^{(n)}\,d\lambda +$$

$$+ \psi^{(n)}\,d\theta) + i\int_0^{\lambda,\theta} (-\,\Phi^{(n)}\,d\theta + \psi^{(n)}\,d\lambda), \tag{11.1.2}$$

where

$$\Phi^{(n)} + i\,\psi^{(n)} = g^{(n)}; \tag{11.1.2a}$$

however, it is possible to use numerical integration. Since $g^{(n)}(Z)$ is an analytic function of a complex variable, the value of the integral $\int_0^{\bar{Z}} g^{(n)}(\bar{Z})\,d\bar{Z}$ is independent of the path of integration. A very convenient path consists of the interval $[(0,0), (\lambda,0)]$ of the negative real axis, followed by the interval $[(\lambda,0),(\lambda,\theta)]$ along a line parallel to the θ-axis, provided this path lies within the image of the flow in the pseudo-logarithmic plane. If this is not the case, an obvious adjustment must be made. Dividing each of these intervals into s-subintervals $\Delta\lambda_k$, $\Delta\theta_k$, gives approximately:

$$g^{(n+1)}(\bar{Z}) = \sum_{k=1}^{s} [(\Phi_k^{(n)} + i\,\psi_k^{(n)})\,\Delta\lambda_k + (\psi_k^{(n)} - i\,\Phi_k^{(n)})\,\Delta\theta_k]. \tag{11.1.3}$$

In many instances it is more convenient to expand $g(\bar{Z})$ into an infinite series and then integrate, term by term. However, as a rule, the function $g(\bar{Z})$ has singularities, and several series developments are needed in order to cover the domain in which the function is to be considered. For the case of function (11.1.1), the integral $\int g\,d\bar{Z}$ can be computed in closed form. Then

$$\int_0^{\bar{Z}} g(\bar{Z})\,d\bar{Z} = (1 - 2\exp\bar{Z})^{1/2} + \log[\exp\bar{Z} - 1 + (1 - 2\exp\bar{Z})^{1/2}] - \bar{Z} - i(1 + \pi/2),$$

$$\tag{11.1.4}$$

and

$$T_0(\lambda, \theta) = Im\,[g(\bar{Z})] = 2^{-3/2}(-A_0 + B_0^{1/2})^{1/2}(1 - B^{-1/2}), \qquad (11.1.5)$$

where

$$A_0 = (1 - 2\cos\theta\,\exp\lambda), \qquad (11.1.5\,\mathrm{a})$$

$$B_0 = (1 - 4\cos\theta\,\exp\lambda + 4\exp(2\,\lambda)), \qquad (11.1.5\,\mathrm{b})$$

$$T_1(\lambda, \theta) = Im\left[\int_0^{\bar{Z}} g(\bar{Z})d\bar{Z}\right] = 2^{-1/2}(-A_0 + B_0^{1/2})^{1/2} - \frac{\theta}{2} -$$

$$- (1 + \pi/2) + 1/2\tan^{-1}(C_1 D_1{}^{-1}), \qquad (11.1.6)$$

$$C_1 = 2^{1/2}\sin\theta\,\exp\lambda + [-A_0 B_0^{1/2}]^{1/2}, \qquad (11.1.6\,\mathrm{a})$$

$$D_1 = 2^{1/2}(\cos\theta\,\exp\lambda - 1) - [A_0 + B_0^{1/2}]^{1/2}. \qquad (11.1.6\,\mathrm{b})$$

One may easily calculate the values of $T_0(\lambda, \theta)$ and $T_1(\lambda, \theta)$ for various values of (λ, θ)[150]. The functions g, $T_0(\lambda, \theta)$ and $T_1(\lambda, \theta)$ can be represented by the following series:

For $-0.691 < \lambda < 0$:

$$g = i\,2^{1/2}\left[\exp(\bar{Z}/2) - (3/4)\exp(-\bar{Z}/2) - \frac{5}{32}\exp\left(-\frac{3\bar{Z}}{2}\right) -\right.$$

$$\left.- \frac{7}{128}\exp\left(\frac{5\bar{Z}}{2}\right) - \frac{45}{2048}\exp(-7\bar{Z}/2) - \ldots\right], \qquad (11.1.7)$$

$$T_0 = 2^{1/2}\left[\cos\left(\frac{\theta}{2}\right)\exp(\lambda/2) - (3/4)\cos\left(\frac{\theta}{2}\right)\exp\left(-\frac{\lambda}{2}\right) -\right.$$

$$- (5/32)\cos(3\,\theta/2)\exp(-3\,\lambda/2) - (7/128)\cos(5\,\theta/2)\exp\cdot$$

$$\left.\cdot(-5\,\lambda/2) - (45/2048)\cos(7\,\theta/2)\exp(-7\,\lambda/2) + \ldots\right], \qquad (11.1.8)$$

$$T_1 = 2^{1/2}[2(\cos(\theta/2)\exp(\lambda/2) - 1) +$$

$$+ (3/4)\cdot(2/1)(\cos(\theta/2)\exp(-\lambda/2) - 1) + (5/32)\cdot(2/3)\cdot$$

$$\cdot(\cos(3\,\theta/2)\exp(-3\,\lambda/2) - 1) + (7/128)\cdot(2/5)(\cos(5\,\theta/2)\cdot$$

$$\cdot\exp(-5\,\lambda/2) - 1) + (45/2048)\cdot(2/7)(\cos(7\,\theta/2)\exp\cdot(-7\,\lambda/2) - 1) + \ldots].$$

$$(11.1.9)$$

For $\lambda < -0.691$:

$$g = -2[1 + (1/2)\exp(2\,\bar{Z}) + \exp(3\,\bar{Z}) + (15/8)\exp(4\,\bar{Z}) +$$

$$+ (7/2)\exp(5\,\bar{Z}) + \ldots], \qquad (11.1.10)$$

$$T_0 = -2[(1/2)\sin(2\,\theta)\exp(2\,\lambda) + \sin(3\,\theta)\exp(3\,\lambda) +$$

$$+ (15/8)\sin(4\,\theta)\exp(4\,\lambda) + (7/2)\sin(5\,\theta)\exp(5\,\lambda) + \ldots], \qquad (11.1.11)$$

$$T_1 = -2[\theta + (1/4)\sin(2\,\theta)\exp(2\,\lambda) + (1/3)\sin(3\,\theta)\exp(3\,\lambda) +$$

$$+ (15/32)\sin(4\,\theta)\exp(4\,\lambda) + (7/10)\sin(5\,\theta)\exp(5\,\lambda) + \ldots]. \qquad (11.1.12)$$

Step II

The construction of the flow in the pseudo-logarithmic plane can be performed conveniently on specially scaled graph paper. One Cartesian coordinate axis is taken as the θ-axis, the other as the M-axis. In addition to the values of M, scales showing the corresponding values of $\lambda(M)$ and q/a_0 should be indicated. Such scales may be prepared once and for all, for each given value of k (see diagrams Ia and Ib[150] where these scales are drawn for $k = -0.5$ and $k = 1.4$). The values found for $Im\, g^{(n)}$ are entered on this paper, and the lines

$$Im\, g^{(n)}\,(\lambda - i\,\theta) = \text{constant},\qquad (11.1.13)$$

are drawn. In this way there are obtained by graphical interpolation the values of the functions

$$T_n\,(M,\,\theta) = Im\,\{g^{(n)}\,[\lambda\,(M) - i\,\theta]\}.\qquad (11.1.14)$$

The function $\psi\,(M,\,\theta)$ is determined by formula (1.9.16). In applying this formula, it is necessary to choose a definite value for the arbitrary constant α. In general, it is possible to choose α so that the terms in the series (1.9.16) become small for large value of M. In [150] α was set equal to 0.1.

In order to obtain the function $\psi\,(M,\,\theta)$ it is necessary to evaluate the products

$$1/2\,Q^{(1)}\,(M)\,T_1\,(M,\,\theta),\,(3/4)\,Q^{(2)}\,(M)\,T_2\,(M,\,\theta),\,\ldots\qquad (11.1.15)$$

The values of the functions $Q^{(n)}\,(M)$ can be tabulated once and for all (for a given value of k). The above products can be evaluated graphically, say, by means of a simple nomogram (see figure 4[148]). Adding a finite number of the terms in the series (1.9.2a) and multiplying by $H\,(M)$ (which function may be tabulated once and for all, see tables Ia, Ib, in [150]) gives an approximate expression for $\psi\,(M,\,\theta)$. Finally, the lines $\psi\,(M,\,\theta) = \text{constant}$ (streamlines) may be drawn. In [150] one can find the lines $\psi\,(M,\,\theta) = \text{constant}$, computed for the function (11.1.1) and for $k = -0.5$ and $k = 1.4$. For $k = -0.5$ the series has been replaced by

$$\psi_1\,(M,\,\theta) = \psi\,(M,\,\theta;\,-0.5) = Im\,\{H\,(2\,\lambda - 0.2)\cdot$$
$$\cdot\,[T_0\,(M,\,\theta) + (1/2)\,Q^{(1)}\,(2\,\lambda - 0.2)\,T_1\,(M,\,\theta)]\},\qquad (11.1.16)$$

with $\lambda = \lambda\,(M)$.

For $k = 1.4$, the series has been replaced by

$$\psi_2\,(M,\,\theta) = \psi\,(M,\,\theta;\,1.4) = Im\,\{H\,(2\,\lambda - 0.2)\,[T_0\,(M,\,\theta) +$$
$$+ (1/2)\,Q^{(1)}\,(2\,\lambda - 0.2)\,T_1\,(M,\,\theta) + (3/4)\,Q^{(2)}\cdot(2\,\lambda - 0.2)\,T_2\,(M,\,\theta)]\},\quad (11.1.17)$$

again with $\lambda = \lambda\,(M)$.

Step III

The flow is first transferred from the pseudo-logarithmic plane to the hodograph plane. This may be done by using (for the hodograph plane) polar coordinate graph paper. The polar coordinates in the hodograph plane are q/a_0 and θ. Since the values of q/a_0 and θ are known, the transition usually presents no difficulties (see figure 8 and 9 in [150]). In order to obtain the streamlines in the physical plane it is necessary to compute the integrals (1.9.18a) and (1.9.18b) along the streamlines in the hodograph plane.

The partial derivatives of ψ entering in the integrands of (1.9.18) can be expressed as follows:

$$\psi_q = R^{(0)}\,Im\,g_{\bar{z}} + R^{(1)}\,Im\,g + (1/2)\,R^{(2)}\,Im\,g^{(1)} + (3/4)\,R^{(3)}\,Im\,g^{(2)} + \ldots,\qquad (11.1.18)$$

where

$$R^{(0)} = H\,\frac{d\lambda}{dq},\qquad R^{(1)} = H_q + (1/2)\,H Q^{(1)}\,\frac{d\lambda}{dq},\,\ldots,\qquad (11.1.19)$$

$$R^{(n)} = \left[\frac{(2\,n - 2)!}{2^{2n-2}\,(n-1)!}\,(H Q^{(n-1)})_\lambda + \frac{2\,n\,(2\,n-1)}{4\,n}\,H Q^{(n)}\right]\frac{d\lambda}{dq},$$
$$n = 2,\,3,\,\ldots,\qquad (11.1.20)$$

and

$$\psi_\theta = H\,[Re\,g_{\bar{z}} + (1/2)\,Q^{(1)}\,Re\,g + (3/4)\,Q^{(2)}\,Re\,g^{(1)} + (15/8)\,Q^{(3)}\,Re\,g^{(2)} + \ldots].\qquad (11.1.21)$$

For the case under consideration the values of $R^{(0)}, \ldots R^{(4)}$ for $k = 1.4$ are given in table I of [150].

The integration is to be performed graphically. When a number of streamlines in the physical plane are drawn, a rather complete picture of the flow is obtained. In fact, for each point in the physical plane for which the corresponding point in the hodograph plane is known, the values of the speed, pressure, local Mach number, and so forth, also are known.

For the function (11.1.1) it happens that a part of the streamline $\psi = 0$ forms a closed contour. The streamline starts at $-\infty$ and divides itself into two branches at the first stagnation point. The two parts come together at the second stagnation point. Thus, a flow around an obstacle has been obtained. The boundary of the obstacle for $k = -0.5$ resembles a rhombus possessing curved lines as its four sides and rounded corners.

2. Use of Punch-card Machines

Step I

A method using the punch-card machines applied to the function (11.1.1) will be briefly outlined following[149]. The above function possesses singularities (branch points of the second order) at the points

$$\bar{Z} = -\log 2 + i\varkappa\pi, \quad \varkappa = 0, \pm 1, \pm 2, \ldots, \qquad (11.2.1)$$

only. By classical results in the theory of functions, $g(\bar{Z})$ can be expanded in series in powers of $\mathfrak{z}^{1/2}$,

$$\mathfrak{z} = \bar{Z} + \log 2, \qquad (11.2.2)$$

which series will converge for $|\mathfrak{z}| < \pi$ and therefore will represent g in a large part of the domain, $D_1 + D_2$, which is of interest. Namely, the domain D in the (λ, θ)-plane, in which the function $g(\bar{Z})$ is to be considered, can be divided into two distinct parts D_1 and D_2, defined as the subdomains in which $\lambda_0 < \lambda < 0$ and $\lambda < \lambda_0$, respectively. The number λ_0 is a preassigned number which can be altered to suit the case, although, in general, it will lie somewhere between $\lambda = -0.4$ and $\lambda = -0.1$, corresponding to local Mach numbers $M = 0.65$ and $M = 0.85$, respectively. The choice of λ_0 will depend upon the conditions in each case; for the

Table 11.1. Coefficients $A_{n,m}$

n \ m	0	1	2	3	4	5
0	-0.50000	0.625000	0.119797	0.024741	0.003871	0.004286
-1	0.25000	0.312500	0.179695	0.061852	0.013548	0.019287
1	$-1.$	0.416667	0.047919	0.007069	0.000860	
2	-0.66660	0.166667	0.013691	0.001571	0.000156	
3	-0.26667	0.047619	0.003042	0.000286	0.000024	
4	-0.076190	0.010582	0.000553	0.000044	0.000003	

Table 11.2. Coefficients $C_{n,m}$

n \ m	0	1	2	3	4	5
1	0.5708	0.	0	0	0	0
2	-0.0817	0.5708	0	0	0	0
3	0.0124	-0.0817	0.2854	0	0	0
4	-0.0016	0.0124	-0.0409	0.0951	0	0

most part, λ_0 must be larger than the maximal λ-coordinate of the regular points of $g\,(\bar{Z})$ in D. We note that in the example under consideration, ψ is determined not merely in D_2, but also in D_1, by the method described in the present section. A formal computation yields:

$$g \equiv g^{(0)} = (1/2)\,[(1 - \exp \mathfrak{z})^{1/2} + (1 - \exp \mathfrak{z})^{-1/2}] = i \sum_{n=0}^{\infty} A_{0,\,m}\,\mathfrak{z}^{m-1/2}, \qquad (11.2.3)$$

$$g^{(-1)} = \frac{dg^{(0)}}{d\bar{Z}} = \frac{dg^{(0)}}{d\mathfrak{z}} = i \sum_{m=0}^{\infty} A_{-1,\,m}\,\mathfrak{z}^{m-3/2}, \qquad (11.2.4)$$

$$g^{(n)} = i\left[\sum_{n=0}^{\infty} A_{n,\,m}\,\mathfrak{z}^{n+m-1/2}\right] + i\sum_{m=0}^{n-1} C_{n,\,m}\,\mathfrak{z}^{m}. \qquad (11.2.5)$$

The values of $A_{n,\,m}$ and $C_{n,\,m}$ are given in tables 11.1 and 11.2, respectively. By writing

$$g^{(n)} = S^{(n)} + i\,T^{(n)}, \qquad n = -1,\,0,\,1,\,2,\,\ldots, \qquad (11.2.6)$$

there is obtained

$$S^{(n)} = -\sum_{m=0}^{\infty} A_{n,\,m}\,\varrho^{n+m-1/2}\sin\,[(n+m-1/2)\,\varphi] - \sum_{m=0}^{n-1} C_{n,\,m}\,\varrho^{m}\sin m\,\varphi, \qquad (11.2.7)$$

$$T^{(n)} = \sum_{m=0}^{\infty} A_{n,\,m}\,\varrho^{n+m-1/2}\cos\,[(n+m-1/2)\,\varphi] + \sum_{n=0}^{m-1} C_{n,\,m}\,\varrho^{m}\cos m\,\varphi, \qquad (11.2.8)$$

where

$$\mathfrak{z} = \varrho \exp\,(i\,\varphi). \qquad (11.2.9)$$

The evaluation of the $S^{(n)}$ and $T^{(n)}$ on a punch card machine proceeds as follows:

The values of $\varrho^{k/2}$, $k = \pm 1,\ \pm 2,\ \pm 3,\ \ldots,\ \varrho^{1/2} = 0.1,\ 0.2,\ \ldots$, of $\cos\,(k\,\varphi/2)$, and of $\sin\,(k\,\varphi/2)$, $k = +1,\ \pm 2,\ \ldots \mu$, $\varphi = 0°,\ 30°,\ 60°,\ \ldots,\ 330°$, can easily be computed (see tables 7 and 8 in [149]) and entered on three sets of punch cards A, B, C, respectively. By using set A, two new sets, D and E are then prepared. On every punch card of the set D, the values of $A^{+}_{n,\,m}\,\varrho^{n+m-1/2}$ and of $C^{+}_{n,\,m}\,\varrho^{m}$ for a fixed n and fixed ϱ are entered, say, $A^{+}_{n,\,0}\,\varrho^{n-1/2}$ are punched in columns 1 to 6, $A^{+}_{n,\,1}\,\varrho^{n+1/2}$ in columns 7 to 12, $A^{+}_{n,\,2}\,\varrho^{n+3/2}$ in columns 13 to 18, and so forth. Here $A^{+}_{n,\,m}$ denotes $A_{n,\,m}$ if $A_{m,\,n}$ is positive, and 0 if $A_{n,\,m}$ is zero or negative; $C^{+}_{n,\,m}$ has an analogous meaning. In a similar manner the values of $A^{-}_{n,\,m}\,\varrho^{n+m-1/2}$ and of $C^{-}_{n,\,m}\,\varrho^{m}$ are entered on the cards of set E. Again $A^{-}_{n,\,m} = 0$ if $A_{n,\,m} > 0$, and equals $-A_{n,\,m}$ if $A_{n,\,m} \leqq 0$; the same holds for $C^{-}_{n,\,m}$. By using the sets C and D

$$\sum_{m=0}^{\mu} A^{+}_{n,\,m}\,\varrho^{n+m-1/2}\sin\,[(n+m-1/2)\,\varphi] + \sum_{m=0}^{n-1} C^{+}_{n,\,m}\,\varrho^{m}\sin m\,\varphi, \qquad (11.2.10)$$

is evaluated, and by using the sets C and E there may be computed

$$\sum_{m=0}^{\mu} A^{-}_{n,\,m}\,\varrho^{n+m-1/2}\sin\,[(n+m-1/2)\,\varphi] + \sum_{m=0}^{n-1} C^{-}_{n,\,m}\,\varrho^{m}\sin m\,\varphi. \qquad (11.2.11)$$

By subtracting (11.2.10) from (11.2.11) $S^{(n)}$ (11.2.7) is obtained. Similarly, $T^{(n)}$ can be determined. By interpolation, the values of $S^{(n)}\,(\lambda,\,\theta)$ and $T^{(n)}\,(\lambda,\,\theta)$ may be determined at intermediate points. Note that $\varrho \exp\,(i\,\varphi) = (\lambda - i\,\theta) + \log 2$, which yields the relation between $(\varrho,\,\varphi)$ and $(\lambda,\,\theta)$. Alternately, the expressions (11.2.7) and (11.2.8) may be evaluated by adding on cards of the sets B and C an extra column, say, column 7, in which nothing is punched if the corresponding sine or cosine is positive and say, 1 is punched if it is negative. In columns 1 to 6 the absolute value of the sine or cosine is entered.

Analogously, on cards of the set D an additional column is provided in which 1 or nothing is punched according to the sign of $A_{n,m}$ or $C_{n,m}$.

Step II

The second stage of the method is then to obtain the values of the stream function and its derivatives in the (λ, θ)-plane—that is, to evaluate the expressions

$$\psi(\lambda, \theta) = L^{(0)}(2\lambda) T^{(0)}(\lambda, \theta) + L^{(1)}(2\lambda) T^{(1)}(\lambda, \theta) + \ldots +$$

$$+ L^{(n)}(2\lambda) T^{(n)}(\lambda, \theta) + \ldots, \tag{11.2.12}$$

$$L^{(0)}(2\lambda) = H(2\lambda), \tag{11.2.12a}$$

$$L^{(1)}(2\lambda) = (1/2) H(2\lambda) Q^{(1)}(2\lambda), \ldots, \tag{11.2.12b}$$

$$L^{(n)}(2\lambda) = \frac{(2n)!}{2^n n!} H(2\lambda) Q^{(n)}(2\lambda), \tag{11.2.12c}$$

$$\psi_q(\lambda, \theta) = R^{(0)}(2\lambda) \operatorname{Im} g_{\bar{Z}} + R^{(1)}(2\lambda) T^{(0)}(\lambda, \theta) + \ldots + R^{(n)}(2\lambda) T^{(n-1)}(\lambda, \theta) + \ldots, \tag{11.2.13}$$

$$\psi_\theta(\lambda, \theta) = L^{(0)}(2\lambda) \operatorname{Re} g_{\bar{Z}} + L^{(1)}(2\lambda) S^{(0)}(\lambda, \theta) + \ldots +$$

$$+ L^{(n)}(2\lambda) S^{(n-1)}(\lambda, \theta) + \ldots \tag{11.2.14}$$

Since the $L^{(s)}(2\lambda)$, $s = 0, 1, 2, \ldots$ are independent of g, they can be entered on master cards once and for all, for different values of s and different values of λ. For instance, on master card No. 1 in columns 1 to 6, the value of $L^{(0)}(2\lambda)$ for a fixed value of λ, say λ_0, is entered, while nothing is punched in column 7 if $L^{(0)}$ is positive; in columns 8 to 14 the absolute value of $L^{(1)}(2\lambda)$ is entered, and in column 15 the number 1 is punched, if $L^{(1)}$ is negative, and so forth. Similarly, on master card No. 2 the corresponding values of $L^{(s)}(2\lambda^{(1)})$ are punched, and so forth.

From previous computations the values of $T^{(\varkappa)}(\lambda, \theta)$, $\varkappa = 0, 1, 2, \ldots$ for $\lambda = \lambda^{(0)}$ and $\theta = \theta^{(0)}$, $\theta^{(1)}$, $\theta^{(2)}$, and so forth, for $\lambda = \lambda^{(1)}$, $\theta = \theta^{(0)}$, $\theta^{(1)}$, $\theta^{(2)}$, and so forth, are obtained; both sets of cards, that is, the $L^{(s)}$ and $T^{(s)}$ are then put into the multiplier, which then yields the values of $\psi(\lambda, \theta)$ for the set of points $(\lambda^{(0)}, \theta^{(0)})$, $(\lambda^{(0)}, \theta^{(1)})$, $\ldots$ $(\lambda^{(1)}, \theta^{(0)})$, $(\lambda^{(1)}, \theta^{(1)})$, $\ldots$, $(\lambda^{(2)}, \theta^{(0)})$, $(\lambda^{(2)}, \theta^{(1)})$, $\ldots$. And ψ_q and ψ_θ may be obtained in similar fashion. The values of ψ, ψ_q, ψ_θ obtained for the case under consideration are given in [149], Tables 11 and 12.

Now, the values of $\psi(\lambda^{(0)}, \theta^{(\varkappa)})$, $\varkappa = 0, 1, 2, \ldots$, $\psi_q(\lambda^{(0)}, \theta^{(\varkappa)})$, $\psi_\theta(\lambda^{(0)}, \theta^{(\varkappa)})$, may be plotted on graph paper along the abscissa of which the values of θ are given. By using this diagram, the values of θ can be determined for $\psi(\lambda^{(0)}, \theta) = $ constant, say, 0, ± 0.1, ± 0.2, and so forth. The values of $\psi_q(\lambda^{(0)}, \theta)$ and of $\psi_\theta(\lambda^{(0)}, \theta)$ corresponding to $\psi(\lambda^{(0)}, \theta) = $ constant may then be determined. This procedure is then repeated for different values of λ.

Step III

To every value $\lambda^{(\varkappa)}$, $\varkappa = 0, 1, 2, \ldots$, the values of $\theta^{(\varkappa\tau)}$ are determined for which

$$\psi(\lambda^{(n)}, \theta^{(n\tau)}) = \tau = \text{constant}, \tag{11.2.15}$$

as well as the corresponding values of $\psi_\lambda(\lambda, \theta)$ and $\psi_\theta(\lambda, \theta)$. Tables (or figures) of q^2, $1 - M^2$, $(\varrho q^2)^{-1}$, $d\lambda/dq$, can be prepared, which, since these quantities are functions of λ alone have to be computed only once. The image of a streamline (11.2.15) in the physical plane is given by the integrals (1.9.18a) and (1.9.18b) taken along the streamline in the hodograph plane. The integrals (1.9.18a) and (1.9.18b) should be approximated by the sums

$$x = x\,(q_{(l)}) = \sum_{s=0}^{s=l} \Delta x_s\,(\tau), \qquad (11.2.16a)$$

$$y = y\,(q_{(l)}) = \sum_{s=0}^{s=l} \Delta y_s\,(\tau), \quad l = 1, 2, 3, \ldots \qquad (11.2.16b)$$

$$\Delta x_s = \varrho_0\,(\varrho_s\,q_s{}^2)^{-1}\,[\psi_\theta{}^{(s\,\tau)\,2}\,(1 - M_s{}^2) + q_s{}^2\,\psi_q{}^{(s\,\tau)\,2}]\cos\theta^{(s\,\tau)}\,[\psi_\theta{}^{(s\,\tau)}]^{-1}\,\Delta q_s, \qquad (11.2.17a)$$

$$\Delta y_s = \varrho_0\,(\varrho_s\,q_s{}^2)^{-1}\,[\psi_\theta{}^{(s\,\tau)\,2}\,(1 - M_s{}^2) + q_s{}^2\,\psi_q{}^{(s\,\tau)\,2}]\sin\theta^{(s\,\tau)}\,[\psi_\theta{}^{(s\,\tau)}]^{-1}\,\Delta q_s, \qquad (11.2.17b)$$

$$\Delta q_s = q_{s+1} - q_s. \qquad (11.2.17c)$$

By using tables for squares and the reciprocal, the values of $\psi_\theta{}^{(s\,\tau)\,2}$, $\psi_q{}^{(s\,\tau)\,2}$ and $[\psi_\theta{}^{s)\,\tau)}]^{-1}$ may be determined, together with the previously described tables for $(\varrho\,q^2)^{-1}$, $1 - M^2$, and so forth. The quantity

$$\varrho_0\,(\varrho_s\,q_s{}^2)^{-1}\,[\psi_\theta{}^{(s\,\tau)\,2}\,(1 - M_s)^2 + q_s{}^2\,\psi_q{}^{(s\,\tau)\,2}]\,[\psi_\theta{}^{(s\,\tau)}]^{-1}\,\Delta q_s, \qquad (11.2.18)$$

is determined with the use of punch cards. The expression (11.2.18) is then multiplied by $\cos\theta^{(s\,\tau)}$ and $\sin\theta^{(s\,\tau)}$ to yield the first and second terms of equations (11.2.17), respectively.

Since the sine and cosine may vary in sign, an extra column must be provided with each term of the product. The cards are put in the multiplier which is set for progressive totaling the values (11.2.16), which correspond to (11.2.15) then resulting.

3. Flow Past an Oval-shaped Obstacle

This type of flow is treated in [160]. Let there be given a circle $|z| = R$, and suppose that it is desired to determine the symmetric flow past this obstacle having velocity $(1/2)\,U$ at infinity, where U is a positive real number. Then it is easily proven that the so-called "complex potential" $w = \Phi + i\,\psi$ of the flow is given by:

$$w = \Phi + i\,\psi = (1/2)\,U\,(z + R^2/z), \qquad (11.3.1)$$

where, as usual, Φ and ψ are the potential and stream-functions, respectively, and dw/dz is the conjugate of the velocity vector (not the velocity vector itself). Now consider the mapping:

$$\mathfrak{z} = (1/2)\,(z + z^{-1}). \qquad (11.3.2)$$

If, as we shall assume, $R > 1$, this mapping transforms the exterior of the aforementioned circle into the exterior of the ellipse. We note that the thickness ratio of the ellipse is equal to $(R^2 - 1)/(R^2 + 1)$. It is easy to show that if (11.3.2) is solved for z as function of $\mathfrak{z}$, and if this substitution is performed in (11.3.1), then $w = \Phi + i\,\psi$ yields a symmetric flow past the ellipse with velocity U (not $(1/2)\,U$, as previously) at infinity. Furthermore, setting $dw/d\mathfrak{z} = \dfrac{dw}{dz}\bigg/\dfrac{d\mathfrak{z}}{dz} = \bar{q}$ (i. e., the conjugate of the velocity vector), we can easily show that w may be expressed as a function of $\bar{q}$ in the following manner:

$$w = (1/2)\,U\left[\left(\frac{\bar{q} - R^2\,U}{\bar{q} - U}\right)^{1/2} + R^2\left(\frac{\bar{q} - U}{\bar{q} - R^2\,U}\right)^{1/2}\right]. \qquad (11.3.3)$$

---

[160] S. Bergman and B. Epstein: Determination of a compressible fluid flow past an oval-shaped obstacle. J. Math. Physics **26**, 195—222 (1948).

Finally, if we set $\bar{q} = \exp \bar{Z}^+$, $\bar{Z}^+ = \lambda^+ - i\,\theta$, we obtain the following expression for the complex potential in the plane of the complex variable $\bar{Z}^+$ (logarithmic plane):

$$w = g^{[0]}\,(\bar{Z}^+) = (1/2)\,U\left[\left(\frac{\exp \bar{Z}^+ - R^2\,U}{\exp \bar{Z}^+ - U}\right)^{1/2} + R^2\left(\frac{\exp \bar{Z}^+ - U}{\exp \bar{Z}^+ - R^2\,U}\right)^{1/2}\right]. \qquad (11.3.4)$$

There are two parameters in the flow, namely U, the speed at infinity, and R^2, which determine the thickness ratio of the ellipse. We have chosen $R^2 = 1.2$, corresponding to a thickness ratio of $1/11$, while U was chosen equal to 0.765; the value of λ at infinity for the compressible flow $\psi = H\,\psi^*$, ψ^* being given by $(1.9.2)$, determined by applying the integral operator to the function $g^{[0]}\,(\bar{Z})$ given by $(11.3.4)$, will therefore be equal to $\alpha = \log\,(0.765) = -0.26785$, corresponding to Mach number 0.6. It should be stressed that the flow of the compressible fluid is situated on a two-sheeted Riemann surface whose two sheets corresponding to the upper and lower halves of the physical plane, respectively, are connected along the interval $-\infty < \lambda^+ < \alpha$, $\theta = 0$. The point $\bar{Z}^+ = \alpha$ is seen to be a branch point of $g^{[0]}\,(\bar{Z})$.

The function $g^{[0]}\,(\bar{Z})$ was evaluated for a number of values of $\bar{Z}$ lying within and slightly outside the streamline $\psi = 0$, (see Table IV.0 in [160]). Then the function

$$g^{[1]}\,(\bar{Z}) = \int_{\alpha}^{\bar{Z}} g^{[0]}\,(\bar{Z}')\,d\bar{Z}',\,\bar{Z}' = \lambda' - i\,\theta, \qquad (11.3.5)$$

which can also be expressed in terms of elementary functions, was evaluated for the same set of values of $\bar{Z}$ (see Table IV.1 in [160]). We have namely:

$$g^{[1]}\,(\bar{Z}) = (1/2)\,U\,(1 + R^2)\,\lg\,\{[1 + R^2 - 2\exp\,(\bar{Z} - \alpha) - 2\,S]/[R^2 - 1]\} +$$

$$+\,U\,R\,\lg\,\{[2\,R\,S + 2\,R^2 - (1 + R^2)\cdot\exp\,(\bar{Z} - \alpha)]/[R^2 - 1]\} - R\,U\,(\bar{Z} - \alpha), \qquad (11.3.6)$$

where

$$S^2 = \exp\,[2\,(\bar{Z} - \alpha)] - (1 + R^2)\exp\,(\bar{Z} - \alpha) + R^2. \qquad (11.3.6\,\mathrm{a})$$

The functions $g^{[2]}\,(\bar{Z})$, $g^{[3]}\,(\bar{Z})$, $g^{[4]}\,(\bar{Z})$, which are defined by the recursion formula:

$$g^{[n+1]}\,(\bar{Z}) = \int_{\alpha}^{\bar{Z}} g^{[n]}\,(\bar{Z}')\,d\bar{Z}', \qquad (11.3.7)$$

were also evaluated for the same set of values of $\bar{Z}$. However, since it was not possible to express the functions in closed form, a graphical method was employed to evaluate them. The values of $g^{[n+1]}\,(\bar{Z})$ were determined in the following manner: The point $\bar{Z}$ was joined to the point α by a polygonal path consisting of a vertical segment from $\bar{Z}$ to the λ-axis and a horizontal segment to the point α. The complex integration $(11.3.7)$ was thus reduced to two real integrations, for on the vertical portion, $d\bar{Z}' = i\,d\theta'$, while on the horizontal portion, $d\bar{Z}' = d\lambda'$. The integrals were estimated by plotting curves on graph paper and counting squares. In this manner the functions $g^{[2]}$, $g^{[3]}$, $g^{[4]}$, successively were determined (see Tables IV.2–IV.4 in [160]). Next for each of the values of $\bar{Z}$ under consideration the corresponding values of $H\,(2\,\lambda)$, $Q^{(n)}\,(2\,\lambda)$, $n = 1, 2, 3, 4$, were determined (see Table V in [160]). Then the imaginary part of the series $(1.9.2)$ multiplied by H $(1.9.7)$ which yields the stream-function of a compressible flow (at least in a certain domain of convergence) was approximated by taking the first four terms of the indicated summation. The stream-function given by $(1.9.2)$ does not yield a flow past the same obstacle as in the incompressible case, but rather past a more or less distorted obstacle. In this manner the approximate values for the stream-function were obtained (see Table VI in [160]). Finally, by using formulas (1.8.1) for the transition to the physical plane, the form of the obstacle and the streamlines may be determined approximately (see Tables VII.1, VII.2, in [160]). The point corresponding to $q = 0.4571$, $\theta = 0.348$ on the zero streamline was arbitrarily chosen as the origin of the physical plane. The position of the image of the point $q = 0.4571$, $\theta = 0.237$ (on the streamline 0.01) was then determined by using formula (1.8.1) along the line $q = 0.4571$). Then the images of the other points on the streamline $\psi = 0.01$ were determined by performing the integrations indicated in (1.8.1) along the streamline.

4. Transonic Flow

As an illustration of the theory presented in section 4.9, an example is briefly outlined below[155].

The complex potential of an incompressible flow around a circle is given in the physical plane (t-plane) by the relation

$$w = \Phi + i\,\psi = q_0\,(t + R/t^2), \tag{11.4.1}$$

where q_0 is the dimensionless speed at infinity. The function

$$\xi = (1/2)\,(t + 1/t), \tag{11.4.2}$$

maps the exterior of the circle onto the exterior of the ellipse and therefore the function $g\,(\xi) = w\,[(t\,(\xi)]$ will represent the complex potential of an incompressible flow around an ellipse in the physical plane. This potential, when transformed to the hodograph plane will furnish, upon expansion about $\eta_0 = \log q_0$, a set of $\{b_\nu\}$ for the incompressible case, which, as a first approximation will also be used for the compressible flow. This transformation is carried out as follows: The conjugate to the velocity vector $\bar{q}$ is given by

$$\bar{q} = dw/d\xi = (dw/dt)/(d\xi/dt), \tag{11.4.3}$$

or

$$\bar{q} = [(1/2)\,q_0\,(1 - R^2/t^2)]/[(1/2)\,q_0\,(1 - 1/t^2)], \tag{11.4.3a}$$

or

$$t^2 = (\bar{q} - q_0\,R^2)/(\bar{q} - q_0). \tag{11.4.3b}$$

Therefore:

$$w\,\{\xi\,[t\,(\bar{q})]\} = (1/2)\,q_0\,\{[(\bar{q} - q_0\,R^2)/(\bar{q} - q_0)]^{1/2} + R^2\,[(\bar{q} - q_0)/(\bar{q} - q_0\,R^2)]^{1/2}\}. \tag{11.4.4}$$

Since the complex potential in the logarithmic plane is needed, $\bar{z}$ is set $= \tilde{\eta} - i\,\theta$, with $\tilde{\eta} = \int^{\bar{q}} d\bar{q}/\bar{q} = \log \bar{q}$ and $\eta_0 = \log q_0$. Thus is obtained

$$\bar{q}\,(\bar{z}) = (1/2)\,Re\,(\exp \eta_0)\,\{(A\,B^{-1})^{1/2} + (B\,A^{-1})^{1/2}\}, \tag{11.4.5}$$

$$A = [1 - R^{-2}\exp\,(\bar{z} - \eta_0)], \tag{11.4.5a}$$

$$B = [1 - \exp\,(\bar{z} - \eta_0)], \tag{11.4.5b}$$

which can be developed in a power series of the form

$$g\,(\bar{z}) = \sum_{\nu=0}^{\infty} b_\nu\,(\bar{z} - \eta_0)^{\nu - 1/2}. \tag{11.4.6}$$

As an example, let us take $q_0 = 0.98$ (or $\eta_0 = -0.02$) and $R^2 = 1.2$, which yields a thickness ratio for the elliptic profile 1:11. With these numerical values the set of $\{b_\nu\}$ listed in Table 11.4.1 is obtained.

Table 11.4.1. The Series Coefficients of Function g

ν	0	1	2	3	4	5
b_ν	$0.24010\,i$	$-2.10086\,i$	$-4.85941\,i$	$-8.17015\,i$	$-82.86923\,i$	$-389.55\,i$

It is observed that the radius of convergence of this series is $|\mathfrak{z} - \eta_0| = \log R^2 = 0.18$ about the center $\mathfrak{z} = \eta_0$. As was mentioned in previous sections, however, by employing suitable summation methods, the series may be used outside its domain of convergence.

Using these values for $\{b_\nu\}$ and the table of $\{\psi^{(\nu)}\}$ (see section 4.9) yields the expression

$$\psi = \sum i\,b_\nu\,\psi^{(\nu)}\,(\eta,\,\theta;\,-0.02), \qquad (11.4.7)$$

which is equated to zero. If this is solved by a graphical method, a number of points are obtained on the streamline

$$\psi\,(\eta,\,\theta;\,-0.02) = 0, \qquad (11.4.8)$$

which are given in Table 11.4.2.

Table 11.4.2. Points on Streamline $\psi\,(\eta,\,\theta;\,-0.02) = 0$

η	0.032	0.029	0.025	0.014	0.	-0.014
θ	0	0.08	0.124	0.151	0.170	0.190

The next task is to obtain the image of this streamline in the physical plane. To do this the values of $\{X^{(\nu)}\}$ and $\{Y^{(\nu)}\}$, $\nu = 0, 1, \ldots 5$, (4.9.22) may be found corresponding to the set $(\eta,\,\theta)$. By forming the expressions

$$x = \sum i\,b_\nu\,X^{(\nu)}\,(\eta,\,\theta;\,-0.02), \qquad (11.4.9\,\mathrm{a})$$

$$y = \sum i\,b_\nu\,Y^{(\nu)}\,(\eta,\,\theta;\,-0.02), \qquad (11.4.9\,\mathrm{b})$$

see (4.9.21), a set of points may be obtained lying on the streamline $\psi = 0$ in the physical plane (see Table VIII in [155]).

Part XII

Errata in Previous Papers

A few errors and misprints were found here and there in previous papers. They will be corrected below, the numbers in brackets referring to the list of references at the end of the book.

Page	Equation	Should read
148		
48	(129)	$\beta = \theta_0,$
66	second (175)	$\ldots \dfrac{\pi^2\,\nu^2\,b_\nu}{4\,L^2} = 0.$
149		
37	(43)	$F\,(2\,\lambda) = \dfrac{1}{64}\,(k+1)\,[\ldots],$
49	(74)	$\leqq \dfrac{c\,(k+1)!}{\beta^{k+2}}.$

Continued on page 126

[153] Appendix III, p. 61: A more correct representation of the proof of a fundamental theorem on supersonic flows is given in N. A. C. A., T. N. No. 1875, Appendix B, pp. 27—31.

The whole proof in Appendix III, pp. 61—70, should be derived for $E^{(n)}\,(2\,\Lambda) = E^{(n)}\,(\tau)$, and not for $E^{(n)}\,(\Lambda)$. The formulas given on pp. 68, 69 and 70 should refer to $\underline{E}^{(n)}\,(2\,\Lambda) = \underline{E}^{(n)}\,(\tau)$ and not $\underline{E}^{(n)}\,(\Lambda)$.

Page	Equation	Should read
153		
67	(156)	$\Lambda^* = \Lambda - \dfrac{a}{2},$
67	Figure 10 is incorrect	Figure 3, Bergman, N. A. C. A., No. 1875.
69	Above (168)	$\Lambda = \dfrac{a}{2}.$
154		
17	(19)	$+ \int\limits_{\eta_0}^{\eta} h_Y(Y) \ldots + \int\limits_{\delta_0}^{\delta} g_X(X) \ldots,$
21	below (31)	$F_2\,[2\,\Lambda - \pi\,(h^{-1} - 1)] = \ldots,$
25	(41)	$\int\limits_{\eta_0}^{\eta} dY \int\limits_{\delta_0}^{\delta} F_1\,(X + Y)\,dX = \ldots,$
29	(60)	$\ldots + \tilde{H}^{(n+1)}(\tau),$
29	(61)	$0 \ll \hat{\tilde{H}}^{(n+1)}(\tau),$
29	(64)	$\dfrac{M}{(\tau_0 - \varepsilon - \tau)^2},$
30	(67)	$\tilde{E}^{(n+1)}(\tau) = C_n\,M\,\tau_0{}^2 \int\limits_{a}^{\tau} \cdot$ $\cdot \left[\dfrac{n\,(n+1) + M}{(\tau_0 - \varepsilon - \tau)^{n+2}}\right] d\tau +$ $+ \dfrac{C_{n+1}\,M\,\tau_0{}^2}{(\tau_0 - \varepsilon - a)^{n+1}} = \dfrac{C_{n+1}\,M\,\tau_0{}^2}{(\tau_0 - \varepsilon - \tau)^{n+1}},$
30	(68)	$C_{n+1} = C_n\left(n + \dfrac{M}{(n+1)}\right),$
31	(73)	$V^{(1)}(\xi', \eta') \ll$ $M_2\left[1 + \dfrac{2!\,M\,M^*\,M_2\,\tau_0{}^2\,|\xi'|}{(\tau_0 - \varepsilon - \tau)} + \ldots \right.$ $\left. \dfrac{(n+1)!\,M\,M^*\,M_2\,\tau_0{}^2\,|\xi'|^n}{n!\,(\tau_0 - \varepsilon - \tau)^n} + \ldots \right].$
156		
470	(4.5)	$\dfrac{\bar{q} - \bar{q}_0}{m} + \ldots$
471	(4.6)	$\ldots + \text{Power series in } (\bar{q} - \bar{q}_0).$
472	Below definition 4.1	$\psi = \text{arc tan}\,[(\theta - \theta_0)/(\tilde{\lambda} - \tilde{\lambda}_0)].$
472	Above (5.1)	$\ldots$ of (2.12b) and (2.12a), respectively.
155		
17	footnote	$\ldots (1 - M^2)^{1/2}\,dq,$
18	(12)	$l\,(\eta) = [2/(k-1)]^{(2-k)/(k+1)} \cdot \{(-2\,\eta) - [(2\,k + 5)/$ $(2\,k + 2)] \ldots\},$
20	(17)	$S_0 = 2^{(2k+1)/(6k-6)}\,3^{-1/6} \ldots,$
21	(27b)	$[\ldots \sin\theta + \psi_\eta \cos\theta]\,d\eta,$
22	6th line from top	$|X| > 1,$
22	(B)	$\ldots - 2\,c\,(-\eta)^{3/2}\,\eta_0 - \eta_0{}^2 = 0,$
22	(28a)	$\ldots (1/6,\ 5/6,\ \nu + 1/2,\ \ldots) = F\,(1/6,\ 5/6,\ \nu + 1/2,\ \ldots),$
23	(28b)	$\ldots (1/6,\ 5/6,\ \nu + 1/2,\ \ldots) = \ldots,$
23	below (28b)	$A_\nu = \Gamma\,(-2/3) \cdot \Gamma\,(\nu + 1/2)/\Gamma\left(-\dfrac{1}{3} - \nu\right)\Gamma\,(5/6).$

Page	Equation	Should read
157		
874	(4.14)	$E^+ = (\ldots) F (\ldots) + b_3 S_0 \ldots,$
876	First line from top	$.. (4.9) \ldots,$
876	5th line from bottom	$\ldots$ of $(2.12).,$
882	(5.29)	$\ldots s^{(3/2)n} s^{\nu}/(\ldots)\ldots,$
882	(5.30)	$\ldots \displaystyle\sum_{n=0}^{\infty} \sum_{\nu=0}^{\infty} \ldots .$

Additional Contributions

Below the author gives some recent contributions to Chapter VII, "Review of Other Methods".

Henrici[161] (1952) had shown that the Bergmann's integral operator of the first kind can be interpreted as the solution of a certain hyperbolic equation by Riemann's method.

One may mention the papers by Gilbarg[162], Loewner[163], Sauer[164], Longhorn[165], Bers[166], Bergman[167], Manwell[168], Gilbard and Serrin[169], and many others.

Diaz and Ludford[170] compare the methods of Bergman and Le Roux[171] for the case of linear hyperbolic equations in two variables. They seem to show that the theorem which is the basis for Bergman's method is a restatement of results of Le Roux. In some applications Le Roux's method may be simpler.

[161] P. Henrici: Bergman's Integraloperator erster Art und Riemannsche Function. ZAMP, III, 3, 228—232 (1952).

[162] D. Gilbarg: Comparison methods in the theory of subsonic flows. J. Rational Mech. Analysis 2, 2, 233—251 (1953).

[163] C. Loewner: Conservation laws in compressible fluid flow and associated mappings. J. Rational Mech. Analysis 2, 3, 537—561 (1953). Some bounds for the critical free stream Mach number of a compressible flow around an obstacle. "Studies in mathematics and mechanics" presented to R. v. Mises by friends, colleagues and pupiles, New York, Academic Press. 177—183 (1954).

[164] R. Sauer: Hyperbolic problem in gas dynamics with more than two independent variables (in German). ZAMM 33, 10/11, 331—336 (1953).

[165] A. L. Longhorn: Subsonic compressible flow past bluff bodies. Aero. Quart. 5, 2, 144—162 (1954).

[166] L. Bers: Existence and uniqueness of a subsonic flow past a given profile. Comm. pure appl. math. 7, 3, 441—504 (1954).

[167] S. Bergman: A method for solving boundary value problems of mathematical physics on punch card machines. J. Assn. Comput. Mach. 1, 3, 101—104 (1954). Tables for the determination of fundamental solutions of equations in the theory of compressible fluids. Mathematical tables and other aids to computation. National Acad. Sci., Nation. Res. Council, U. S. A., 9, 49, 8—14 (Jan. 1955).

[168] A. R. Manwell: A new singularity of transonic plane flows. Quart. Appl. Math. 12, 4, 343—349 (1955).

[169] D. Gilbarg and J. Serrin: Uniqueness of axially symmetric subsonic flow past a finite body. J. Rat. Mech. Analysis, 4, 1, 169—175 (1955).

[170] J. B. Diaz and G. S. S. Ludford: On two methods of generating solutions of linear partial differential equations by means of definite integrals. Quart. Appl. Math. 12, 4, 422—427 (1955).

[171] J. Le Roux: Sur les intégrales des équations linéares aux dérivées partielles du second ordre à deux variables indépendantes. Ann. Ec. Norm. Sup. (3), 12, 227—316 (1895).

Part XIII

Generalization of Bergman's Method to Diabatic Flow

In this part the author jointly with Dr. H. A. Hassan presents the generalization of Bergman's method to diabatic flow. This part originated from a chapter of the dissertation, done by Dr. H. A. Hassan under the direction of the senior author, as a partial fulfillment of the requirements toward the Ph. D. degree at the University of Illinois. The example refers to a very limited case.

A diabatic flow can be thought of as a flow of an inviscid, non-heat conducting fluid with heat addition or subtraction by means of sources or sinks distributed in the flow domain. Bergman's method is extended here to diabatic flows in the elliptic domain. Starting from the generalized Cauchy-Riemann conditions, the stream function equation is derived. This equation is transformed to a form similar to that in isentropic flow. Following Bergman, v. Mises and Schiffer, the stream function is represented as an infinite series. A proof of convergence along the lines of the work done by v. Mises and Schiffer is presented and the domain of convergence is determined.

The fundamental law used in this generalization is the first law of thermodynamics:

$$q = dQ = c_v\, dT + p\, dv, \qquad v = \varrho^{-1},$$

where the symbol c_v denotes the specific heat of the fluid. It is assumed that c_v is a constant value. M. Planck opposed the following example by Clausius who denoted the quantity q by dQ to indicate that it is infinitely small. According to Planck this notation has frequently given rise to misunderstanding, for dQ has been repeatedly regarded as the differential of a known finite quantity Q. M. Planck wrote:

"This faulty reasoning may be illustrated by the following example. If we choose T and v as independent variables, we have:

$$dQ\,(T, v) = du + p\, dv = u,_T\, dT + (u,_v + p)\, dv; \qquad u = c_v\, T.$$

On the other hand:

$$dQ\,(T, v) = Q,_T\, dT + Q,_v\, dv.$$

Accordingly, since dT and dv are independent of one another:

$$Q,_T = u,_T; \qquad Q,_v = u,_v + p.$$

Differentiating the first with respect to v and the second with respect to T, we have:

$$Q,_{Tv} = u,_{Tv} = u,_{vT} + p,_T.$$

Therefore $p,_T = 0$, which is certainly not true."

Analogous demonstrations are given in many books. The usual interpretation of the results obtained above is the following: In general, $Q,_{Tv} \neq Q,_{vT}$; hence dQ is not a perfect differential or $\oint dQ\,(T, v) \neq 0$, and consequently there does not exist a function Q whose differential dQ satisfies the first law of thermodynamics. If we assume that $\oint dQ\,(T, v) = 0$, with $\oint c_v\, dT = 0$, then $\oint p\, dv = 0$, which is satisfied if and only if $p = p\,(v)$ (barotropic flow). But this reasoning of Planck is not complete.

In another of his works, the senior author proved the following lemma:

If in a certain region $E\,[x, y]$ the functions $T\,(x, y)$ and $v\,(x, y)$ are functions of bounded variation and the function $p\,(x, y)$ is continuous in the equality

$$dQ\,(x, y) = c_v\, dT\,(x, y) + p\,(x, y)\, dv\,(x, y),$$

with c_v being a constant, then the Riemann-Stieltjes integral $Q(x, y) = \int dQ(x, y)$ always exists on E.

The proof follows immediately from the fundamental theorems of the theory of real functions. The conditions appearing in the Lemma, referring to T, p, and v, are very light and one may believe that they are satisfied in a predominant number of cases occurring in fluid dynamics. Finding a solution of the stream-function equation (13.4.14) is not easy. But the general approach to solving the problem maybe well illustrated on a very limited case like $Q = Q(q)$ only. Any other approach may obviously follow the outlined pattern.

The total differential of a function $z(x, y)$ is defined by:

$$dz = z,_x\, dx + z,_y\, dy.$$

For the existence of a total differential at (x_0, y_0) it is necessary that the function be continuous with respect to (x, y) at (x_0, y_0).

1. Basic Equations

The diabatic flow equations are based upon the hydrodynamic equations for an inviscid, non-heat conducting compressible perfect fluid and the first law of thermodynamics. With the use of Cartesian tensor notation, these equations in a steady flow have the following form:

The equation of motion:

$$\varrho\, u_j u_{i,\,j} + p,_i = 0, \quad (i, j = 1, 2) \tag{13.1.1}$$

the continuity equation and equation of state:

$$(\varrho\, u_i),_i = 0; \quad p = R\, \varrho\, T. \tag{13.1.2}$$

The first law of thermodynamics $dQ = dE + dW$, (dQ = heat introduced; E = internal energy, W = work done by the fluid) reduces for a perfect gas to the following form: $dW = p\, d(\varrho^{-1})$, $dE = c_v\, dT$ and by expressing dp from the equation of state; and using $R = c_p - c_v$,

$$c_p\, dT = dQ + \varrho^{-1}\, dp. \tag{13.1.3}$$

From eq. (13.1.3) the energy equation can be written in the form:

$$u_i (c_p\, T),_i = u_i Q,_i + \varrho^{-1} u_i p,_i. \tag{13.1.4}$$

Expressing the total differentials in eq. (13.1.3) as $dT = T,_i\, dx_i$, etc. and, since x_i, x_j are independent for $i \neq j$, one finds

$$c_p\, T,_i = Q,_i + \varrho^{-1} p,_i. \tag{13.1.4a}$$

Dividing equation (13.1.3) by T and defining the entropy $dS = T^{-1} dQ$, one finds for constant c_p,

$$c_p\, T^{-1} dT = dS + R\, p^{-1} dp. \tag{13.1.5}$$

Integration of equation (13.1.5) gives:

$$p\, p_0^{-1} = (\varrho\, \varrho_0^{-1})^\gamma \exp\left[c_v^{-1} (S - S_0)\right], \quad \gamma = c_p/c_v, \tag{13.1.6}$$

which can be written in the form:

$$p = C\, \varrho^\gamma \exp(c_v^{-1} S); \quad S = \int T^{-1} dQ + \text{const.} \tag{13.1.7}$$

This is the pressure-density-entropy relationship. Following previous authors, the local "isentropic" velocity of sound is defined as:

$$a^2 = (\partial p/\partial \varrho)_S = \gamma\, \varrho^{-1} p = \gamma\, R\, T. \tag{13.1.8}$$

2. The Stream Function Equation

In a steady, two-dimensional motion, the stream function is defined as:

$$\varrho\, u = \psi,_y; \quad -\varrho\, v = \psi,_x; \quad \varrho^2 q^2 = \psi^2,_x + \psi^2,_y; \quad q^2 = u^2 + v^2. \quad (13.2.1)$$

Differentiating the third of equations (13.2.1) first with respect to x and multiplying the result by $(-v)$, then with respect to y and multiplying the result by u, using the first of equations (13.2.1) and adding the products furnish the result:

$$u^2\, \psi,_{yy} - 2\, u\, v\, \psi,_{xy} + v^2\, \psi,_{xx} = q^2\, (u\, \varrho,_y - v\, \varrho,_x) + \varrho\, q\, (u\, q,_y - v\, q,_x). \quad (13.2.2)$$

Eliminating T in equation (13.1.4a) by means of the equation of state, eliminating $p,_i$ in the same equation by means of equation (13.1.1) and, using equation (13.1.8) one finds

$$- (\gamma - 1)^{-1}\, \varrho^{-1}\, a^2\, \varrho,_i = Q,_i + (\gamma - 1)^{-1}\, u_j\, u_{i,\,j}. \quad (13.2.3)$$

Putting $i = x, y$ respectively, in equation (13.2.3), i. e., writing $\ldots \varrho,_x = Q,_x + \ldots$, etc., using the last of equations (13.2.1) and multiplying the first by v, the second by u and subtracting give the formula:

$$u\, \varrho,_y - v\, \varrho,_x = \varrho\, a^{-2}\, [(\gamma - 1)\, (v\, Q,_x - u\, Q,_y) - q\, (u\, q,_y - v\, q,_x) - q^2\, \omega], \quad (13.2.4)$$

where the symbol $\omega = v,_x - u,_y$, denotes the vorticity. Calculating u and v from the first of equations (13.2.1), differentiating these values with respect to y and x, respectively, and subtracting, one finds

$$u\, \varrho,_y - v\, \varrho,_x = \varrho\, \omega + \psi,_{xx} + \psi,_{yy}. \quad (13.2.5)$$

Combining equations (13.2.4) and (13.2.5), gives

$$a^2\, (\varrho\, \omega + \psi,_{xx} + \psi,_{yy}) = \varrho\, [(\gamma - 1)\, (v\, Q,_x - u\, Q,_y) - q\, (u\, q,_y - v\, q,_x) - q^2\, \omega]. \quad (13.2.6)$$

Eliminating $(u\, \varrho,_y - v\, \varrho,_x)$ between equations (13.2.2) and (13.2.5) and adding the result to equation (13.2.6), one obtains

$$(a^2 - u^2)\, \psi,_{xx} - 2\, u\, v\, \psi,_{xy} + (a^2 - v^2)\, \psi,_{yy} = \varrho\, [(\gamma - 1)\, (v\, Q,_x - u\, Q,_y) - a^2\, \omega]. \quad (13.2.7)$$

This is one of possible forms of the stream function equation.

By using equation (13.1.1) to eliminate $p,_i$ from equation (13.1.4) one finds:

$$u_i\, (c_p\, T),_i + u_i\, u_j\, u_{i,\,j} = u_i\, Q,_i. \quad (13.2.8)$$

But since:

$$q\, \partial/\partial s = u_i\, \partial/\partial x_i, \quad (13.2.9)$$

where $\{s, n\}$ represent the orthogonal coordinates system running along a streamline (s — tangential, n — normal); hence equation (13.2.8) furnishes:

$$\partial/\partial s\, (c_p\, T + {}^1/_2\, q^2 - Q) = 0; \quad (13.2.10)$$

$$c_p\, T + {}^1/_2\, q^2 - Q = c_p\, T_0 - Q_0 = \text{const.}, \quad (13.2.10\,a)$$

where the subscript '0' refers to the stagnation conditions. Calculating $c_p\, T$ from equation (13.2.10a), i. e.,

$$c_p\, T = Q - Q_0 - {}^1/_2\, u_j\, u_j + c_p\, T_0, \quad (13.2.11)$$

and inserting this value into equation (13.1.4a) one obtains

$$(c_p\, T_0),_i - Q_0,_i = u_j\, (u_{j,\,i} - u_{i,\,j}). \quad (13.2.12)$$

Let $i = x, y$ in equation (13.2.12), multiply the results by v and $(-u)$, respectively, and add the products, thus obtaining:

$$u\, Q_0,_y - v\, Q_0,_x + v\, (c_p\, T_0),_x - u\, (c_p\, T_0),_y = q^2\, \omega. \quad (13.2.13)$$

Denoting the angle of the inclination of the streamline to the horizontal axis by θ furnishes the well-known relation:

$$\partial/\partial n = -\sin\theta\,\partial/\partial x + \cos\theta\,\partial/\partial y, \tag{13.2.14}$$

or

$$q\,\partial/\partial n = -v\,\partial/\partial x + u\,\partial/\partial y. \tag{13.2.14a}$$

Of course:

$$d\psi/dn = \varrho\,q. \tag{13.2.15}$$

With the use of equations (13.2.14a) and (13.2.15), equation (13.2.13) reduces to:

$$\omega = \varrho\,[dQ_0/d\psi - dh_0/d\psi], \quad h_0 = c_p\,T_0. \tag{13.2.16}$$

Inserting equation (13.2.16) into equation (13.2.7) furnishes another form of stream function equation:

$$(a^2 - u^2)\,\psi,_{xx} - 2\,u\,v\,\psi,_{xy} + (a^2 - v^2)\,\psi,_{yy} =$$
$$= \varrho^2\,[a^2\,(dh_0/d\psi - dQ_0/d\psi) - (\gamma - 1)\,q^2\,dQ/d\psi], \tag{13.2.17}$$

with

$$\varrho^2 = q^{-2}\,[(\psi,_x)^2 + (\psi,_y)^2]. \tag{13.2.17a}$$

3. The "Generalized" Potential Equation

The concept of a function analogous to the velocity potential was used by v. Krzywoblocki. The equation $v,_x - u,_y = \omega$ is fulfilled if one puts:

$$u = \varphi,_x + g; \quad v = \varphi,_y + g; \quad g,_x - g,_y = \omega, \tag{13.3.1}$$

where the function $g = g(x, y)$ is defined by the last of equations (13.3.1); the condition is that for an irrotational flow the trivial solution of (13.3.1), i. e., $g \equiv 0$, must be assumed.

The first two equations (13.3.1) are squared and added, thus giving an expression for q^2, differentiating it with respect to x, and y, respectively; multiplying the first expression so obtained by u and the second by v and adding the products, one obtains

$$q\,(u\,q,_x + v\,q,_y) = u^2\,\varphi,_{xx} + 2\,u\,v\,\varphi,_{xy} + v^2\,\varphi,_{yy} + (u + v)\,(u\,g,_x + v\,g,_y). \tag{13.3.2}$$

Letting $i = x, y$, respectively, in equation (13.2.3); multiplying the first by u and the second by v and adding, and finally eliminating the expression $\varrho^{-1}\,(u\,\varrho,_x + v\,\varrho,_y)$ by substituting the first two equations (13.3.1) into the continuity equation; one finds

$$a^2\,(\varphi,_{xx} + \varphi,_{yy} + g,_x + g,_y) = (\gamma - 1)\,(u\,Q,_x + v\,Q,_y) + q\,(u\,q,_x + v\,q,_y). \tag{13.3.3}$$

Eliminating the last term in equation (13.3.3) by using (13.3.2), and arranging terms, give the equation:

$$(a^2 - u^2)\,\varphi,_{xx} - 2\,u\,v\,\varphi,_{xy} + (a^2 - v^2)\,\varphi,_{yy} = (\gamma - 1)\,(u\,Q,_x + v\,Q,_y) -$$
$$- (a^2 - u^2)\,g,_x + u\,v\,(g,_x + g,_y) - (a^2 - v^2)\,g,_y. \tag{13.3.4}$$

This is one of possible forms of the generalized potential equation of diabatic flow. Other forms are listed below. Using equations (13.3.1) and the new function $u\,Q,_x + v\,Q,_y = DQ/Dt = \overline{Q}$ in equation (13.3.4) furnishes the formula:

$$(a^2 - u^2)\,u,_x - u\,v\,(v,_x + u,_y) + (a^2 - v^2)\,v,_y = (\gamma - 1)\,\overline{Q}, \tag{13.3.5}$$

where $\overline{Q}$ is the rate of the convection of the total energy introduced. Consider the function Q as function of u, v, i. e., $Q = Q\,[u\,(x, y), v\,(x, y)]$, then:

$$Q,_x = Q,_u\,u,_x + Q,_v\,v,_x; \quad Q,_y = Q,_u\,u,_y + Q,_v\,v,_y. \tag{13.3.6}$$

Inserting equations (13.3.1) and (13.3.6) in equation (13.3.4) gives:

$$[a^2 - u^2 - (\gamma - 1)\,u\,Q,_u]\,u,_x - [u\,v + (\gamma - 1)\,v\,Q,_u]\,u,_y -$$

$$- [u\,v + (\gamma - 1)\,u\,Q,_v]\,v,_x + [a^2 - v^2 - (\gamma - 1)\,v\,Q,_v]\,v,_y = 0. \qquad (13.3.7)$$

In the case of an irrotational flow: $u,_x = \varphi,_{xx};\ u,_y = v,_x = \varphi,_{xy};\ v,_y = \varphi,_{yy}.$
For a rotational flow these derivatives should be derived from equation (13.3.1).

4. Hodograph Transformation

Letting $W = f(x, y) = F(u, v)$ be an arbitrary function, then:

$$F,_u = f,_x\,x,_u + f,_y\,y,_u; \qquad F,_v = f,_x\,x,_v + f,_y\,y,_v, \qquad (13.4.1)$$

and

$$f,_x = J_1^{-1}(F,_u\,y,_v - F,_v\,y,_u); \qquad f,_y = J_1^{-1}(F,_u\,x,_v - F,_v\,x,_u), \qquad (13.4.2)$$

where the Jacobian of the transformation, J_1, is defined as:

$$J_1 = x,_u\,y,_v - x,_v\,y,_u. \qquad (13.4.3)$$

Applying equations (13.4.2) to u, v, respectively, i. e., $u,_x = J_1^{-1}\,y,_v;\, v,_x = -J_1^{-1}\,y,_u,$
etc., and inserting the resulting expressions into equation (13.3.5) and into the
equation defining the vorticity, one obtains the following system of equations:

$$(a^2 - u^2)\,y,_v + u\,v\,(x,_v + y,_u) + (a^2 - v^2)\,x,_u = (\gamma - 1)\,J_1\,\overline{Q}; \qquad (13.4.4a)$$

$$x,_v - y,_u = J_1\,\omega. \qquad (13.4.4b)$$

The expressions for the total differentials $d\varphi,\ d\psi$ are given by

$$d\varphi = \varphi,_u\,du + \varphi,_v\,dv = \varphi,_x\,dx + \varphi,_y\,dy = (u - g)\,dx + (v - g)\,dy; \qquad (13.4.5a)$$

$$d\psi = \psi,_u\,du + \psi,_v\,dv = \psi,_x\,dx + \psi,_y\,dy = \varrho\,(u\,dy - v\,dx). \qquad (13.4.5b)$$

Treating equations (13.4.5a, b) as an algebraic system in two unknowns dx and dy
and calculating dx and dy one can obtain the partial derivatives $x,_u;\ x,_v;$ etc.:

$$x,_u = [\varrho\,u\,\varphi,_u - (v - g)\,\psi,_u]\,E^{-1}; \qquad x,_v = [\varrho\,u\,\varphi,_v - (v - g)\,\psi,_v]\,E^{-1}; \qquad (13.4.6a)$$

$$y,_u = [\varrho\,v\,\varphi,_u + (u - g)\,\psi,_u]\,E^{-1}; \qquad y,_v = [\varrho\,v\,\varphi,_v + (u - g)\,\psi,_v]\,E^{-1}; \qquad (13.4.6b)$$

$$E = \varrho\,[q^2 - g\,(u + v)]. \qquad (13.4.6c)$$

Substituting these expressions into the system (13.4.4a, b) furnishes the system:

$$\varrho\,a^2\,(u\,\varphi,_u + v\,\varphi,_v) + (q^2 - a^2)\,(v\,\psi,_u - u\,\psi,_v) - g\,\psi,_u\,(u\,v + v^2 - a^2) +$$

$$+ g\,\psi,_v\,(u\,v + u^2 - a^2) = (\gamma - 1)\,\varrho\,[q^2 - g\,(u + v)]\,J_1\,\overline{Q}; \qquad (13.4.7a)$$

$$(u - g)\,\psi,_u + (v - g)\,\psi,_v + \varrho\,(v\,\varphi,_u - u\,\varphi,_v) = -\varrho\,[q^2 - g\,(u + v)]\,J_1\,\omega. \qquad (13.4.7b)$$

Now:

$$u = q \cos \theta; \qquad v = q \sin \theta; \qquad D = \frac{\partial(u, v)}{\partial(q, \theta)} = q. \qquad (13.4.8)$$

Then:

$$\varphi,_q = \varphi,_u\,u,_q + \varphi,_v\,v,_q; \qquad \varphi,_\theta = \varphi,_u\,u,_\theta + \varphi,_v\,v,_\theta, \qquad (13.4.9)$$

calculating the derivatives $u,_q,\ u,_\theta,\ v,_\theta$, etc., by use of equations (13.4.8) and inserting
these values into equations (13.4.9); treating the system, so obtained, as an algebraic
system one can calculate the quantities $\varphi,_u$ and $\varphi,_v$. Application of the same procedure
to the function $\psi = \psi(u, v) = \psi(q, \theta)$ furnishes the following results

$$\varphi,_u = \varphi,_q \cos \theta - q^{-1} \varphi,_\theta \sin \theta; \qquad \varphi,_v = \varphi,_q \sin \theta + q^{-1} \varphi,_\theta \cos \theta; \qquad (13.4.10a)$$

$$\psi,_u = \psi,_q \cos \theta - q^{-1} \psi,_\theta \sin \theta; \qquad \psi,_v = \psi,_q \sin \theta + q^{-1} \psi,_\theta \cos \theta. \qquad (13.4.10b)$$

The following discussion will be restricted to an irrotational flow only, i. e., $\omega \equiv g \equiv 0$.

Use of equations (13.4.10a, b) in equations (13.4.7a, b) furnishes a system of equations corresponding to Cauchy-Riemann equations in incompressible flow:

$$\varphi,_q = (\varrho\, q)^{-1} (q^2\, a^{-2} - 1)\, \psi,_\theta + (\gamma - 1)\, q\, a^{-2} J_1 \overline{Q};\qquad (13.4.11a)$$

$$\varphi,_\theta = q\, \varrho^{-1}\, \psi,_q. \qquad (13.4.11b)$$

Identification of f and F with Q in equations (13.4.2) and the use of equations (13.4.6a, b) together with relations similar to those in equations (13.4.10) for $Q,_u$ and $Q,_v$, yield the following equalities for $\overline{Q}$:

$$\overline{Q} = (\varrho\, J_1)^{-1} (Q,_u\, \psi,_v - Q,_v\, \psi,_u);\qquad \overline{Q} = (\varrho\, q\, J_1)^{-1} (\psi,_\theta\, Q,_q - \psi,_q\, Q,_\theta). \quad (13.4.12)$$

Inserting the second equation (13.4.12) into the system (13.4.11a, b) furnishes the system:

$$\varphi,_q = b_1\, \psi,_\theta + c_1\, \psi,_q;\qquad \varphi,_\theta = a_1\, \psi,_q;\qquad (13.4.13a)$$

$$a_1 = \varrho^{-1} q;\qquad b_1 = (\varrho\, q)^{-1} [q^2\, a^{-2} - 1 + (\gamma - 1)\, q\, a^{-2} Q,_q];\qquad (13\ 4.13b)$$

$$c_1 = -(\gamma - 1)\, (\varrho\, a^2)^{-1} Q,_\theta. \qquad (13.4.13c)$$

Differentiation of the first of equations (13.4.13a) with respect to θ, and the second one with respect to q, gives the stream function equation in the hodograph plane:

$$a_1\, \psi,_{qq} - c_1\, \psi,_{q\theta} - b_1\, \psi,_{\theta\theta} + a_{1,q}\, \psi,_q - b_{1,\theta}\, \psi,_\theta - c_{1,\theta}\, \psi,_q = 0. \quad (13.4.14)$$

5. General Considerations

The total differentials given in equation (13.1.3), considered as function of q, θ, can be written as

$$dT = T,_q\, dq + T,_\theta\, d\theta, \ldots, \text{etc.} \qquad (13.5.1a)$$

hence, equation (13.1.3), after collecting terms, gives

$$(c_p\, T,_q - Q,_q - \varrho^{-1} p,_q)\, dq + (c_p\, T,_\theta - Q,_\theta - \varrho^{-1} p,_\theta)\, d\theta = 0. \qquad (13.5.1b)$$

Since q, θ are two independent variables, equation (13.5.1b) gives

$$c_p\, T,_q = Q,_q + \varrho^{-1} p,_q;\qquad c_p\, T,_\theta = Q,_\theta + \varrho^{-1} p,_\theta. \qquad (13.5.1c)$$

Similar reasoning applied to Bernoulli's equation (13.2.10a) gives

$$q + c_p\, T,_q - Q,_q = 0,\qquad c_p\, T,_\theta - Q,_\theta = 0. \qquad (13.5.2)$$

Comparison of equations (13.5.1c) and (13.5.2) yields along a streamline

$$p,_q = dp/dq = -\varrho\, q;\qquad p,_\theta = 0, \qquad (13.5.3)$$

hence, $p = p(q)$. From equation (13.5.3) it is clear that $\varrho = \varrho(q)$, from the equation of state $T = T(q)$, from equation (13.5.2) $Q = Q(q)$ and therefore, it follows from equation (13.1.7) $S = S(q)$. Briefly, all the dependent variables are functions of the velocity only. A more general proof of this result is given below.

From equation (13.5.2) along a streamline

$$dT/dq = c_p^{-1} (dQ/dq - q), \qquad (13.5.4)$$

and from the equation of state with the use of equations (13.5.3) and (13.5.4)

$$d\varrho/dq = -\varrho\, a^{-2} [q + (\gamma - 1)\, dQ/dq]. \qquad (13.5.5)$$

The above results show that, in equations (13.4.13a, b, c) $c_1 \equiv 0$, $a_1 = a_1(q)$ and $b_1 = b_1(q)$. Hence, the stream function equation (13.4.14) reduces to the form

$$a_1(q)\, \psi,_{qq}(q, \theta) - b_1(q)\, \psi,_{\theta\theta}(q, \theta) + (da_1/dq)\, \psi,_q(q, \theta) = 0. \qquad (13.5.6)$$

The results given above are, in fact, a consequence of Kelvin's Circulation Theorem. A proof of this statement is given below.

The circulation Γ is given by

$$\Gamma = \oint u_i \, dx_i \tag{13.5.7}$$

hence,

$$\frac{D\Gamma}{Dt} = \oint (Du_i/Dt) \, dx_i + \oint u_i \, D \, (dx_i)/Dt =$$

$$= - \oint q^{-1} \, (\partial p/\partial x_i) \, dx_i + \oint u_i \, du_i \quad \text{(by using eq. 13.1.1)}$$

$$= - \oint dp/\varrho + {}^1/_2 \oint d \, (q^2). \tag{13.5.8}$$

$\Gamma = 0$ implies that $\omega = 0$ by applying Stokes theorem to equation (13.5.7). Hence, equation (13.5.8) gives, for $\omega = 0$, $\oint dp/\varrho = 0$ since $d(q^2)$ is an exact differential. The vanishing of $\oint dp/\varrho$ implies that dp/ϱ is an exact differential and therefore, $p = p \, (\varrho)$. Hence, equation (13.1.6) gives $S = S \, (\varrho)$, the equation of state gives $T = T \, (\varrho)$, the First Law of Thermodynamics gives $Q = Q \, (\varrho)$ and finally, Bernoulli's equation gives $\varrho = \varrho \, (q)$. This shows that $Q = Q \, (\varrho \, (q)) = F \, (q)$ or $q = q \, (Q)$ along a streamline.

Equations (13.2.16) and (13.2.10a) show that the condition of irrotationality implies that the constant of integration in Bernoulli's equation (eq. 13.2.10a) is independent of the stream function. Since the form of the functional relationship $Q = Q \, (x, y)$ was not used in the proof given above; it follows that any function $Q = Q \, (x, y)$ transforms into $Q = Q \, (q)$ along a streamline in the hodograph plane. Hence $Q \, (x, y)$ is arbitrary. $Q = Q \, (q)$ is valid in rotational flow as well.

$p = p \, (\varrho)$ also follows from Silberstein's theorem which states that "an irrotational flow of an inviscid non-heat conducting fluid subject to conservative forces is either baratropic, isochoric or isobaric". This theorem was given in his paper published in the Bulletin Internationale de l'Academie des Sciences de Cracovie Comptes Rendus in 1896.

The above outlined relation $Q = Q \, (q)$ may be subject to a radical change when the boundary value problem is considered. In this case one has to satisfy the usual geometric boundary conditions and the "a priori" given heat distribution $Q = Q \, (x, y)$, which due to $\omega = 0$ must result in $Q = Q \, (q)$ only. The question whether this can be always satisfied seems to be a difficult one and is beyond the scope of the present work. Additional remarks on that are at the end of the book.

6. Generalized Chaplygin's Equation and Its Transformation

The basic equations are (for $Q = Q \, (q)$ and $\omega = o$):

$$\varphi,_\theta = \varrho^{-1} q \, \psi,_q; \quad \varphi,_q = \varrho^{-1} q \, h^2 \, \psi,_\theta, \tag{13.6.1}$$

where

$$h^2 = q^{-2} \, [a^{-2} \, q^2 - 1 + (\gamma - 1) \, q \, a^{-2} \, (dQ/dq)] = - \bar{h}^2. \tag{13.6.2}$$

Only the elliptic region will be discussed; in this region $h^2 < 0$ and therefore $\bar{h}^2 > 0$.

Eliminating φ from eqs. (13.6.1) by differentiating the first partially with respect to q and the second partially with respect to θ and using eq. (13.6.2), one finds

$$\psi,_{\theta\theta} + \varrho \, q^{-1} \, \bar{h}^2 \, (\varrho^{-1} \, q \, \psi,_q),_q = 0. \tag{13.6.3}$$

This is the generalized Chaplygin's equation.

The dimensionless variable λ is introduced by

$$\lambda = - \int_q^{q_*} \bar{h} \, dq, \quad \text{where} \quad q_* = q \text{ for } h^2 = 0. \tag{13.6.4}$$

Since eq. (13.6.4) defines an indefinite integral

$$(d\lambda/dq) = \bar{h}, \tag{13.6.5}$$

and therefore

$$\varrho^{-1}\, q\, \psi,_q = \varrho^{-1}\, q\, \psi,_\lambda\, d\lambda/dq = \varrho^{-1}\, q\, \bar{h}\, \psi,_\lambda. \tag{13.6.6}$$

Introducing z by

$$z = \varrho^{-1}\, q\, \bar{h}, \tag{13.6.7}$$

then, using eqs. (13.6.5), (13.6.6) and (13.6.7), eq. (13.6.3) reduces to

$$\psi,_{\theta\theta} + z^{-1}\,(z\,\psi,_\lambda),_\lambda = 0. \tag{13.6.8}$$

Letting

$$\psi^* = z^{1/2}\,\psi, \tag{13.6.9}$$

and substituting into eq. (13.6.8), one finds

$$\Delta^2\,\psi^* + f\,(\lambda)\,\psi^* = 0, \tag{13.6.10}$$

where

$$f\,(\lambda) = -\,z^{-1/2}\,(d^2\,(z^{1/2})/d\lambda^2). \tag{13.6.11}$$

Eq. (13.6.10) is formally similar to the corresponding equation in isentropic subsonic flows. Because of this formal relationship the convergence proofs which do not make use of the functional dependence of f on λ remain the same and such proofs will not be included below[*]. In diabatic flow, in contrast to isentropic flow, $f\,(\lambda)$ depends on the heat distribution, therefore, while in isentropic flow $f\,(\lambda)$ can be calculated once and for all, in diabatic flow this is not the case. In the following discussion Q and its derivatives up to the required order are assumed bounded.

7. Computation of f for Diabatic Flow

The pressure-density-entropy relationship is given by eq. (13.1.7), i. e.

$$p = c\,\varrho^\gamma \exp\,(c_v^{-1}\,S), \tag{13.7.1}$$

and the velocity of sound a is defined by eq. (13.1.8) as

$$a^2 = \gamma\,p\,\varrho^{-1} = \gamma\,c\,\varrho^{\gamma-1} \exp\,(c_v^{-1}\,S). \tag{13.7.2}$$

From eqs. (13.1.8) and (13.2.10a), the energy equation can be written in the form

$$\tfrac{1}{2}\,(\gamma - 1)\,q^2 + a^2 - (\gamma - 1)\,Q = \tfrac{1}{2}\,(\gamma - 1)\,q^2_{\text{MAX}} - Q/_{T=0} = a^2_0 = \text{const.} \tag{13.7.3}$$

(See Section 13.14.)

Letting

$$m = a^{-2}\,q^2, \tag{13.7.4}$$

then eq. (13.7.3) reduces to

$$q^2\,[1 + \tfrac{1}{2}\,(\gamma - 1)\,m] = m\,[a_0^2 + (\gamma - 1)\,(Q - Q_0)]; \tag{13.7.5}$$

differentiation of eq. (13.7.5) with respect to m gives

$$d\,(q^2)/dm = m^{-1}\,q^2\,[1 + (\gamma - 1)\,m^2\,q^{-2}\,(dQ/dm)]\,[1 + \tfrac{1}{2}\,(\gamma - 1)\,m]^{-1}. \tag{13.7.6}$$

Similarly, differentiating eq. (13.7.2) after taking logarithms, one finds

$$(\gamma - 1)\,d\,(\log \varrho)/dm + c_v^{-1}\,(dS/dm) = a^{-2}\,d\,(a^2)/dm =$$
$$= (\gamma - 1)\,a^{-2}\,[(dQ/dm) - \tfrac{1}{2}\,(d\,[q^2]/dm)], \tag{13.7.7}$$

but, from the definition of S and eq. (13.1.8),

$$c_v^{-1}\,(dS/dm) = (c_v\,T)^{-1}\,(dQ/dm) = \gamma\,(\gamma - 1)\,a^{-2}\,(dQ/dm), \tag{13.7.8}$$

---

[*] For more details on this part see Ref. R. v. Mises and M. Schiffer.

therefore, substitution of eqs. (13.7.6) and (13.7.8) into eq. (13.7.7) gives

$$d\,(\log \varrho)/dm = -\,{}^1\!/_2\,[1 + (\gamma - 1)\,m^2\,q^{-2}\,(dQ/dm)]\,[1 + {}^1\!/_2\,(\gamma - 1)\,m]^{-1} -$$
$$-\,(\gamma - 1)\,m\,q^{-2}\,(dQ/dm). \tag{13.7.9}$$

From eq. (13.6.11)

$$f = -\,z^{-\,{}^1\!/_2}\,(d^2\,(z^{\,{}^1\!/_2})/d\lambda^2) = -\,({}^1\!/_2)^2\,(d\,(\log z)/d\lambda)^2 - \{d\,[{}^1\!/_2\,(d\,(\log z)/d\lambda)]/d\lambda\} =$$
$$= -\,n^2 - (dn/d\lambda), \tag{13.7.10}$$

where

$$z = \varrho^{-1}\,[1 - m - (\gamma - 1)\,q\,a^{-2}\,(dQ/dq)]^{{}^1\!/_2}, \tag{13.7.11}$$

and

$$dQ/dq = 2\,q\,(dQ/dm)\,(dq^2/dm)^{-1}. \tag{13.7.12}$$

Differentiating eq. (13.7.11) after intriducing eq. (13.7.12) and taking logarithms, one finds

$$d\,(\log z)/dm = -\,d\,(\log \varrho)/dm - {}^1\!/_2\,\{1 - m\,[1 +$$
$$+ 2\,(\gamma - 1)\,(dQ/dm)\,(d\,(q^2)/dm)^{-1}]\}^{-1}\,\{1 + 2\,(\gamma - 1)\,(d\,[m\,(d\,(q^2)/dm)^{-1}\,(dQ/dm)]/dm)\}.$$

$$\tag{13.7.13}$$

From eq. (13.6.6)

$$d\lambda/dq = \bar{h} = q^{-1}\,\{1 - m\,[1 + 2\,(\gamma - 1)\,(d\,(q^2)/dm)^{-1}\,(dQ/dm)]\}^{{}^1\!/_2}, \tag{13.7.14}$$

hence,

$$d\lambda/d\,(q^2) = (d\lambda/dq)\,(d\,(q^2)/dq)^{-1} =$$
$$= (2\,q^2)^{-1}\,\{1 - m\,[1 + 2\,(\gamma - 1)\,(d\,(q^2)/dm)^{-1}\,(dQ/dm)]\}^{{}^1\!/_2}. \tag{13.7.15}$$

Similarly, from eq. (13.7.10)

$$n = {}^1\!/_2\,(d\,(\log z)/d\lambda) = {}^1\!/_2\,[d\,(\log z)/dm]\,(d\,(q^2)/dm)^{-1}\,(d\lambda/d\,(q^2))^{-1}. \tag{13.7.16}$$

Substitution of eqs. (13.7.6), (13.7.13) and (13.7.15) into eq. (13.7.16), and arranging terms, gives

$$n = -\,{}^1\!/_4\,(\gamma + 1)\,m^2\,(1 - m)^{-3/2}\,\bar{n}, \tag{13.7.17}$$

where n is given by

$$\bar{n} = \{1 + (\gamma - 1)\,m^2\,q^{-2}\,(dQ/dm)\}^{-1}\,\{1 - 2\,(\gamma - 1)\,m\,(1 - m)^{-1}\,(dQ/dm)\,(d\,(q^2)/dm)^{-1}\}^{3/2}$$
$$\{1 - 2\,m\,(1 - m)\,q^{-2}\,h_1{}^2\,(dQ/dm) - 8\,h_1{}^2\,q^{-2}\,(dQ/dm)\,[1 + {}^1\!/_2\,(\gamma - 1)\,m]\,[1 - m\,\{1 +$$
$$+ (\gamma - 1)\,(dQ/dm)\,(d\,(q^2)/dm)^{-1}\}] + 4\,h_1{}^2\,m^{-1}\,[d\,[m\,(d\,(q^2)/dm)^{-1}\,(dQ/dm)]/dm]\},$$

$$\tag{13.7.18}$$

and

$$h_1{}^2 = (\gamma - 1)\,(\gamma + 1)^{-1}. \tag{13.7.19}$$

It can be seen that for $Q = 0$, $\bar{n}$ reduces to 1.

Representation of n as given by eq. (13.7.17) enables the reader to see how the term is modified in case of diabatic flow. As $m \to 1$ the term

$$\{1 - 2\,(\gamma - 1)\,m\,(1 - m)^{-1}\,(dQ/dm)\,(d\,(q^2)/dm)^{-1}\}^{-3/2}$$

approaches zero in such a manner that $\lim\limits_{m \to 1} n$ exists. This statement can be proved by considering

$$\lim_{m \to 1}\,(1 - m)^{-3/2}\,\{1 - 2\,(\gamma - 1)\,m\,(1 - m)^{-1}\,(dQ/dm)\,(d\,(q^2)/dm)^{-1}\}^{-3/2} =$$
$$= \lim_{m \to 1}\,\{1 - m - 2\,(\gamma - 1)\,m\,(dQ/dm)\,(d\,(q^2)/dm)^{-1}\}^{-3/2} =$$
$$= \{-\,2\,(\gamma - 1)\,(dQ/dm)\,(d\,(q^2)/dm)^{-1}\}^{-3/2}_{m=1}, \tag{13.7.20}$$

and this is finite since in diabatic flow the sonic line is different from the parabolic line.

Computation of $(dn/d\lambda)$ by differentiating eq. (13.7.17) and using

$$(dn/d\lambda) = (dn/dm)\,(d\,(q^2)/dm)^{-1}\,(d\lambda/d\,(q^2))^{-1}, \tag{13.7.21}$$

gives

$$(dn/d\lambda) = -\,^1/_4\,(\gamma+1)\,(1-m)^{-3}\,m^2\,\{1-2\,(\gamma-1)\,m\,(1-m)^{-1}\,(d\,(q^2)/dm)^{-1}\,(dQ/dm)\}^{-1/_2}.$$
$$.\{1+(\gamma-1)\,m^2\,q^{-2}\,(dQ/dm)\}^{-1}\,\{\overline{n}\,[4-(3-2\,\gamma)\,m-\,^1/_2\,(\gamma-1)\,m^2]+$$
$$+\,2\,m\,(1-m)\,[1+\,^1/_2\,(\gamma-1)\,m]\,(d\overline{n}/dm)\}. \tag{13.7.22}$$

f can be calculated by substituting eqs. (13.7.17), (13.7.18) and (13.7.22) into eq. (13.7.10).

The expression for λ is obtained as follows:

$$(d\lambda/dm) = (d\lambda/d\,(q^2))\,(d\,(q^2)/dm), \tag{13.7.23}$$

or by using eqs. (13.7.6) and (13.7.15)

$$(d\lambda/dm) = \,^1/_2\,m^{-1}\,\{1+\,^1/_2\,(\gamma-1)\,m\}^{-1}\,\{1-m-2\,(\gamma-1)\,m\,(d\,(q^2)/dm)^{-1}\,(dQ/dm)\}^{1/_2}.$$
$$.\{1+(\gamma-1)\,m^2\,q^{-2}\,(dQ/dm)\}, \tag{13.7.24}$$

hence

$$\lambda = \,^1/_2\!\int_{m^*}^{m} m^{-1}\,\{1+\,^1/_2\,(\gamma-1)\,m\}^{-1}\,\{1-m-2\,(\gamma-1)\,m\,(d\,(q^2)/dm)^{-1}\,(dQ/dm)\}.$$
$$.\{1+(\gamma-1)\,m^2\,q^{-2}\,(dQ/dm)\}\,dm, \tag{13.7.25}$$

where m^* corresponds to q_*, i. e., m^* satisfies the equation

$$1-m-2\,(\gamma-1)\,m\,(d\,(q^2)/dm)^{-1}\,(dQ/dm) = 0. \tag{13.7.26}$$

To integrate eq. (13.7.25), one may put

$$\{1-m\,[1+2\,(\gamma-1)\,(d\,(q^2)/dm)^{-1}\,(dQ/dm)]\}^{1/_2} = \{1-z\}^{1/_2}, \tag{13.7.27}$$

in the elliptic region $0 \leqslant z < 1$, and by expanding $\{1-z\}^{1/_2}$ by the binomial theorem, the integration may be carried out term by term. In the hyperbolic region $z > 1$, and the term $\{1-z\}^{1/_2}$ has to be written as $i\,z^{1/_2}\,(1-z^{-1})^{1/_2}$ before applying the binomial theorem; this shows that λ is imaginary in the hyperbolic region.

On integrating the expression for λ the following cases may arise:

I. $Q\,(m)$ is given in a simple form, the integration may be carried out in a closed form or it may reduce to an elliptic or a hyperelliptic integral.

II. $Q\,(m)$ may be a complicated function of m, it may be convenient to express the function in a Fourier Series in terms of (m/m^*) first and then carry out the integration; in case this cannot be done, expansion of $\sin\,(m/m^*)$ and $\cos\,(m/m^*)$ may be used.

8. Method of Generating Stream Functions

The method of generating stream functions employed here is exactly the same as the method used in isentropic flows.

Concequently, a few results will be cited here without proof.

Following Bergman, ψ^* is assumed to be represented as

$$\psi^* = \sum_{n=0}^{\infty} g_n\,G_n, \tag{13.8.1}$$

where the G_n's are functions of λ and satisfy the following recurrence formula

$$G_0 = 1; \quad (dG_{n+1}/d\lambda) = (d^2G_n/d\lambda^2) + f\,G_n, \tag{13.8.2}$$

and the g_n's are harmonic functions of λ and θ and satisfy $2\,g_{n,\lambda} = -\,g_{n-1}$, ($g_0$ is an arbitrary harmonic function).

$$\tag{13.8.3}$$

The solution of eq. (13.8.3) is given by

$$g_n(\lambda, \theta) = (-1)^n [2^n (n-1)!]^{-1} \, Re \left\{ \int_0^\xi \varphi_0(t) (\xi - t)^{n-1} \, dt \right\}, \quad n > 0, \qquad (13.8.4)$$

where

$$g_0(\lambda, \theta) = Re \{\varphi_0(\xi)\}; \quad \xi = \lambda + i\,\theta. \qquad (13.8.5)$$

The series defined in eq. (13.8.1), with the conditions given by eqs. (13.8.2) and (13.8.3), satisfies eq. (13.6.10) in its domain of convergence.

9. The Domain of Convergence of the Development for ψ^*

Introducing eq. (13.8.4) into eq. (13.8.1), one finds

$$\psi^*(\lambda, \theta) = g_0(\lambda, \theta) + \sum_{n=1}^\infty (-1)^n [2^n (n-1)!]^{-1} G_n \, Re \left\{ \int_0^\xi \varphi_0(t) (\xi - t)^{n-1} \, dt \right\}. \qquad (13.9.1)$$

Assuming uniform convergence for a certain domain of t, which will be determined later, summation and integration can be interchanged and eq. (13.9.1) can be written in the form

$$\psi^*(\lambda, \theta) = g_0(\lambda, \theta) + Re \left\{ \int_0^\xi \varphi_0(t) \, U(t, \lambda, \theta) \, dt \right\}, \qquad (13.9.2)$$

where

$$U(t, \lambda, \theta) = \sum_{n=1}^\infty (-1)^n [2^n (n-1)!]^{-1} G_n (\xi - t)^{n-1}. \qquad (13.9.3)$$

This transformation of the series for $\psi^*(\lambda, \theta)$ is valid only where eq. (13.9.3) converges uniformly.

The domain of uniform convergence of $U(t, \lambda, \theta)$ will be studied by considering the uniform convergence of a series that dominates $U(t, \lambda, \theta)$. A first step in this direction will be to find functions $Q^{(n)}(\lambda)$ such that each $Q^{(n)}(\lambda)$ dominates $G_n(\lambda)$. This means that for $-\infty < \lambda < 0$.

$$|G_n| \leqslant Q^{(n)}, \text{ and } |d^k G_n/d\lambda^k| \leqslant d^k Q^n/d\lambda^k \text{ for all } k, \qquad (13.9.4)$$

or, in short

$$G_n << Q^{(n)}. \qquad (13.9.5)$$

These functions will be picked similar to those used in isentropic flow, i. e., they satisfy the following recurrence formula

$$(dQ^{(n+1)}/d\lambda) = (d^2 Q^{(n)}/d\lambda^2) + c\,(\varepsilon - \lambda)^{-2} Q^{(n)}; \quad Q^{(0)} = 1, \quad Q^{(n)}(-\infty) = 0. \qquad (13.9.6)$$

A theorem similar to the one proved for isentropic flows will be proved here, namely: For each differential equation of the type of eq. (13.6.10) where $f(\lambda)$ can be dominated by an expression of the form $c\,(\varepsilon - \lambda)^{-2}$, i. e.

$$f \ll F = c\,(\varepsilon - \lambda)^{-2}; \quad c > 0; \quad \varepsilon < 0; \qquad (13.9.7)$$

the corresponding series $U(t, \lambda, \theta)$ is uniformly convergent in the domain

$$|(\xi - t)\,[2\,(\varepsilon - \lambda)]^{-1}| \leqq a < 1, \qquad (13.9.8)$$

and a suitable constant c can be determined for diabatic flows.

The proof of the first part, i. e. the condition given by eq. (13.9.8) is exactly the same as the one given for isentropic flows and therefore, it will not be reproduced. In order to prove that a suitable constant c can be found, Q and its derivatives up to the required order are assumed to be bounded and the following dependent variable will be introduced by

$$T_1^2 = m^* - m. \qquad (13.9.9)$$

The parabolic line is defined by $m = m^*$ or, from eq. (13.9.9), by $T_1 = 0$. Since m^* satisfies eq. (13.7.26), it can be seen by using eq. (13.9.9) that $T_1 = 0$ satisfies the equation

$$1 - m^* + T_1{}^2 - 2 (\gamma - 1) (m^* - T_1{}^2) (dQ/dT_1) (d (q^2)/dT_1)^{-1} = 0. \quad (13.9.10)$$

From eq. (13.7.10)

$$f = - n^2 - (dn/d\lambda), \quad (13.9.11)$$

where n and $(dn/d\lambda)$ are defined by eqs. (13.7.17), (13.7.18) and (13.7.22). It is noticed from eqs. (13.7.17) and (13.7.18) that n has the factor

$$(1 - m)^{-3/2} \{1 - 2 (\gamma - 1) m (1 - m)^{-1} (dQ/dm) (d (q^2)/dm)^{-1}\}^{-3/2} =$$
$$= \{1 - m - 2 (\gamma - 1) m (dQ/dm) (d (q^2)/dm)^{-1}\}^{-3/2} =$$
$$= \{1 - m^* + T_1{}^2 - 2 (\gamma - 1) (m^* - T_1{}^2) (dQ/dT_1) (d (q^2)/dT_1)^{-1}\}^{-3/2}. \quad (13.9.12)$$

By considering eq. (13.9.10) one concludes that n is infinite at $T_1 = 0$. Exactly similar reasoning of eqs. (13.7.18) and (13.7.22) leads to the conclusion that $(dn/d\lambda)$ is infinite at $T_1 = 0$.

The factor $\{1 + (\gamma - 1) m^2 q^{-2} (dQ/dm)\}$ in eq. (13.7.19) which is proportional to $d (q^2)/dm$ [see eq. (13.7.6)] cannot vanish. Vanishing of $d (q^2)/dm$ for value of m and q which are finite and different from zero gives finite values for dQ/dm. Therefore, the term z defined by eq. (13.7.27) approaches infinity. This is a contradiction. This together with the assumptions on Q shows that f is regular everywhere except at

$$T_1 = 0 \quad (m = m^* \quad \text{or} \quad \lambda = 0).$$

The expression for λ was given by eq. (13.7.25). It will be proved below that λ is infinite at $q = 0$ ($m = 0$ or $T = \pm (m^*)^{1/2}$). By using eq. (13.7.27), λ can be expressed as

$$\lambda (m) = {}^1\!/_2 \int_{m^*}^{m} m^{-1} \{1 + {}^1\!/_2 (\gamma - 1) m\}^{-1} \{1 - z\}^{1/2} \{1 + (\gamma - 1) m^2 q^{-2} (dQ/dm)\} dm. \quad (13.9.13)$$

In the elliptic region both z and $[1 + (\gamma - 1) m^2 q^{-2} (dQ/dm)$ are bounded since (dQ/dm) is assumed bounded. However m^{-1} is infinite at $m = 0$. The value of λ at $m = 0$ can be evaluated as follows:

$$\lambda (0) = {}^1\!/_2 \int_{m^*}^{0} (\quad) dm = {}^1\!/_2 \left\{ \int_{m^*}^{\delta} (\quad) dm + \int_{\delta}^{0} (\quad) dm \right\}, \quad (13.9.14)$$

where

$$0 < \delta < m^*. \quad (13.9.15)$$

The integrand in the first integral is bounded and since the region of integration is finite the contribution of the first integral is a finite number. On the other hand, in the second integral $0 \leqslant z \leqslant b < 1$ on $[0, \delta]$, hence on expanding $\{1 - z\}^{1/2}$ by the binomial theorem the integrand can be expressed as $(1/m)$ + terms which are bounded on $[0, \delta]$; on carrying out the integration one finds that the contribution of the second integral is $- \infty$.

$\lambda (- 2/(\gamma - 1))$ can be evaluated by re-writing eq. (13.9.13) as

$$\lambda (- 2/(\gamma - 1)) = {}^1\!/_2 \left\{ \int_{m^*}^{\delta} (\quad) dm + \int_{\delta}^{-2/(\gamma-1)} (\quad) dm \right\}. \quad (13.9.16)$$

Proceeding as above, the second integrand can be written as $1/(1 + [\gamma - 1/2] m)$ + + terms in m which are bounded on $[- 2/(\gamma - 1), \delta]$. Therefore, $\lambda (- 2/(\gamma - 1)) = = - \infty$. However, since $m < 0$, M is imaginary.

From eqs. (13.7.24) and (13.9.9)

$$(d\lambda/dT_1) = -\,2\,(\gamma + 1)^{-1}\,T_1\,(m^* - T_1{}^2)^{-1}\,[1 - h_1{}^2\,(1 - m^*) - h_1{}^2\,T_1{}^2]^{-1}\,.$$

$$.\{1 - {}^1\!/_2\,(\gamma - 1)\,T_1{}^{-1}\,(m^* - T_1{}^2)^2\,q^{-2}\,(dQ/dT_1)\}\,.$$

$$.\{1 - m^* + T_1{}^2 - 2\,(\gamma - 1)\,(m^* - T_1{}^2)\,(dQ/dT_1)\,(d\,(q^2)/dT_1)^{-1}\}^{1/2}.\qquad (13.9.17)$$

Eq. (13.9.17) and the above remarks about λ show that every branch of the function $T_1\,(\lambda)$ is regular at each point λ in the half plane $Re\,\{\lambda - \varepsilon\} < 0$, $\varepsilon < 0$. Only real values of $T_1\,(\lambda)$ for real λ are chosen.

Applying the monodromy theorem one concludes that $T_1\,(\lambda)$ is regular and uniform in the entire half plane $Re\,\{\lambda - \varepsilon\} < 0$. Therefore for a fixed negative λ, T_1 is bounded and different from zero on a circle $|\tau - \lambda| = \varepsilon - \lambda$ and $f\,(T_1)$ is also bounded

$$f\,(T_1) < K. \qquad (13.9.18)$$

Letting $f\,(T_1) = \varphi\,(\lambda)$ and applying Cauchy's formula for the n'th derivative of φ, one finds

$$|(d^n\,f/d\lambda^n| \leqslant (d^n\,F/d\lambda^n), \qquad (13.9.19)$$

if

$$c = K\,A^2\,(n + 1)^{-1} \quad \text{and} \quad \varepsilon > \lambda \geqslant -\,A. \qquad (13.9.20)$$

This proves the theorem.

Since $U\,(t,\,\lambda,\,\theta)$ has to be integrated from zero to ξ, it is required at least that

$$|\xi| < 2\,(\varepsilon - \lambda) \quad \text{or} \quad \lambda^2 + \theta^2 < 4\,(\varepsilon - \lambda)^2; \qquad (13.9.21)$$

this determines the interior of a hyperbola in the ξ plane, and, for $\varepsilon \to 0$ the hyperbola approaches two straight lines $\theta = \pm\,3^{1/2}\,|\lambda|$. Hence the series defined by eq. (13.8.1) converges uniformly and absolutely in every closed subdomain of this angular domain as long as φ_0 is bounded in the subdomain. This domain of uniform convergence is formally the same as the one obtained for isentropic flows.

10. Discussion of the Domain of Convergence of the Series Representing ψ^*

A discussion similar to the one given by R. v. Mises and M. Schiffer shows that there is a little chance of improving the conditions of convergence outside the angular domain $|\theta| = 3^{1/2}\,|\lambda|$. However one may improve the convergence either by approximating the function $f\,(\lambda)$ by a polynomial in e^λ or by using Borel's exponential or integral methods as suggested by R. v. Mises and M. Schiffer.

11. Behavior of G_n Near $m = m^*$

The recurrence formula for the G_n's (eq. 13.8.2) can be written in the form

$$G_{n+1} = dG_n/d\lambda + \int^{\lambda} f\,G_n\,d\lambda, \qquad G_0 = 1. \qquad (13.11.1)$$

This equation shows that G_n depends on an arbitrary constant of integration. This constant will be fixed by assuming that the G_n's vanish at $\lambda = -\,\infty$ $(m = 0)$. (This is possible because $f = 0$ for $m = 0$). However, at $\lambda = 0$ $(m = m^*)$ the G_n's are infinite. In this section a possible representation of the G_n's near $m = m^*$, in terms of T_1, will be given.

It can be seen from eq. (13.7.26) and the subsequent remarks about z that

$$1 - m - 2\,(\gamma - 1)\,m\,(dQ/dm)\,(d\,(q^2)/dm)^{-1} =$$

$$= 1 - m^* + T_1{}^2 - 2\,(\gamma - 1)\,(m^* - T_1{}^2)\,(dQ/dT_1)\,(d(q^2)/dT_1)^{-1} = T_1{}^2\,\Phi\,(T_1), \qquad (13.11.2)$$

where $\Phi(0) \neq 0$. For simplicity, the term $q^{-2}\,(dQ/dm)$ will be denoted by

$$q^{-2}\,(dQ/dm) = -\,(2\,T_1\,q^2)^{-1}\,(dQ/dT_1) = \chi\,(T_1). \tag{13.11.3}$$

From eqs. (13.7.17), (13.7.18), (13.11.2) and (13.11.3)

$$n = -\,{}^1\!/_4\,(\gamma+1)\,(m^* - T_1{}^2)^2\,T_1{}^{-3}\,\Phi^{-3/2}\{1 + (\gamma-1)\,(m^* - T_1{}^2)^2\,\chi\}^{-1}.$$
$$.\{1 - H\,(T_1)\}, \tag{13.11.4}$$

where,

$$H\,(T_1) = -\,\{2\,h_1{}^2\,(1 - m^* + T_1{}^2)\,(m^* - T_1{}^2)\,\chi + [(\gamma+1)\,T_1\,(m^* -$$
$$-\,T_1{}^2)]^{-1}\,d\,(T_1{}^2 - T_1{}^2\,\Phi)/dT_1 + 4\,(\gamma-1)\,\chi\,[1 - h_1{}^2\,(1 - m^* +$$
$$+\,T_1{}^2)]\,[1 - m^* + T_1{}^2 + T_1{}^2\,\Phi]\}. \tag{13.11.5}$$

Therefore, eqs. (13.9.17), (13.11.2), (13.11.3) and (13.11.4) give

$$(dn/d\lambda) = (dn/dT_1)\,(d\lambda/dT_1)^{-1} = 2^{-3}\,(\gamma+1)^2\,T_1{}^{-6}\,(m^* - T_1{}^2)\,\Phi^{-3}\,R\,(T_1)\,[1 -$$
$$-\,h_1{}^2\,(1 - m^* + T_1{}^2)]\,[1 + (\gamma-1)\,(m^* - T_1{}^2)\,\chi]^{-3}, \tag{13.11.6}$$

where

$$R\,(T_1) = -\,\{4\,T_1{}^2\,\Phi + 3\,(m^* - T_1{}^2)\,\Phi + {}^3\!/_2\,T_1\,(m^* - T_1{}^2)\,(d\Phi/dT_1)\}.$$
$$.\{1 + (\gamma-1)\,(m^* - T_1{}^2)^2\,\chi\}\,\{1 + H\} + (\gamma-1)\,(m^* - T_1{}^2)\,T_1\,\Phi\,[-\,4\,T_1\,\chi +$$
$$+\,(m^* - T_1{}^2)\,(d\chi/dT_1)]\,[1 + H] +$$
$$+\,T_1\,(m^* - T_1{}^2)\,\Phi\,\{1 + (\gamma-1)\,(m^* - T_1{}^2)^2\,\chi\}\,(dH/dT_1). \tag{13.11.7}$$

Hence, eqs. (13.7.10), (13.11.4) and (13.11.6) give

$$f = -\,n^2 - (dn/d\lambda) = -\,4^{-2}\,(\gamma+1)^2\,(m^* - T_1{}^2)^2\,T_1{}^{-6}\,\Phi^{-3}\,\{1 + (\gamma-1)\,(m^* -$$
$$-\,T_1{}^2)^2\,\chi\}^{-3}\,\{(m^* - T_1{}^2)\,(1 + H)^2\,[1 + (\gamma-1)\,(m^* - T_1{}^2)^2\,\chi] + 2\,[1 - h^2{}_1\,(1 - m^* +$$
$$+\,T_1{}^2)\,R\,(T_1)]\}. \tag{13.11.8}$$

The following representations will be assumed

$$G_n = c_n\,T_1{}^{-3n}\,[1 + b_n\,T_1{}^2 + \ldots] = c_n\,T^{-3n}\,r_n\,(T_1), \quad G_0 = 1, \tag{13.11.9}$$

where

$$r_n\,(T_1) = 1 + b_n\,T_1{}^2 + \ldots, \quad c_0 = 1, \quad r_0 = 1, \quad b_0 = 0; \tag{13.11.10}$$

$$B = f\,(d\lambda/dT_1) = -\,2^{-3}\,(\gamma+1)\,T_1{}^{-4}\sum_{\nu=0}^{\infty}\beta_\nu\,T_1{}^{2\nu}, \tag{13.11.11}$$

and

$$A = (d\lambda/dT_1) = -\,2^{-1}\,(\gamma-1)\,T_1{}^{-2}\sum_{\nu=0}^{\infty}\alpha_\nu\,T_1{}^{2\nu}. \tag{13.11.12}$$

Eq. (13.9.17) gives $\lambda = \lambda\,(T_1)$ and since $f\,(\lambda)$ is a known function of λ, the coefficients β_ν, α_ν are determined uniquely by carrying out the expansions denoted in eqs. (13.11.11) and (13.11.12).

Eqs. (13.11.1), (13.11.11) and (13.11.12) give

$$G_{n+1} = A\,(dG_n/dT_1) + \int_{m^{*1/2}}^{T_1} B\,G_n\,dT_1. \tag{13.11.13}$$

Substitution of eqs. (13.11.9), (13.11.11), (13.11.12) into eq. (13.11.13) gives

$$c_n\,T_1{}^{-3(n+1)}\,[1 + b_{n+1}\,T_1{}^2 + \ldots] =$$
$$= -\,(c_n/2)\,(\gamma+1)\,T_1{}^{-3(n+1)}\,\{-\,3\,n\,[1 + b_n\,T_1{}^2 + \ldots] + [2\,b_n\,T_1{}^2 + \ldots]\}.$$

$$.\left\{\sum_{\nu=0}^{\infty}\alpha_\nu\,T_1{}^{2\nu}\right\} - 2^{-3}\,(\gamma+1)\,c_n\int_{m^{*1/2}}^{T_1} x^{-(3n+4)}\left[\sum_{\nu=0}^{\infty}\beta_\nu\,x^{2\nu}\right][1 + b_n\,x^2 + \ldots]\,dx. \tag{13.11.14}$$

Carrying out the integration and comparing factors of equal powers of T_1, one finds

$$(c_{n+1}/c_n) = {}^3/_2\, n\, (\gamma + 1)\, \{\alpha_0 + \beta_0\, [36\, n\, (n + 1)]^{-1}\}. \qquad (13.11.15)$$

Eq. (13.11.15) gives the following recurrence formula for the c_n's

$$c_{n+1} = [3\, (\gamma + 1)/2]^n\, n!\, \prod_{\nu=1}^{n} \{\alpha_0 + \beta_0\, [36\, \nu\, (\nu + 1)]^{-1}\}. \qquad (13.11.16)$$

Since $c_0 = 1$, c_1 can be determined from eq. (13.11.15) as

$$c_1 = \beta_0\, (\gamma + 1)/24. \qquad (13.11.17)$$

Similarly

$$(c_{n+1}/c_n)\, b_{n+1} = 2^{-3}\, (\gamma + 1)\, \{4\, \alpha_0\, (3\, n - 2) + \beta_0\, (3\, n + 1)^{-1}\}\, b_n +$$
$$+ 2^{-3}\, (\gamma - 1)\, \{12\, n\, \alpha_1 + \beta_1\, (2\, n + 1)^{-1}\}. \qquad (13.11.18)$$

(c_{n+1}/c_n) is given by eq. (13.11.15) and, since $b_0 = 0$, the above equation gives the recurrence relation for the b_n's.

12. Recurrence Formula for the Functions $r_n\,(T_1)$

Substitution of eq. (13.11.9) into eq. (13.11.13) gives the following recurrence relation for the r_n's

$$c_{n+1}\, r_{n+1}\, (T_1) = A\, c_n\, \{T_1{}^3\, (dr_n/dT_1) - 3\, n\, T_1{}^2\, r_n\} +$$
$$+ c_n\, T_1{}^{3n+3} \int_{m^{*1/2}}^{T_1} B\, x^{-3n}\, r_n\, (x)\, dx, \qquad (13.12.1)$$

or, by using eqs. (13.11.11) and (13.11.12)

$$(c_{n+1}/c_n)\, r_{n+1}\, (T_1) = - {}^1/_2\, (\gamma + 1) \sum_{\nu=0}^{\infty} \alpha_\nu\, T_1{}^{2\nu}\, (T_1\, (dr_n/dT_1) - 3\, n\, r_n) -$$
$$- 2^{-3/2}\, (\gamma + 1)\, T_1{}^{3n+3} \int_{m^{*1/2}}^{T_1} x^{-(3n+4)} \left[\sum_{\nu=0}^{\infty} \beta_\nu\, x^{2\nu} \right] r_n\, (x)\, dx. \qquad (13.12.2)$$

Since $c_0 = 1$ and $r_0 = 1$, it can be seen from eqs. (13.12.1) and (13.12.2) that

$$c_1\, r_1\, (T_1) = T_1{}^3 \int_{m^{*1/2}}^{T_1} B\, (x)\, dx = - 2^{-3}\, (\gamma + 1) \sum_{\nu=0}^{\infty} (\beta_\nu/2\, \nu - 3)\, T_1{}^{2\nu} + \text{const.}\, T_1{}^3. \qquad (13.12.3)$$

Similarly, from eqs. (13.12.1) and (13.12.3)

$$(c_2/c_1)\, r_2\, (T_1) = A\, \{T_1{}^3\, (dr_1/dT_1) - 3\, T_1{}^2\, r_1\} + T_1{}^6 \int_{m^{*1/2}}^{T_1} x^{-3}\, B\, r_1\, (x)\, dx =$$
$$= A\, \{T_1{}^3\, (dr_1/dT_1) - 3\, T_1{}^2\, r_1\} + (T_1{}^6/2\, c_1) \left\{ \int_{m^{*1/2}}^{T_1} B\, (x)\, dx \right\}^2. \qquad (13.12.4)$$

From eqs. (13.12.2), (13.12.3) and (13.12.4), it can be seen that the expression for $r_3\,(T_1)$ has a term of the form

$$T_1{}^9 \int_{m^{*1/2}}^{T_1} x^{-10} \left[\sum_{\nu=0}^{\infty} d_\nu\, x^\nu \right] dx = T_1{}^9 .$$

$$\cdot \left\{ d_9 \log T_1 + \sum_{\nu=0}^{8} (d_\nu/\nu - 9) T_1{}^2 + \sum_{\nu=10}^{\infty} (d_\nu/\nu - 9)\, T_1{}^2 + \text{const.} \right\}. \qquad (13.12.5)$$

Hence, it is clear from eqs. (13.12.3), (13.12.4) and (13.12.5) that, while $r_1\,(T_1)$, $r_2\,(T_1)$ are free of logarithmic terms, $r_3\,(T_1)$ contains a logarithmic term. However, it can be seen from eq. (13.12.5) that $r_3 (T_1)$ has a regular development at the point $T_1 = 0$ up to the ninth order. Further application of eq. (13.12.3) will preserve this term and produces other terms of the form $T^\nu (\log T)^\mu$. In general, $r_n (T_1)$ can be represented as

$$r_n (T_1) = P_0^{(n)} (T_1) + P_1^{(n)} (T_1) \log T_1 + \ldots + P_{n-2}^{(n)} (T_1) (\log T_1)^{n-2}, \quad n > 2, \quad (13.12.6)$$

where the $P^{(n)}$'s are regular powers of T_1.

13. Discussion and Conclusions

As was pointed out in the introduction, the main reason for assuming Q, and not $\overline{Q}$, given is that this renders the governing equations in the hodograph plane linear. This is apparent on considering the two equations (13.4.11a) and (13.4.13a): equation (13.4.11a) is non-linear due to the presence of the Jacobian of the transformation, whereas equation (13.4.13a) is linear. In addition to the fact that it is more convenient to deal with linear equations, certain techniques used in the theory of isentropic flow can be applied in the theory of diabatic flow with great success. This is to be expected because in both cases one deals with linear equations of the same character. However, the practical side of the problem is more involved. This will be apparent on discussing Bergman's linear operator method.

Bergman's linear operator method is extended here to diabatic flows in the elliptic domain. It is clear from equation (13.6.3) that if $\bar{h}^2 > 0$, where

$$\bar{h}^2 = q^{-2} \{1 - a^{-2} [q^2 + (\gamma - 1) q (dQ/dq)]\}, \qquad (13.13.1)$$

the equation is elliptic. The quantity $a^{-2} [q^2 + (\gamma - 1) q (dQ/dq)]$ varies from 0 to 1 and, the algebraic equation

$$a^{-2} [q^2 + (\gamma - 1) q (dQ/dq)] = 1$$

or, by using equation (13.7.3)

$$a_0^2 + (\gamma - 1) Q - (\gamma - 1) q^2/2 = q^2 + (\gamma - 1) q (dQ/dq) \qquad (13.13.2)$$

determines the value of q_*. Due to the restriction imposed on the given function $Q (q)$, i. e. Q and its derivatives up to the required order are assumed bounded, it follows from Weierstrass' approximation theorem that $Q (q)$ can be approximated uniformly by a polynomial of order n, say. Hence, in this case, equation (13.13.2) represents an algebraic equation of order n in q and, q_* is the smallest positive root since values of $q > q^*$ are in the hyperbolic domain.

The region of uniform convergence of the series development for ψ^* [eq. (13.8.1)] is formally the same as the one for isentropic flows in the $\{\lambda, \theta\}$ plane. This is because the two equations are of the same type and the dominating series in both cases are identical in structure.

In practical problems the boundary conditions are given in the physical plane. This leads to an extremely complicated problem in the hodograph plane since the boundary curve depends on the solution itself. The method of determining the solution is that of successive approximations with the first approximation being the known solution of the corresponding problem in incompressible flow.

In diabatic flows two distinct cases may arise.

1. Q is given as $Q (q)$: In this case λ and $f (\lambda)$ can be calculated as outlined in Section (XIII.7). The functions G_n can be calculated from equation (13.8.2) and the solution of the problem for an incompressible fluid furnishes $\varphi_0 (z)$. When the expression for ψ is known one can transform it back to the physical plane by a method which will be explained below. This gives a profile which does not coincide with

the original profile in the physical plane and consequently, $\varphi_0\,(z)$ has to be modified and the process repeated.

If it is assumed that $\psi\,(q, \theta)$ is known in the hodograph plane, the method employed in passing to the physical plane is as follows:

Solving for dx, dy in terms of $d\psi, d\varphi$ from equations (13.4.5a) and (13.4.5b) and using equation (13.4.8), one finds

$$dx = q^{-1} \cos \theta\, d\varphi - (\varrho\, q)^{-1} \sin \theta\, d\psi; \qquad (13.13.3a)$$

$$dy = q^{-1} \sin \theta\, d\varphi - (q\, \varrho)^{-1} \cos \theta\, d\psi, \qquad (13.13.3b)$$

where, [using equations (13.1.1) and (13.1.2)]

$$d\varphi = \varphi,_q\, dq + \varphi,_\theta\, d\theta = \varrho^{-1} q\, (- \bar{h}^2\, \psi,_\theta\, dq + \psi,_q\, d\theta), \qquad (13.13.4a)$$

and,

$$d\psi = \psi,_q\, dq + \psi,_\theta\, d\theta. \qquad (13.13.4b)$$

Inserting equations (13.13.4a) and (13.13.4b) into equations (13.13.3a) and (13.13.3b) and integrating, one finds

$$x = \int \varrho^{-1} \{- [\cos \theta\, \bar{h}^2\, \psi,_\theta + q^{-1} \sin \theta\, \psi,_q]\, dq + [\cos \theta\, \psi,_q - q^{-1} \sin \theta\, \psi,_\theta]\, d\theta\};$$
$$(13.13.5a)$$

$$y = \int \varrho^{-1} \{- [\sin \theta\, \bar{h}^2\, \psi,_\theta - q^{-1} \cos \theta\, \psi,_q]\, dq + [\sin \theta\, \psi,_q + q^{-1} \cos \theta\, \psi,_\theta]\, d\theta\}.$$
$$(13.13.5b)$$

Along a streamline ψ is constant, hence equation (13.13.4b) gives

$$d\theta = - \psi,_\theta^{-1}\, \psi,_q\, dq. \qquad (13.13.6)$$

Substituting equation (13.13.6) into equations (13.13.5a) and (13.13.5b), one obtains the following parametric representation for the streamlines in the physical plane,

$$x = - \int \varrho^{-1} \cos \theta\, [\bar{h}^2\, \psi,_\theta + \psi^2,_q]\, \psi,_\theta^{-1}\, dq, \qquad (13.13.7a)$$

$$y = - \int \varrho^{-1} \sin \theta\, [\bar{h}^2\, \psi,_\theta + \psi^2,_q]\, \psi,_\theta^{-1}\, dq. \qquad (13.13.7b)$$

It should be noted that the path of integration in equations (13.13.5a) and (13.13.5b) is not essential since the generalized Cauchy-Riemann equations are the integrability conditions for these two integrals.

2. Q is given as $Q\,(x, y)$: This is the situation which is likely to be encountered in practical applications. In this case, x, y are calculated from equations (13.13.5a) and (13.13.5b) using the incompressible flow solution ($\bar{h}^2 = q^{-2}$ since $a^2 \to \infty$). This gives the expression for $Q\,(q)$ and one may proceed as in 1. The new expression for $\psi\,(q, \theta)$ is used in calculating $Q\,(q)$ from equations (13.13.5a) and (13.13.5b). This procedure is repeated until $Q\,(q)$ shows reasonable behavior. It is inseparably connected with the entire limiting process of calculating $\psi\,(q, \theta)$ from the stream function equation in the hodograph plane.

As Bergman pointed out, it is necessary to use high speed computing machines to calculate the function $G_n\,(\lambda)$ and $g_n\,(\lambda, \theta)$. However, from a practical viewpoint, there is a major difference between isentropic flows and diabatic flows: In isentropic flows the $G_n\,(\lambda)$'s are calculated once and for all, while the $g_n\,(\lambda, \theta)$'s, which depend on the boundary conditions, must be calculated in each particular case. On the other hand, in diabatic flows the functions $G_n\,(\lambda)$ should be calculated a number of times due to the fact that $Q\,(x, y)$ is not known as a function of q and may be determined by the iteration procedure outlined above. This, together with the fact that the functions $G_n\,(\lambda)$ are more complex in diabatic flows, shows that the numerical work is more involved. Finally, one may consider the most general and difficult case, $Q = Q(q, \theta)$, not discussed here.

14. Discussion of Q

Concentrating the attention on a thermodynamic system which consists of a streamline and letting q' represent the heat added per unit mass and per unit length along the streamline, then the First Law of Thermodynamics [see equation (13.1.3)] applied to this system can be written as (zero denotes container's conditions):

$$dQ = q'\, ds = c_p\, dT - \varrho^{-1}\, dp, \qquad (13.14.1)$$

where 's' is the running coordinate along the streamline. Hence

$$Q = \int dQ = \int_0^s q'\, ds + Q_0. \qquad (13.14.2)$$

Q is therefore the total heat added above a certain reference point. Choosing this reference point at stagnation then, $Q = 0$ for $q = 0$, $(s = 0)$ and therefore $Q_0 = 0$.

The generalized Bernoulli's equation [equation (13.2.10a)] was given as

$$\tfrac{1}{2} q^2 + c_p\, T - Q = \text{const.} \qquad (13.14.3)$$

For $q = 0$, $Q = 0$ and therefore, equation (13.14.3) reduces to

$$\tfrac{1}{2} q^2 + c_p\, T - Q = c_p\, T_0. \qquad (13.14.4)$$

The quantity q' may be related to the rate of heat added or subtracted per unit mass, $\overline{Q}$. Equation (13.14.1) gives

$$\partial Q/\partial s = q', \qquad (13.14.5)$$

but, $\overline{Q} = DQ/Dt = q\, \partial Q/\partial s$ and therefore,

$$q' = \overline{Q}/q. \qquad (13.14.6)$$

Isentropic flow is defined as $dQ = q'\, ds \equiv 0$ and therefore $Q \equiv 0$. In this case equation (13.14.4) reduces to the familiar Bernoulli's equation in isentropic flow.

An interesting situation arises on considering the physical significance of $Q = c$ (const.). A function q' such that, $q' = 0$ for all $s \neq s_0$, $0 \leqslant s_0 \leqslant s$, and $Q = \int_0^s q'\, ds = c$ may exist and $Q = c$ is justified mathematically if one considers $Q = \int dQ$ as a Lebesgue-Stieltjes integral. This situation may simulate the presence of a flame front in the flow domain. Another possibility would be a Dirac delta function.

The above outlined relation $Q = Q(q)$ may be subject to a radical change when the boundary value problem is considered. In this case one has to satisfy the usual geometric boundary conditions and the "a priori" given heat distribution $Q = Q(x, y)$, which must result in $Q = Q(q)$ only. The question whether this can be always satisfied seems to be a difficult one and is beyond the scope of the present work[1]. The most general case $Q(q, \theta)$ requires a special attention.

---

[1] Chapter XIII was reproduced in part from the paper: M. Z. v. Krzywoblocki and H. A. Hassan: Generalization of Bergman's Linear Integral Method to Diabatic Flow. Journal of the Society for Industrial and Applied Mathematics, Vol. 5, No. 2, June 1957, pp. 47—65.

Part XIV

Integral Operators in the Case of Certain Differential Equations in Three Variables

1. Some Examples

Below we quote Section 5 from Bergman's paper[1].

The next step in developing this approach is the generalization of these methods to the case of differential equations and systems of differential equations in three and more variables. The first step in this direction is to consider the harmonic functions in three variables, i. e., solutions of the equation

$$\Delta_3\,\psi \equiv \frac{\partial^2\psi}{\partial x_1{}^2} + \frac{\partial^2\psi}{\partial x_2{}^2} + \frac{\partial^2\psi}{\partial x_3{}^2} = 0, \tag{14.1.1}$$

and harmonic vectors, i. e., vectors $\vec{q} = \{H_1, H_2, H_3\}$ which satisfy the system of equations

$$\operatorname{curl}\vec{q} = 0, \quad \operatorname{div}\vec{q} = 0. \tag{14.1.2}$$

These vectors can be considered as a generalization of analytic functions of one complex variable. Various investigations similar to those discussed before are based on a formula which transforms analytic functions of two complex variables into harmonic functions ψ and harmonic vectors. If we denote by x_k, $k = 1, 2, 3$, the coordinates of the (real) three dimensional space and we write

$$Z = (i\,x_2 + x_3)/2; \quad Z^* = (i\,x_2 - x_3)/2, \tag{14.1.3}$$

then the operator

$$B_3\,(f, L, X_0) = (2\,\pi\,i)^{-1} \int_L f\,(u, \zeta)\,\zeta^{-1}\,d\zeta, \quad (x_1, x_2, x_3)\,\varepsilon\,V\,(X_0), \tag{14.1.4}$$

where L is a closed curve in the complex ζ plane and $V\,(X_0)$ a sufficiently small neighborhood of the point $X_0 = (x_1{}^0, x_2{}^0, x_3{}^0)$ transforms an analytic function of $u = x_1 + Z\,\zeta + Z^*\,\zeta^{-1}$ and ζ into a harmonic function of three variables (x_1, x_2, x_3) which is regular in the neighborhood of the point X_0. That is to say harmonic functions H of three variables can be represented in the form:

$$H\,(X) = B_3\,(f, L, X_0), \quad X\,\varepsilon\,V\,(X_0). \tag{14.1.5}$$

For simplicity's sake we shall assume in the following that X_0 is the origin, i. e., $x_1{}^0 = x_2{}^0 = x_3{}^0 = 0$.

In the representation (14.1.4) which generalized the well known formula of Whittaker the variables u and ζ play a different role. It is therefore useful to go a step further and represent f as an integral of a function of two variables. In this way we obtain the representation

$$H\,(X) = C_3\,(\chi); \quad C_3\,(\chi) \equiv (\pi\,i)^{-1} \int_L \int_{T=0}^{1} u^{1/2}\,\zeta^{-1}\,dT\,d\zeta\,d\,[u^{1/2}\,\chi\,(u\,\zeta^{-1}\,T^2,$$

$$u\,\zeta\,(1 - T^2))]/du, \tag{14.1.6}$$

where L is a conveniently chosen closed curve in the ζ-plane, e. g., the unit circle $|\zeta| = 1$, χ is a function of the form

$$\chi\,(Z, Z^*) = (Z\,Z^*)^{1/2}\,\chi_1\,(Z, Z^*) + \chi_2\,(Z, Z^*), \tag{14.1.7}$$

where χ_1 and χ_2 are analytic functions of two complex variables which are regular in the neighborhood of the point $Z = 0$, $Z^* = 0$. (14.1.7) represents a generalization

---

[1] S. Bergman: New Methods for Solving Boundary Value Problems. Z. angew. Math. Mechan. **36**, No. 5/6, 182—191 (May/June 1956).

to the case of harmonic functions of three variables of the formula (14.1.8) for harmonic functions of two variables. While in the case of (real) harmonic functions

$$h\,(x_1,\,x_2) = H\,(z,\,z^*) = \frac{1}{2}\,[g\,(z) + \bar{g}\,(z^*)];$$

$$z = x_1 + i\,x_2; \qquad z^* = x_1 - i\,x_2; \tag{14.1.8}$$

the corresponding "associate" function $g\,(z)$ is given by

$$g\,(z) = 2\,H\,(z,\,0) - \bar{g}\,(0); \tag{14.1.9}$$

in the case of harmonic functions

$$h\,(x_1,\,x_2,\,x_3) \equiv H\,(x_1,\,Z,\,Z^*) \tag{14.1.10}$$

of three variables for the associate function we have

$$\chi\,(Z,\,Z^*) = H\,(2\,(Z\,Z^*)^{1/2},\,Z,\,Z^*). \tag{14.1.11}$$

Similarly for the components of harmonic vectors

$$\vec{q} = \{H_1,\,H_2 + Re\,(g\,(x_2 + i\,x_3)),\,H_3 - Im\,(g\,(x_2 + i\,x_3))\}, \tag{14.1.12}$$

we obtain the representation in terms of an analytic function $f\,(u,\,\zeta)$ of two variables and an analytic function g of one variable:

$$\vec{q} = R_3\,(f,\,L,\,X_0,\,g) \equiv \left\{B_3\,(f,\,L,\,X_0),\,B_3\left[\frac{i}{2}\,(\zeta + \zeta^{-1})\,f\,(u,\,\zeta),\,L,\,X_0\right] + \right.$$

$$\left. + Re\,(g\,(x_2 + i\,x_3)),\,B_3\left[\frac{1}{2}\,(\zeta + \zeta^{-1})\,f\,(u,\,\zeta),\,L,\,X_0\right] - Im\,(g\,(x_2 + i\,x_3))\right\}. \tag{14.1.13}$$

We proceed to the description of operators generating solutions of the differential equations

$$T\,(\Phi) \equiv \Delta_3\Phi + C\,(r^2)\,\Phi = 0, \tag{14.1.14}$$

where $C\,(r^2)$ is an entire function of $r^2 = x^2 + y^2 + z^2$. To every differential equation one can determine a function $B\,(r^2,\,\sigma^2)$ defined for $0 \leq r^2 < \infty$, $0 \leq \sigma^2 \leq 1$, such that the operator

$$\Phi\,(X) = p_3\,[H\,(X)] = H\,(X) + \int\limits_0^1 B\,(r^2,\,\sigma^2)\,H\,(\sigma^2\,X)\,d\sigma, \tag{14.1.15}$$

yields solutions of (14.1.14). Here H is an arbitrary harmonic function of three variables which is regular at the origin.

Combining (14.1.6) and (14.1.15) we obtain the representation

$$\Phi\,(x,\,Z,\,Z^*) = p_3\,[C_3\,(\chi\,(Z,\,Z^*))], \tag{14.1.16}$$

of solutions of (14.1.14) in terms of analytic functions of two complex variables.

The inverse operator, i. e., the operator yielding $\chi\,(Z,\,Z^*)$ in terms of Φ,

$$\chi\,(Z,\,Z^*) = \Phi\,(2\,(Z,\,Z^*)^{1/2},\,Z,\,Z^*), \tag{14.1.17}$$

is independent of the coefficient C of the equation (14.1.14).

Similar integral operators can be derived in the case of the equations of the form:

$$\Delta_3\Phi + A\,(r^2)\left(x_1\,\frac{\partial\Phi}{\partial x_1} + x_2\,\frac{\partial\Phi}{\partial x_2} + x_3\,\frac{\partial\Phi}{\partial x_3}\right) + C\,(r^2)\,\Phi = 0, \tag{14.1.18}$$

where $A\,(r^2)$ and $C\,(r^2)$ are entire functions of r^2 (see [2]). In [3] operators generating solutions of equations

$$\partial^s\,\psi/\partial t^s + \partial^2\psi/\partial y^2 + \partial^2\psi/\partial z^2 + F\,(y,\,z)\,\psi = 0, \qquad s = 1,\,2,$$

are considered. Here F is an entire function.

Similar integral operators can be derived for certain systems of linear partial differential equations, e. g., in [4] functions $\psi = \psi (x_1, x_2, x_3, x_4)$ satisfying the system

$$\partial^2 \psi / \partial x_1{}^2 + \partial^2 \psi / \partial x_2{}^2 = F (x_1, x_2)\, \psi; \quad \partial^2 \psi / \partial x_3{}^2 + \partial^2 \psi / \partial x_4{}^2 = G (x_3, x_4)\, \psi,$$

are considered. Here F is an entire function of two variables (x_1, x_2) while G is an entire function of (x_3, x_4). In [4] the following representation of ψ in terms of two functions of two complex variables f and g is derived:

$$\psi (x_1, x_2, x_3, x_4) = Re \left\{ \int_{t=-1}^{1} \int_{\tau=-1}^{1} E_1 (z, \bar{z}, t)\, E_2 (w, \bar{w}, \tau)\, [f (z (1 - t^2), w (1 - \tau^2)) + \right.$$

$$\left. + g (z (1 - t^2), \bar{w} (1 - \tau^2))]\, (1 - t^2)^{-1/2}\, (1 - \tau^2)^{-1/2}\, dt\, d\tau \right\}.$$

Here $z = x_1 + i\, x_2$, $w = x_3 + i\, x_4$, $\bar{z} = x_1 - i\, x_2$, $\bar{w} = x_3 - i\, x_4$. f and g are two arbitrary analytic functions of two complex variables which are regular at the origin.

Part XV

Some SEAC Computations of Subsonic Fluid Flows by Bergman's Method of Integral Operators*

By

Philip Davis and Philip Rabinowitz

1. Introduction

This chapter presents the results of some computations of two-dimensional irrotational subsonic compressible flow patterns which were carried out in the pseudo-logarithmic plane by means of Bergman's method of integral operators. This is a method of superposition of particular solutions using integral operators to generate these solutions. A total of eleven particular solutions to the stream equation were obtained in the region $.615 < M < .711$, $0 < \theta < .2$. With these solutions at hand, and having computed by the same means certain fundamental singular solutions, a boundary value problem was set up in the pseudologarithmic plane which corresponds in the physical plane to a flow in a channel bounded by straight line segments and free boundaries. This boundary value problem was solved numerically by obtaining that linear combination of particular solutions which rendered the integral square error along the boundary a minimum. The computations were carried out on the National Bureau of Standards Eastern Automatic Computer (SEAC).

We shall first review the basic equations of the fluid flow. This will serve principally to introduce working definitions and to develop the boundary value problem. We shall then discuss the computational aspects of the problem; the specific computations which were made, the numerical methods employed, the difficulties encountered, the quality of the results, and the overall experience gathered in carrying out these computations.

---

[2] S. Bergman: Classes of Solutions of Linear Partial Differential Equations in Three Variables. Duke Math. J. 13, 216—249, 419—458 (1946).

[3] S. Bergman: On Singularities of Solutions of Certain Differential Equations in Three Variables. Transactions of the Amer. Mathem. Soc., Vol. 85, No. 2, pp. 462—488, July, 1957.

[4] S. Bergman and M. Schiffer: Properties of Solutions of a System of Partial Differential Equations. Studies in Mathematics and Mechanics presented to R. v. Mises, pp. 79—87. New York: Acad. Press. 1954.

* This work was completed in 1953 under the sponsorship of Air Resarch and Development Command, USAF.

2. Basic Equations of Flow

As the equations of flow are well known and can be found in many places, we shall merely summarize them here. The principal references for the theoretical portions of this paper are those of Bergman.

Let ϱ and p be the density and the pressure respectively of the fluid, let q be its speed and θ the angle which the velocity vector makes with some initial direction. The equation of state of the fluid is assumed to be:

$$p = \sigma \varrho^k. \tag{15.2.1}$$

(In the numerical computations which were carried out a value of $k = 1.4$ corresponding to that of air was adopted.) If the units of mass and velocity are chosen in such a way that $\varrho = 1$ when $q = 0$ and the sound velocity $dp/d\varrho = 1$, then we have:

$$\varrho = \left(1 - \frac{k-1}{2}\, q^2\right)^{1/(k-1)}. \tag{15.2.2}$$

The equations of continuity and irrotationality guarantee the existence of stream and potential functions ψ and Φ such that:

$$\Phi_\theta = \frac{q}{\varrho}\, \psi_q, \qquad \Phi_q = -\,\frac{1-M_2}{\varrho\, q}\, \psi_\theta, \tag{15.2.3}$$

where M is the Mach number given by:

$$M = q \Big/ \left(1 - \frac{(k-1)}{2}\, q^2\right)^{1/2}. \tag{15.2.4}$$

The pseudo-logarithmic plane is introduced by means of the variables λ and θ where λ is that function of M (or of q) which is given by:

$$\lambda = \frac{1}{2}\, \log\left[\frac{1 - (1 - M^2)^{1/2}}{1 + (1 - M^2)^{1/2}} \left(\frac{1 + h\,(1 - M^2)^{1/2}}{1 - h\,(1 - M^2)^{1/2}}\right)^{1/h}\right], \tag{15.2.5}$$

where:

$$h = \sqrt{\frac{k-1}{k+1}}. \tag{15.2.6}$$

In this plane, the interval $0 < M < 1$ corresponds to the interval $-\infty < \lambda < 0$. The equations (15.2.3) become:

$$\Phi_\theta = \psi_\lambda\, l^{1/2}; \qquad \Phi_\lambda = -\,\psi_\theta\, l^{1/2}, \tag{15.2.7}$$

where the function l is defined by:

$$l = (1 - M^2)/\varrho^2. \tag{15.2.8}$$

When the system (15.2.7) is solved for ψ and Φ respectively there results:

$$\psi_{\lambda\lambda} + \psi_{\theta\theta} + \left(l^{-1/2}\frac{d}{d\lambda}\,(l^{1/2})\right)\psi_\lambda = 0, \tag{15.2.9}$$

and

$$\Phi_{\lambda\lambda} + \Phi_{\theta\theta} + \left(l^{1/2}\frac{d}{d\lambda}\,(l^{-1/2})\right)\Phi_\lambda = 0. \tag{15.2.10}$$

It is now convenient to reduce equations (15.2.9) and (15.2.10) to normal form. This is accomplished in the usual way by modifying the dependent variables. Accordingly, we introduce as modified stream and potential functions the quantities:

$$\psi^* = \psi/H, \tag{15.2.11}$$

$$\Phi^* = H\, \Phi, \tag{15.2.12}$$

where:

$$H = (1 - M^2)^{-1/4}\left(1 + \frac{k-1}{2}\, M^2\right)^{-\frac{1}{2(k-1)}}. \tag{15.2.13}$$

This change of variable now reduces equations (15.2.9) and (15.2.10) to:

$$\frac{\partial^2 \psi^*}{\partial \lambda^2} + \frac{\partial^2 \psi^*}{\partial \theta^2} + 4\,F(\lambda)\,\psi^* = 0, \tag{15.2.14}$$

and

$$\frac{\partial^2 \Phi^*}{\partial \lambda^2} + \frac{\partial^2 \Phi^*}{\partial \theta^2} + 4\,P(\lambda)\,\Phi^* = 0, \tag{15.2.15}$$

where the functions $F(\lambda)$ and $P(\lambda)$ are as follows:

$$F = \frac{k+1}{64}\,M^4\left(\frac{-(3k-1)\,M^4 - 4(3-2k)\,M^2 + 16}{(1-M^2)^3}\right), \tag{15.2.16}$$

$$P = \frac{(k+1)}{64}\,M^4\left(\frac{(k-3)\,M^4 + (12-8k)\,M^2 - 16}{(1-M^2)^3}\right). \tag{15.2.17}$$

In the equations which follow, an asterisk (*) used on a stream or a potential function will always denote the modified functions defined by (15.2.11) and (15.2.12).

It should be observed that the functions $F = F(\lambda)$ and $P = P(\lambda)$ are regular analytic functions of the variable λ in the range $-\infty < \lambda < 0$ and at $\lambda = 0$ $(M = 1)$ possess singularities. Thus, for instance, the function $F(\lambda)$ possesses an expansion valid in the neighborhood of $\lambda = 0$ of the form:

$$F(\lambda) = \frac{5}{144}\,\lambda^{-2} + A_{-2}(-\lambda)^{-2/3} + A_0 + A_2(-\lambda)^{2/3} + \ldots, \tag{15.2.18}$$

where $A_{-2},\,A_0,\,A_2,\ldots$ are certain fixed constants which depend solely upon k.

The equations which relate the pseudologarithmic (λ,θ) plane with the physical (x,y) plane are as follows:

$$x = \int \frac{1}{\varrho}\left[\left(-\frac{(1-M^2)^{1/2}\cos\theta}{q}\,\psi_\theta - \frac{\sin\theta}{q}\,\psi_\lambda\right)d\lambda + \left(\frac{(1-M^2)^{1/2}\cos\theta}{q}\,\psi_\lambda - \frac{\sin\theta}{q}\,\psi_\theta\right)d\theta\right];$$

$$y = \int \frac{1}{\varrho}\left[\left(-\frac{(1-M^2)^{1/2}\sin\theta}{q}\,\psi_\theta + \frac{\cos\theta}{q}\,\psi_\lambda\right)d\lambda + \left(\frac{(1-M^2)^{1/2}\sin\theta}{q}\,\psi_\lambda + \frac{\cos\theta}{q}\,\psi_\theta\right)d\theta\right].$$

The bracketed expressions above are perfect differentials and the integrals are therefore independent of the path. The transformation can also be expressed in terms of the potential function Φ.

3. Generation of Particular Solutions of the Differential Equation (15.2.14)

We now describe a method whereby particular solutions of the differential equation (15.2.14) may be generated. Bergman extends the range of variation of the real variables λ, θ into the complex plane, and then introduces as two independent complex variables the quantities:

$$z = \lambda + i\,\theta, \quad z^* = \lambda - i\,\theta. \tag{15.3.19}$$

The object of this transformation is to change the differential equations (15.2.14) and (15.2.15) into equations which are formally of hyperbolic type. Since for an arbitrary function $f(z, z^*)$ we have:

$$\frac{\partial f}{\partial z} = \frac{\partial f}{\partial \lambda}\,\frac{\partial \lambda}{\partial z} + \frac{\partial f}{\partial \theta}\,\frac{\partial \theta}{\partial z}, \tag{15.3.20}$$

and since:

$$\lambda = \frac{1}{2}(z + z^*), \quad \theta = -\frac{i}{2}(z - z^*), \tag{15.3.21}$$

we have:

$$\frac{\partial}{\partial z} = \frac{1}{2}\left(\frac{\partial}{\partial \lambda} - i\,\frac{\partial}{\partial \theta}\right); \quad \frac{\partial}{\partial z^*} = \frac{1}{2}\left(\frac{\partial}{\partial \lambda} + i\,\frac{\partial}{\partial \theta}\right), \tag{15.3.22}$$

and

$$\frac{\partial}{\partial z\, \partial z^*} = \frac{1}{4}\left(\frac{\partial^2}{\partial \lambda^2} + \frac{\partial^2}{\partial \theta^2}\right) = \frac{1}{4}\,\Delta. \tag{15.3.23}$$

Thus, the equations (15.2.14) and (15.2.15) become:

$$\frac{\partial^2 \psi^*}{\partial z\, \partial z^*} + F\left(\frac{z + z^*}{2}\right)\psi^* = 0, \tag{15.3.24}$$

and

$$\frac{\partial^2 \Phi^*}{\partial z\, \partial z^*} + P\left(\frac{z + z^*}{2}\right)\Phi^* = 0. \tag{15.3.25}$$

The formal hyperbolic character of (15.3.24) suggests the possibility of arriving at particular solutions by iteration, leading to a series of Neumann type:

$$\Psi^* = g\,(z) - \int_{z_0}^{z}\int_{z_0^*}^{z^*} F\left(\frac{z_1 + z_1^*}{2}\right) g\,(z_1)\, dz_1\, dz_1^* +$$

$$+ \int_{z_0}^{z}\int_{z_0^*}^{z^*} F\left(\frac{z_1 + z_1^*}{2}\right)\int_{z_0}^{z_1}\int_{z_0^*}^{z_1^*} F\left(\frac{z_2 + z_2^*}{2}\right) g\,(z_2)\, dz_2\, dz_2^*\, dz_1\, dz_1^* - \ldots \tag{15.3.26}$$

Here $g\,(z)$ is any arbitrary analytic function of a complex variable, while (z_0, z_0^*) is an arbitrary fixed point lying in the region of analyticity of $g\,(z)$ and $F\left(\frac{z + z^*}{2}\right)$. If we introduce the operator:

$$\overline{\Theta}\,(h\,(z, z^*)) = \int_{z_0}^{z}\int_{z_0^*}^{z^*} F\left(\frac{z_1 + z_1^*}{2}\right) h\,(z_1, z_1^*)\, dz_1\, dz_1^*, \tag{15.3.27}$$

then we may rewrite (15.3.26) symbolically in the form

$$\psi^* = g - \overline{\Theta}\,g + \overline{\Theta}^2\, g - \ldots = (I - \overline{\Theta} + \overline{\Theta}^2 - \ldots)\,g = \frac{I}{I + \overline{\Theta}}\,g. \tag{15.3.28}$$

Thus, in view of the fact that $\dfrac{\partial^2}{\partial z\, \partial z^*}\,\overline{\Theta}\,(h) = F\,h$, it is clear that the series (15.3.26) satisfies (15.3.24) formally. Now, let B be any closed region in the z, z^* space lying in the common region of analyticity of F and g, and set:

$$\max_{z,\, z^* \,\in\, B} \left| F\left(\frac{z + z^*}{2}\right)\right| = M; \qquad \max_{z \,\in\, B} |g\,(z)| = m. \tag{15.3.29}$$

Then, we have,

$$|\overline{\Theta}^n\,(g)| \leqslant m\,\frac{|z - z_0|^n\,|z^* - z_0|^n\, M^n}{n!\, n!}. \tag{15.3.30}$$

Thus, the series (15.3.26) must converge uniformly and absolutely in every such closed bounded region. Furthermore the error E committed in terminating the series with the nth term is at most

$$|E| \leqslant m\sum_{j = n+1}^{\infty} \frac{(M\,|z - z_0|\,|z^* - z_0^*|)^j}{(j!)^2}. \tag{15.3.31}$$

For purposes of numerical estimation it is convenient to use the equivalent form

$$|E| \leqslant m\left(I_0\,(2\,t^{1/2}) - \sum_{j=0}^{n} \frac{t^j}{(j!)^2}\right), \tag{15.3.31'}$$

where $t = M\,|z - z_0|\,|z^* - z_0^*|$. The function I_0 is the so-called modified Bessel-Clifford function and will be found tabulated in [1].

<hr>

[1] U. S. Department of Commerce, Nat. Bureau of Standards: Tables of Bessel-Clifford functions of order zero and one. Appl. Math. Series, No. 28.

The operator defined by (15.3.26) has been called by Bergman the integral operator of the first kind. We shall designate it by $\bar{\bar{P}}\,(g)$. Various representations of this operator have been obtained by him and their properties have been studied in considerable detail. This operator is related to the Riemann formula in the case of real hyperbolic differential equations. We should observe, however, that if ψ^* is a complex solution of (15.3.24) then $Re\,\psi^*\,(\lambda + i\,\theta,\ \lambda - i\,\theta)$ and $Im\,\psi^*\,(\lambda + i\,\theta,\ \lambda - i\,\theta)$ where the λ and θ are now taken as real variables, will be real valued solutions of the real equation (15.2.14). When one is dealing with bounded, simply connected, one sheeted regions with piecewise smooth boundaries, then it is known that all solutions may be approximated by the particular solutions generated by selecting successively in (15.3.26) for $g\,(z)$ the functions $g\,(z) = 1, z, z^2, \dots.$

There are other methods for the generation of particular solutions. There is, e. g., the Chaplygin method of separation of variables. Setting $\psi_n^* = e^{i\,n\,\theta}\,a_n\,(\lambda)$, we find that $a_n\,(\lambda)$ must satisfy the ordinary differential equation:

$$a_n''\,(\lambda) + [4\,F\,(\lambda) - n^2]\,a_n\,(\lambda) = 0. \tag{15.3.32}$$

There is also the method of Σ-monogenic functions which has been expounded by Bers and Gelbart.

There is an important theoretical distinction between these two methods and the integral operator of Bergman which consists in the following. In most cases of physical interest we are looking for a solution which possesses a singular behavior for certain finite and non-zero values of the velocity q. The Chaplygin solutions (which may be multivalued if $n \neq$ integer) can be singular only at $\lambda = -\infty$ corresponding to $q = 0$ and cannot therefore exhibit such a singular behavior. But such solutions can be generated by the Bergman integral operator. On the other hand, the importance of this distinction for numerical solutions which necessarily use approximate methods and are frequently local in character is imperfectly understood at the present time.

4. Fluid Flow of Type S

We shall consider only flows which have a very simple character in the pseudo-logarithmic (λ, θ) plane, namely, those which are confined to a bounded, single-sheeted region lying in $-\infty < \lambda < 0$. (The flow around a closed body possesses a pseudograph which lies on a multi-sheeted Riemann Surface.) One way in which such a flow can be realized physically is as follows: In the physical (x, y) plane, consider the flow in a strip bounded by the line $y = y_0 = $ const. on one side, while on the other side it is bounded by a succession of straight line segments alter-

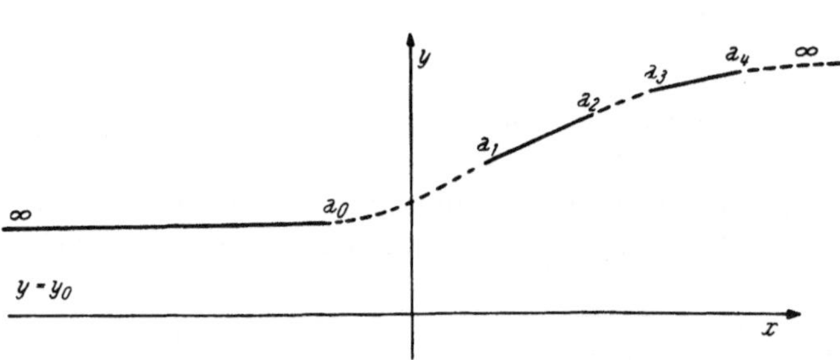

Fig. 1. Flow of Type S. Physical Plane

nating with free boundaries. Thus, in Figure 1, $y = y_0,\ \overline{\infty\,a_0},\ \overline{a_1\,a_2},\ \overline{a_3\,a_4}$ are thought of as being solid confining walls, while the alternate segments $\overline{a_0\,a_1},\ \overline{a_2\,a_3},\ \overline{a_4\,\infty}$ are portions of free boundaries. The fluid enters the channel on the left with a speed at infinity of q_E, and departs on the right with a speed at infinity of q_D. Along the solid walls, the angle of flow θ is constant, while along the free boundaries, the pressure and hence the speed (or the quantity λ) is constant. In view of these facts, the image of the flow in the pseudologarithmic plane looks as follows (Figure 2). It issues from W_1 (with a λ corresponding to q_E) and flows to W_2 (with a λ corresponding to q_D). The bounding line $y = y_0$ corresponds to $\overline{W_1\,W_2}$, while the segments $\overline{\infty\,a_0}$, $\overline{a_0\,a_1}$, $\overline{a_1\,a_2}$, $\overline{a_2\,a_3}$, $\overline{a_3\,a_4}$, $\overline{a_4\,\infty}$ correspond respectively to the sides $\overline{W_1\,A_0}$, $\overline{A_0\,A_1}$, $\overline{A_1\,A_2}$, $\overline{A_2\,A_3}$, $\overline{A_3\,A_4}$, $\overline{A_4\,W_2}$ of the reentrant right angled figure B in Figure 2. Such flows have been called by Bergman flows of type S. The stream function ψ of such flows must be constant along $\overline{W_1\,W_2}$, say 0, while on $\overline{W_1\,A_0\,A_1\,A_2\,A_3\,A_4\,W_2}$, it must also be constant, say, π. In the interior of the polygonal domain B, $\psi^* = \psi\,H^{-1}$ must be a solution of the partial differential equation (15.2.14). Furthermore, the solution ψ must possess singularities at the points W_1 and W_2 which produce jumps of π and $-\pi$ in its functional values at these points. The stream function ψ in its singular behavior at W_1 and W_2, must behave like the image in the (λ, θ) plane of an assumed singular behavior at infinity in the physical plane. The singularity which we shall take in the physical plane will be that due to a source.

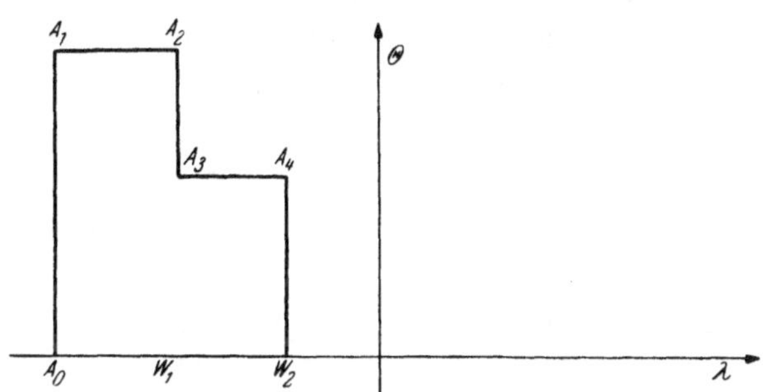

Fig. 2. Flow of Type S. Pseudologarithmic Plane

Thus, if we designated by $\chi^* = \chi\,H^{-1}$ any solution of the differential equation (15.2.14) which at the points W_1 and W_2 possesses appropriate singularities, we can then write:

$$\psi^*\,(\lambda,\,\theta) = \chi^*\,(\lambda,\,\theta) + R^*\,(\lambda,\,\theta), \qquad (15.4.33)$$

where $R^* = RH^{-1}$ designates a solution of (15.2.14) which is perfectly continuous on the boundary b of B. An appropriate boundary value problem for the stream function ψ may therefore be formulated as follows: Having determined the singular function $\chi\,(\lambda,\,\theta)$ and its boundary values $\chi\,(\lambda_b,\,\theta_b)$ for points $(\lambda_b,\,\theta_b)$ on the boundary b of B, one forms the difference $\psi\,(\lambda_b,\,\theta_b) - \chi\,(\lambda_b,\,\theta_b)$ where $\psi\,(\lambda_b,\,\theta_b)$ is taken as that discontinuous function defined by:

$$\left.\begin{aligned}\psi\,(\lambda_b,\,\theta_b) &= 0, \qquad (\lambda_b,\,\theta_b)\,\epsilon\,\overline{W_1\,W_2}, \\ \psi\,(\lambda_b,\,\theta_b) &= \pi, \qquad (\lambda_b,\,\theta_b)\,\epsilon\,\text{comp.}\,\overline{W_1\,W_2}.\end{aligned}\right\} \qquad (15.4.34)$$

This difference is now a continuous function along b. Determine that solution $R^*\,(\lambda,\,\theta)$ of (15.2.14) which on b assumes the values $\psi^*\,(\lambda_b,\,\theta_b) - \chi^*\,(\lambda_b,\,\theta_b)$. Having then obtained $R^*\,(\lambda,\,\theta)$ the stream function ψ^* is given by (15.4.33). In what follows, we shall assume that this boundary value problem possesses a unique solution. In view of the fact that the function $F\,(\lambda)$ is positive, this is not necessarily the case, and the whole question is related to the eigenvalue problem corresponding to (15.2.14).

We turn now to the construction of the singular function χ.

5. Singular Solutions of Type S

By a fundamental singularity of a linear differential equation (15.2.14) is meant a function $S^*\,(\lambda,\,\theta;\,\lambda_0,\,\theta_0)$ of the four indicated real variables which, for fixed $(\lambda_0,\,\theta_0)$ is a solution of the differential equation in $(\lambda,\,\theta)$ and which possesses a logarithmic singularity in the vicinity of $(\lambda_0,\,\theta_0)$. That is, S^* behaves in such a way that we can write:

$$S^*\,(\lambda,\,\theta;\,\lambda_0,\,\theta_0) = \frac{1}{2}\,A\,(\lambda,\,\theta;\,\lambda_0,\,\theta_0)\,\log\{\,(\lambda - \lambda_0)^2 + (\theta - \theta_0)^2\} + B\,(\lambda,\,\theta;\,\lambda_0,\,\theta_0), \quad (15.5.35)$$

where A and B are perfectly regular in the vicinity of (λ_0, θ_0). It may be shown that the images in the (λ, θ) plane of such singular flows as sources, sinks, dipoles and vortices may be obtained from $S^*(\lambda, \theta; \lambda_0, \theta_0)$ and from $T^*(\lambda, \theta; \lambda_0, \theta_0)$ where T^* is defined as the fundamental singularity corresponding to the potential equation (15.2.15). It may be shown that T is the velocity potential function of a flow which has a source at (λ_0, θ_0); hence, the stream function corresponding to T will be a stream function possessing the required type of singularity at (λ_0, θ_0). This stream function has been designated by Bergman by $\psi^{(L,2)}(\lambda, \theta)$.

In order to discuss the computation of the fundamental singularity and of the function $\psi^{(L,2)}$, it is again convenient to return to the variables z, z^*. In terms of these variables a fundamental singularity $T(z, z^*; z_0, z_0^*)$ for the differential equation (15.2.10) is given explicitly by the following series of Neumann type:

$$T(z, z^*; z_0, z_0^*) = \frac{C}{2}\{\log(z - z_0) + \log(z^* - z_0^*)\} + D, \qquad (15.5.36)$$

where:

$$C = C(z, z^*; z_0, z_0^*) =$$

$$= H^{-1}\left[1 - \int_{z_0}^{z}\int_{z_0^*}^{z^*} P(z_1, z_1^*)\, dz_1\, dz_1^* + \int_{z_0}^{z}\int_{z_0^*}^{z^*} P(z_1, z_1^*)\left(\int_{z_0}^{z_1}\int_{z_0^*}^{z_1^*} P(z_2, z_2^*)\, dz_2\, dz_2^*\right) dz_1\, dz_1^* - \ldots\right],$$

$$(15.5.37)$$

and

$$D = D(z, z^*; z_0, z_0^*) =$$

$$= H^{-1}\left[\int_{z_0}^{z}\int_{z_0^*}^{z^*} K(z_1, z_1^*)\, dz_1\, dz_1^* - \int_{z_0}^{z}\int_{z_0^*}^{z^*} P(z_1, z_1^*)\left(\int_{z_0}^{z_1}\int_{z_0^*}^{z_1^*} K(z_2, z_2^*)\, dz_2\, dz_2^*\right) dz_1\, dz_1^* + \ldots\right],$$

$$(15.5.38)$$

where:

$$K(z, z^*) = -\frac{1}{z^* - z_0^*}\frac{\partial(HC)}{\partial z} - \frac{1}{z - z_0}\frac{\partial(HC)}{\partial z^*}. \qquad (15.5.39)$$

The function H is that given by (15.2.13). A real fundamental singularity may be obtained by taking

$$T(\lambda, \theta; \lambda_0, \theta_0) = Re\left(T(z, z^*; z_0, z_0^*)\right) \text{ with } \begin{cases} z = \lambda + i\theta \\ z_0 = \lambda_0 + i\theta_0 \end{cases}, \qquad (15.5.40)$$

and $\begin{cases} z^* = \lambda - i\theta \\ z_0^* = \lambda_0 - i\theta_0 \end{cases}$

An argument similar to (15.3.19) to (15.3.22) shows that:

$$\frac{\partial}{\partial\lambda} = \frac{\partial}{\partial z} + \frac{\partial}{\partial z^*}; \qquad \frac{\partial}{\partial\theta} = i\left(\frac{\partial}{\partial z} - \frac{\partial}{\partial z^*}\right). \qquad (15.5.41)$$

Hence, the system (15.2.7) relating the stream and potential functions becomes:

$$\frac{\partial\Phi}{\partial z} = -i\, l^{1/2}\frac{\partial\psi}{\partial z}; \qquad \frac{\partial\Phi}{\partial z^*} = i\, l^{1/2}\frac{\partial\psi}{\partial z^*}. \qquad (15.5.42)$$

When a potential function is known, the corresponding stream function is obtained in the usual way through the contour integral:

$$\psi = i\int l^{-1/2}\frac{\partial\Phi}{\partial z}\, dz - l^{-1/2}\frac{\partial\Phi}{\partial z^*}\, dz^*. \qquad (15.5.43)$$

Thus

$$\psi^{(L,2)}(z, z^*; z_0, z_0^*) = i\int_{(z_R, z_R^*)}^{(z, z^*)} l^{-1/2}\frac{\partial T}{\partial z}\, dz - l^{-1/2}\frac{\partial T}{\partial z^*}\, dz^*, \qquad (15.5.44)$$

where (z_R, z_R^*) is some arbitrary reference point.

The real form is given by:

$$\psi^{(L,\,2)}\,(\lambda,\,\theta;\,\lambda_0,\,\theta_0) = Re\,[\psi^{(L,\,2)}\,(z,\,\bar{z};\,z_0,\,\overline{z_0})].\qquad(15.5.45)$$

If we consider now the function:

$$\chi\,(\lambda,\,\theta) = a\,\psi^{(L,\,2)}\,(\lambda,\,\theta;\,\lambda_1,\,\theta_1) + b\,\psi^{(L,\,2)}\,(\lambda,\,\theta;\,\lambda_2,\,\theta_2),\qquad(15.5.46)$$

where:

$$W_1 = (\lambda_1,\,\theta_1);\qquad W_2 = (\lambda_2,\,\theta_2),\qquad(15.5.47)$$

and where a and b are two arbitrary constants, then χ is a solution of the stream equation which at the two points W_1 and W_2 possesses the singular character of a source. By an appropriate selection of the constants a and b, the required jumps at the points W_1 and W_2 can be produced.

To evaluate a and b we proceed as follows: As the point $(\lambda,\,\theta)$ turns in a counterclockwise sense around W_1 and W_2, the stream function has an increment of π and $-\pi$ respectively. Hence, χ should have increments of $-\pi$ and π respectively. Now it is known that at any point $(\lambda_0,\,\theta_0)$ the function $\psi^{(L,\,2)}\,(\lambda,\,\theta;\,\lambda_0,\,\theta_0)$ behaves like $-\dfrac{l^{-1/2}(\lambda_0)}{H\,(\lambda_0)}\tan^{-1}\dfrac{\theta-\theta_0}{(\lambda-\lambda_0)}$. Hence, if we take:

$$a = \frac{H\,(\lambda_1)}{l^{-1/2}(\lambda_1)}\,,\qquad b = -\,2\,\frac{H\,(\lambda_2)}{l^{-1/2}(\lambda_2)}\,,\qquad(15.5.48)$$

we shall obtain the required jumps. Thus,

$$\chi\,(\lambda,\,\theta) = \frac{+H\,(\lambda_1)}{l^{-1/2}(\lambda_1)}\,\psi^{(L,\,2)}\,(\lambda,\,\theta;\,\lambda_1,\,\theta_1) - \frac{2\,H\,(\lambda_2)}{l^{-1/2}(\lambda_2)}\,\psi^{(L,\,2)}\,(\lambda,\,\theta;\,\lambda_2,\,\theta_2).\quad(15.5.49)$$

6. Numerical Computation of Particular Solutions and Singular Solutions

The family of particular solutions to be computed are $Re\,\overline{\overline{P}}\,(z^n)$ and $Im\,\overline{\overline{P}}\,(z^n)$ defined essentially by (15.3.26). (Under certain circumstances it may be more convenient to select appropriate linear combinations of these.) The singular solutions are $\psi^{(L,\,2)}\,(\lambda,\,\theta;\,\lambda_1,\,\theta_1)$ and $\psi^{(L,\,2)}\,(\lambda,\,\theta;\,\lambda_2,\,\theta_2)$ defined essentially by (15.5.36) to (15.5.44). In view of the highly complicated structure of the functions F and P when expanded in the z, z^* variables, the series (15.3.26) and (15.5.36) to (15.5.38) must be evaluated by some method of approximate analysis. There are two methods which suggest themselves immediately. They can be described in the following way:

Method A. Compute the functions F and P as functions of a complex argument on a sufficiently dense set of points in the complex plane. With these values at hand, carry out the indicated operations using approximate numerical methods.

Method B. Approximate a real function $F\,(\lambda)$, $P\,(\lambda)$ by a polynomial of some degree N over some interval $[\lambda_1,\,\lambda_2]$. Once this has been done, all the indicated operations can be carried out in closed form prior to the use of computing machines, and the numerical work consists principally of the evaluation of certain polynomials in λ and θ.

These two methods have relative advantages and disadvantages, and it is important to indicate what they are.

There is considerably more numerical and preparatory work associated with Method A than with Method B. The iterated integrals are functions of two complex variables and must be stored. The functions F and P must therefore themselves be computed for complex values and stored. Moreover, in the absence of interpretive subroutines which would enable complex integration to be carried out with the ease of real integration, these iterated integrals must be expressed in real form, and each consists of a great many real iterated integrals. There is consequently a greater opportunity for the build up of error with Method A than with Method B, and these errors are harder to isolate. The numerical work associated with the computation of the singular solutions is particularly troublesome using Method A.

Method B, on the other hand, yields solutions which are somewhat local in character. In order to obtain results which are valid over a wider range of values of λ (or of M), it would be necessary to approximate $F(\lambda)$ and $P(\lambda)$ over this wider range. This means that the degree N of the approximating polynomial becomes larger, and consequently the amount of formal work which must be performed prior to any computation grows tremendously.

Yet, using Method B, the solutions obtained are immediately interpretable as perturbations of order N of the incompressible flows. There is, furthermore, the possibility of patching together local solutions obtained by Method B and arriving in this way at solutions valid over a larger region.

7. Computation of F and P as Functions of a Complex Variable

We designate by F_1 and F_2 the quantities $F_1 = Re\,(F)$, $F_2 = Im\,(F)$. If we introduce the complex variable:

$$T = \tau\, e^{i\sigma}, \qquad T^2 = 1 - M^2, \tag{15.7.50}$$

then from (15.2.16) with $k = 1.4$ there is obtained:

$$\left.\begin{aligned}
F_1 &= 0.51 - \left(0.12\,\tau^2 + \frac{0.21}{\tau^2}\right)\cos 2\,\sigma - \frac{0.63}{\tau^4}\cos 4\,\sigma + \frac{0.45}{\tau^6}\cos 6\,\sigma, \\
F_2 &= \left(\frac{0.21}{\tau^2} - 0.12\,\tau^2\right)\sin 2\,\sigma + \frac{0.63}{\tau^4}\sin 4\,\sigma - \frac{0.45}{\tau^6}\sin 6\,\sigma.
\end{aligned}\right\} \tag{15.7.51}$$

Furthermore, we have from (15.2.5), (15.2.6), and (15.3.19):

$$\left.\begin{aligned}
\lambda &= \frac{1}{2}\log\,(1 + \tau^2 - 2\,\tau\cos\sigma)^{1/2} - \frac{1}{2}\log\,(1 + \tau^2 + 2\,\tau\cos\sigma)^{1/2} + \\
&\quad + \frac{\sqrt{6}}{2}\log\left(1 + \frac{\tau^2}{6} + \frac{2\,\tau}{\sqrt{6}}\cos\sigma\right)^{1/2} - \frac{\sqrt{6}}{2}\log\left(1 + \frac{\tau^2}{6} - \frac{2\,\tau}{\sqrt{6}}\cos\sigma\right)^{1/2}; \\
\theta &= \frac{1}{2}\arctan\left(\frac{-\,\tau\sin\sigma}{1 - \tau\cos\sigma}\right) - \frac{1}{2}\arctan\left(\frac{\tau\sin\sigma}{1 + \tau\cos\sigma}\right) + \\
&\quad + \frac{\sqrt{6}}{2}\arctan\left(\frac{\frac{\tau}{\sqrt{6}}\sin\sigma}{1 + \frac{\tau}{\sqrt{6}}\cos\sigma}\right) - \frac{\sqrt{6}}{2}\arctan\left(\frac{-\frac{\tau}{\sqrt{6}}\sin\sigma}{1 - \frac{\tau}{\sqrt{6}}\cos\sigma}\right).
\end{aligned}\right\} \tag{15.7.52}$$

For given values of λ, θ, the system (15.7.52) may be solved implicitly to produce a corresponding value of σ and τ. When these values of σ and τ are known, F_1 and F_2 may then be computed directly from (15.7.51). The system (15.7.52) was solved by inverse interpolation in two variables using Newton's method of successive approximations. This computation was carried out on the SEAC. The following tabulations are computed: τ, σ, F_1, and F_2 for values of λ in the range $-1.04(.01)-.13$ and for values of θ in the range $0(.04).80$.

The function P may be tabulated as a function of a complex variable in the same manner. For reasons which will be indicated subsequently, this was not carried out in the present set of computations.

8. Estimation of Truncation Errors in the Computation of Particular Solutions

We consider the estimate (15.3.31) in more detail. Let us assume that we are interested in computing solutions over the (real) region $R: -\infty < \lambda_1 < \lambda < \lambda_2 < 0$, $0 < \theta < \theta_1$. We select $\theta_0 = 0$ so that $z_0 = z_0^* = \lambda_0$. The integrations in (15.3.26) must be carried out from z_0 to $z = \lambda + i\theta$ and from z_0^* to $z^* = \lambda - i\theta$ where λ and θ are confined to the rectangle R. It is easily verified that as z and z^* perform such variations independently, the variable $u = \frac{1}{2}(z + z^*)$ need vary only in that rectangle R' which is obtained from R by projection from the point $(\lambda_0, 0)$ and dimunition by a factor of $\frac{1}{2}$. Thus, with $g(z) = Z^N$, we have in (15.3.31):

$$m = (\lambda_1^2 + \theta_1^2)^{N/2}, \quad M = \max_{u \in R'} |F(u)| \left. \begin{matrix} \\ \\ \end{matrix} \right\}$$
$$|z - z_0|, \quad |z^* - z_0^*| < \max_{i=1,2} [(\lambda_i - \lambda_0)^2 + \theta_i^2]^{1/2}. \tag{15.8.53}$$

Let us select, for instance, $\lambda = \lambda_2 = -.13$ (Mach No. $= .734$). An examination of the function F in the complex plane reveals that over the rectangle R', its maximum modulus M occurs at $z = \lambda = -.13$ and equals $.87$. Let $r^2 = (\lambda_1 - \lambda_0)^2 + \theta_1^2$. Then (15.3.31') becomes:

$$|E| < (\lambda_1^2 + \theta_1^2)^{N/2} \left[I_0(2\,t^{1/2}) - \sum_{i=0}^{n} \frac{t^j}{(j!)^2} \right] = (\lambda_1^2 + \theta_1^2)^{N/2}\, E(t, n), \tag{15.8.54}$$

where:

$$t = .87\, r^2 = .87\,[(\lambda_1 - \lambda_0)^2 + \theta_1^2]. \tag{15.8.54'}$$

In view of the fact that the solutions (15.3.26) may be regarded as perturbations from the incompressible solutions z^N, and that the maximum modulus of the quantity z^N over R is precisely $(\lambda_1^2 + \theta_1^2)^{N/2}$, it is clear that the quantity $E(t, n)$ measures the accuracy achieved by retaining in (15.3.26) $n + 1$ terms over and above that provided by the basic z^N. A table of $E(t, n)$ is now given for a few selected values. The allowed range of λ, θ is measured by r as per the above definition, but λ must be $< -.13$.

Table A: $E(t, n)$

r	.34	.76	1.07	1.31	1.52	1.86
n \ t	.1	.5	1.0	1.5	2.0	3.0
1	.002	.07	.3	.7	1.3	3.2
2	.00003	.004	.03	.1	.3	.9
3	.0000002	.0001	.002	.009	.03	.2
4				.0005	.002	.02

9. Computation of Particular Solutions $\bar{\bar{P}}(z^N)$. Method A

The quantities to be computed are $Re\,\psi^*(\lambda + i\theta, \lambda - i\theta)$ and $Im\,\psi^*(\lambda + i\theta, \lambda - i\theta)$ where ψ^* is given by (15.3.26) and $g(z) = z^N$. We consider the first iterated integral in this series. Writing $z_0 = \lambda_0 + i\theta_0$, $z_0^* = \Lambda_0 - i\vartheta_0$, $z = \lambda + i\theta$, $z^* = \Lambda - i\vartheta$, $\lambda_0 = \Lambda_0$, $\theta_0 = \vartheta_0 = 0$, and taking advantage of the fact that the integration is independent of the path of integration, we may carry out the integration process with respect to z as follows: from λ_0 to λ with $dz = d\lambda$ and from λ to $\lambda + i\theta$ with $dz = i\,d\theta$. The integration with respect to z^* is from Λ_0 to Λ with $dz^* = d\Lambda$ and from Λ to $\Lambda - i\vartheta$ with $dz^* = -i\,d\vartheta$. (Λ and ϑ are dummy variables playing the same role as λ and θ.) We therefore obtain:

$$
\begin{aligned}
\int_{z_0}^{z}\int_{z_0^*}^{\bar z} z^N F\Big(\frac{z+z^*}{2}\Big)\,dz\,dz^* &= \int_{\lambda_0}^{\lambda}\int_{\lambda_0}^{\lambda} \lambda^N \Big[F_1\Big(\frac{\lambda+\Lambda}{2}\Big) + i F_2\Big(\frac{\lambda+\Lambda}{2}\Big)\Big]\,d\lambda\,d\Lambda + \\
&+ \int_{\lambda_0}^{\lambda}\int_{0}^{\theta} \lambda^N \Big[F_2\Big(\frac{\lambda+\Lambda-i\vartheta}{2}\Big) - i F_1\Big(\frac{\lambda+\Lambda-i\vartheta}{2}\Big)\Big]\,d\lambda\,d\vartheta + \\
&+ \int_{\lambda_0}^{\lambda}\int_{0}^{\theta} -F_1\Big(\frac{\lambda+i\theta+\Lambda}{2}\Big)\,Im\,(\lambda+i\theta)^N - F_2\Big(\frac{\lambda+i\theta+\Lambda}{2}\Big) \cdot \\
&\qquad\qquad \cdot Re\,(\lambda+i\theta)^N\,d\Lambda\,d\theta + \\
&+ i\int_{\lambda_0}^{\lambda}\int_{0}^{\theta} F_1\Big(\frac{\lambda+i\theta+\Lambda}{2}\Big)\,Re\,(\lambda+i\theta)^N - F_2\Big(\frac{\lambda+i\theta+\Lambda}{2}\Big) \cdot \\
&\qquad\qquad \cdot Im\,(\lambda+i\theta)^N\,d\Lambda\,d\theta + \\
&+ \int_{0}^{\theta}\int_{0}^{\theta} Re\,(\lambda+i\theta)^N F_1\Big(\lambda+i\Big[\frac{\theta-\vartheta}{2}\Big]\Big) - Im\,(\lambda+i\theta)^N \cdot \\
&\qquad\qquad F_2\cdot\Big(\lambda+i\Big[\frac{\theta-\vartheta}{2}\Big]\Big)\,d\theta\,d\vartheta + \\
&+ i\int_{0}^{\theta}\int_{0}^{\theta} Re\,(\lambda+i\theta)^N F_2\Big(\lambda+i\Big[\frac{\theta-\vartheta}{2}\Big]\Big) + Im\,(\lambda+i\theta)^N \cdot \\
&\qquad\qquad F_1\cdot\Big(\lambda+i\Big[\frac{\theta-\vartheta}{2}\Big]\Big)\,d\theta\,d\vartheta,
\end{aligned}
\tag{15.9.54''}
$$

where we have written $F_1(z) = Re\,F(z)$, $F_2(z) = Im\,F(z)$. Each of the above real double integrals must be broken down further for machine computation. Thus, e. g.:

$$
\begin{aligned}
\int_{\lambda_0}^{\lambda}\int_{0}^{\theta} -F_1\Big(\frac{\lambda+i\theta+\Lambda}{2}\Big)\,Im\,(\lambda+i\theta)^N\,d\Lambda\,d\theta &= \\
= \int_{0}^{\theta} Im\,(\lambda+i\theta)^N\,d\theta \int_{\lambda_0}^{\lambda} -F_1\Big(\frac{\lambda+i\theta+\Lambda}{2}\Big)\,d\Lambda. &
\end{aligned}
\tag{15.9.55}
$$

Now:

$$
\begin{aligned}
\int_{\lambda_0}^{\lambda} -F_1\Big(\frac{\lambda+i\theta+\Lambda}{2}\Big)\,d\Lambda &= -2\int_{(\lambda+\lambda_0)/2}^{\lambda} F_1\Big(s+\frac{i\theta}{2}\Big)\,ds = \\
&= -2\Big[\int_{0}^{\lambda} F_1\Big(s+\frac{i\theta}{2}\Big)\,ds - \int_{0}^{(\lambda+\lambda_0)/2} F_1\Big(s+\frac{i\theta}{2}\Big)\,ds\Big].
\end{aligned}
\tag{15.9.56}
$$

To evaluate (15.9.55), the quantity $\int_{0}^{\lambda} F_1\Big(s+\frac{i\theta}{2}\Big)\,ds$ was computed for fixed θ, as a function of λ and then the integration with respect to θ was carried out using (15.9.56).

In the carrying out of these integration on SEAC, certain large scale data handling operations had to be performed. A function $f(x,y)$ of two variables must be stored in blocks, each block corresponding to a fixed value x_i of the variable x and consisting of the values of the function $f(x_i,y)$. Thus, F was tabulated in blocks corresponding to λ_j and in each block the function $F(\lambda_j,\theta)$ was tabulated as a function of θ. When $\int F(\lambda,\theta)\,d\lambda$ was computed, the data had to be shuffled into θ-blocks to give functions tabulated for λ. Then, when the second integration had to be carried out, the results of the first computation which were printed out in θ-blocks had to be rearranged in λ-blocks to permit integration with respect to θ.

The following particular solutions (one iterated integral only) computed by Method A on SEAC are available, $Re\ \psi_N{}^*$ and $Im\ \psi_N{}^*$ for $N = 0, 1, \ldots, 5$ over the range $-1.03(.02)-.13$ for λ and $0(.08).80$ for θ. The total machine time for this computation was approximately 8 hours. This is exclusive of the machine time used for code checking purposes. An examination of the third term in (15.3.26) reveals that the inner iterated integral which must be evaluated numerically is a function of two complex variables or, equivalently, of four real variables, and this must be computed and stored prior to the computation of the outer integral. This fact means that the programming of the quadruple integral will be extremely complicated and, due to the large number of rearrangements of data stored on high access time media which have to be performed, the computation time for the quadruple integral will be very large. For these reasons, it was considered feasible to compute only with the first two terms of (15.3.26).

An upper bound for the truncation error incurred in retaining only two terms of the series (15.3.26) has been given in Table A ($n = 1$). This table reveals that unless $(\lambda - \lambda_0)^2 + \theta^2$ is severely restricted, the percentage error mounts rapidly.

In addition to truncation errors, there are other sources for error in this computation which should be mentioned. There is the error in the computation of F in the complex plane, and the deviation of the computed values from the true analytic continuation of the F on the real line will tend to throw off the values of the particular solutions. There is also the error incurred in replacing the integrals of (15.8.54) by finite numerical rules (in the cases carried out numerically, Simpson's rule was employed with a mesh width of .01 in the λ direction and .04 in the θ direction). These errors are somewhat difficult to estimate, and due to the large number, 24, of single real integrals which must be combined to yield the one complex integral (15.8.54), the pile up of error may amount to more than anticipated. The large number of terms here also increases the opportunity for gross blunders such as wrong signs, faulty coefficients, etc. Local checks carried out on the solutions obtained by Method A were of poor quality, and this method was abandoned in favor of the more flexible method B, now to be described.

10. Local Development of the Solutions $\bar{\bar{P}}(z^N)$. Method B

Assume that in a certain neighborhood of the point λ_0 we have approximated $F(\lambda)$ with sufficient accuracy by a polynomial of the form:

$$F^*(\lambda) = a_0 + a_1(\lambda - \lambda_0) + \ldots + a_M(\lambda - \lambda_0)^M. \qquad (15.10.57)$$

We may then obtain each term of the series (15.3.26) in an explicit form. We select $z_0 = \lambda_0 + i\,\theta_0$, $z_0{}^* = \lambda_0 - i\,\theta_0$, $\theta_0 = 0$, and then set:

$$w = \frac{1}{2}(z - \lambda_0), \qquad w^* = \frac{1}{2}(z^* - \lambda_0). \qquad (15.10.58)$$

The quantity $F^*(\lambda) = F^*\left(\frac{1}{2}(z + z^*)\right)$ becomes:

$$F^* = a_0 + a_1(w + w^*) + \ldots + a_M(w + w^*)^M. \qquad (15.10.59)$$

If we now select for $g(z)$ in (15.3.26) not z^n but what is more convenient (and equivalent from the point of view of the generation of complete sets of solutions), $g(z) = \left(\frac{1}{2}(z - \lambda_0)\right)^N$, then (15.3.26) can be written in the form:

$$\psi_N{}^* = w^N - 4 \int_0^w \int_0^{w^*} w_1{}^N F^* \, dw_1 \, dw_1{}^* + 16 \int_0^w \int_0^{w^*} F^* \, dw_1 \, dw_1{}^* \int_0^{w_1} \int_0^{w_1{}^*} w_2{}^N F^* \, dw_2 \, dw_2{}^* - \ldots$$

$$\tag{15.10.60}$$

For the purposes of the computation, M was selected as $M = 3$. Inserting (15.10.59) in (15.10.60), integrating, and retaining only those terms of the infinite series whose total degree in w, w^* is at most $N + 5$, (the numbers $M = 3$ and $N + 5$ are arbitrary and may be increased at will, but at the expense of more algebra) we obtain:

$$\psi_N{}^* = w^N [A_N + B_N w w^* + C_N w w^{*2} + D_N w w^{*3} + E_N w w^{*4} + F_N w^2 w^* + G_N w^2 w^{*2} +$$
$$+ H_N w^2 w^{*3} + I_N w^3 w^* + J_N w^3 w^{*2} + K_N w^4 w^*] \equiv w^N Q_N, \tag{15.10.61}$$

where:

$$A_N = 1, \; B_N = - \frac{4\,a_0}{N+1}, \; C_N = \frac{-2\,a_1}{N+1},$$

$$D_N = - \frac{4}{3} \frac{a_2}{(N+1)}, \; E_N = - \frac{a_3}{N+1}, \; F_N = - \frac{4\,a_1}{N+2},$$

$$G_N = - \frac{4\,a_2}{N+2} + \frac{8\,a_0{}^2}{(N+1)(N+2)}, \; H_N = - 4\left(\frac{a_3}{N+2} - \frac{2\,a_0\,a_1}{(N+1)(N+2)}\right),$$

$$I_N = - \frac{4\,a_2}{N+3}, \; J_N = - \frac{6\,a_3}{N+3} + \frac{8\,a_0\,a_1}{(N+2)(N+3)} + \frac{8\,a_0\,a_1}{(N+1)(N+3)},$$

$$K_N = - \frac{4\,a_3}{N+4}.$$

$$\tag{15.10.62}$$

The quantity Q_N (which is the bracketed portion of (15.10.61)) measures the deviation of the solution $\psi_N{}^*$, from the solution w^N of the incompressible (Laplace's) equation. Since F^* has been taken as the polynomial (15.10.57) the series (15.10.59) converges for all values of w, yielding solutions for all values of λ, θ. Since however, $F(\lambda)$ is approximated by $F^*(\lambda)$ only over a certain range of λ, the solutions will be valid only for this range. This range may be increased at will by increasing the degree of the polynomial (15.10.57). For a fixed range of acceptable approximation, the range in the θ-direction over which the solutions will be valid may be augmented by increasing the number of terms in the series (15.3.26) or (15.10.60). This results in solutions of the form of (15.10.61) in which the perturbation term Q_n is of the higher order. An increased amount of algebraic preparation is necessary prior to carrying out the computation, but the computation itself presents no added difficulties.

We should remark further that when the real solutions derived from (15.10.61) are expressed in terms of λ and θ, they are found to be polynomials in these variables. The introduction of the complex variables z and z^* may be justified from even a purely formal point of view in that it is a very convenience device for systematically obtaining an infinite set of power series in λ, θ which satisfy the differential equation.

Truncation error estimates may be obtained in § 8, using $M = \max\limits_{u \in R'} |F^*(u)|$. This measures the discrepancy between the solution of $\dfrac{\partial^2 \psi^*}{\partial \lambda^2} + \dfrac{\partial^2 \psi^*}{\partial \theta^2} + 4\, F^*(\lambda)\, \psi^* = 0$ and the truncated series (15.10.60).

11. Development of the Singular Solutions $\psi^{(L,\,2)}(\lambda, \theta)$

The computation of the singular solutions by means of Method A poses very great numerical difficulties. In the first place, we must now deal with two infinite series of iterated integrals (15.5.37) and (15.5.38), and there are, in addition, the partial differentiations and line integrations implied.

Method B can be carried out with a certain amount of ease. The preliminary algebraic work is considerably more involved than that met in the computation of the particular solutions $\bar{\bar{P}}(z^N)$ but can, in principle, be carried out with an arbitrarily high degree of accuracy. For the work of this section, we shall assume that the functions $P(\lambda)$, $H(\lambda)$, and $l^{-\frac{1}{2}}(\lambda)$ are represented over the λ-interval with sufficient accuracy by a constant. Furthermore, we shall use only those terms of (15.5.37) and (15.5.38) which are four fold integrals or lower.

Let therefore:

$$P(\lambda) = p_0, \qquad H(\lambda) = h_0, \qquad l^{-\frac{1}{2}}(\lambda) = l_0, \tag{15.11.63}$$

be a sufficiently good approximation for these respective functions over the interval in question. Write $z = \lambda + i\,\theta$, $z^* = \lambda - i\,\theta$, $z_0 = \lambda_0$, $z_0{}^* = \lambda_0$, $\theta_0 = 0$, $z - z_0 = (\lambda - \lambda_0) + i\,\theta$, $z^* - z_0{}^* = (\lambda - \lambda_0) - i\,\theta$. From (15.5.37), retaining 3 terms, we have:

$$C = \frac{1}{h_0}\left(1 - p_0\,(z - z_0)\,(z^* - z_0{}^*) + \frac{p_0{}^2}{4}\,(z - z_0)^2\,(z^* - z_0{}^*)^2\right). \tag{15.11.64}$$

Using (15.11.64) and (15.5.39), there is obtained:

$$K = 2\,p_0 - p_0{}^2\,(z - z_0)\,(z^* - z_0{}^*). \tag{15.11.65}$$

Using (15.11.65) and (15.5.38), we find that:

$$D = \frac{1}{h_0}\left(2\,p_0\,(z - z_0)\,(z^* - z_0{}^*) - \frac{3}{4}\,p_0{}^2\,(z - z_0)^2\,(z^* - z_0{}^*)^2 + \frac{p_0{}^3}{36}\,(z - z_0)^3\,(z^* - z_0{}^*)^3\right). \tag{15.11.66}$$

From (15.5.36) we have $T(z, z^*; z_0, z_0{}^*) = \dfrac{C}{2}\{\log(z - z_0) + \log(z^* - z_0{}^*)\}$, while from (15.5.44), $\psi^{(L,\,2)}(z, z^*; z_0, z_0{}^*) = i\,l_0\displaystyle\int_{(z_R,\,z_R{}^*)}^{+D\,(z,\,z^*)}\left(\frac{\partial T}{\partial z}\,dz - \frac{\partial T}{\delta z^*}\,dz^*\right)$. Now, for an arbitrary function of two complex variables $f(z, z^*)$ we have:

$$\int_{(z_R,\,z_R{}^*)}^{(z,\,z^*)}(f_z\,dz - f_{z^*}\,dz^*) = f(z, z_R{}^*) - f(z_R, z^*).$$

Hence, selecting $z_R{}^* = z_0{}^* + 1$, $z_R = z_0 + 1$, we obtain:

$$(i\,l_0)^{-1}\,\psi^{(L,\,2)}(z, z^*; z_0, z_0{}^*) = \frac{1}{2}\,C(z, z_0{}^* + 1; z_0, z_0{}^*)\log(z - z_0) + D(z, z_0{}^* + 1, z_0, z_0{}^*) -$$

$$- \frac{1}{2}\,C(z_0 + 1, z^*; z_0, z_0{}^*)\log(z^* - z_0{}^*) - D(z_0 + 1, z^*; z_0, z_0{}^*). \tag{15.11.67}$$

Therefore from (15.5.45):

$$\psi^{(L,\,2)}(\lambda, \theta; \lambda_0, \theta_0) = Re\,[\psi^{(L,\,2)}(z, \bar{z}; z_0, \overline{z_0})] =$$

$$= -2\,l_0\,Im\left(\frac{1}{2}\,C(z, \overline{z_0} + 1; z_0, \overline{z_0})\log(z - z_0) + D(z, \overline{z_0} + 1; z_0, \overline{z_0})\right) =$$

$$= 2\,\frac{l_0}{h_0}\,Im\left[\frac{1}{2}\left(1 - b_0\,(z - z_0) + \frac{b_0{}^2}{4}\,(z - z_0)^2\right)\log(z - z_0) +$$

$$+ \left(2\,p_0\,(z - z_0) - \frac{3}{4}\,p_0{}^2\,(z - z_0)^2 + \frac{p_0{}^3}{36}\,(z - z_0)^3\right)\right]. \tag{15.11.68}$$

Thus, finally:

$$\frac{-h_0}{l_0}\,\psi^{(L,\,2)}(\lambda, \theta; \lambda_0, \theta_0) = Q_1\tan^{-1}\frac{\theta}{\lambda - \lambda_0} + Q_2\log[(\lambda - \lambda_0)^2 + \theta^2] + Q_3, \tag{15.11.69}$$

where:

$$\left.\begin{aligned}
Q_1 &= 1 - p_0\,(\lambda - \lambda_0) + \frac{p_0^2}{4}\,((\lambda - \lambda_0)^2 - \theta^2),\\
Q_2 &= \frac{-p_0}{2}\,\theta + \frac{p_0^2}{4}\,(\lambda - \lambda_0)\,\theta,\\
Q_3 &= 4\,p_0\,\theta - 3\,p_0^2\,(\lambda - \lambda_0)\,\theta + \frac{p_0^3}{18}\,[3\,(\lambda - \lambda_0)^3 - \theta^3].
\end{aligned}\right\} \qquad (15.11.70)$$

12. Computation of Particular Solutions ($\bar{\bar{P}}(z - \lambda_0)^N$). Method B

According to this method, the function $F(\lambda)$ must be approximated by a polynomial in λ over a range including λ_0. The range selected was $-.25 \leqslant \lambda \leqslant -.15$, corresponding to a Mach number range of $.615 \leqslant M \leqslant .711$. The approximation F^* to F (Cf. Eq. (15.10.57)) adopted was:

$$F^*(\lambda) = .5703 + 5.2679\,(\lambda - \lambda_0) + 90.618\,(\lambda - \lambda_0)^2 + 1855.8\,(\lambda - \lambda_0)^3; \qquad (15.12.71)$$

$$\lambda_0 = -.20.$$

The approximation (15.12.71) was obtained by taking the tabulated values of F over the range $-.25(.01)-.15$ and approximating them by linear combinations of 1, λ, λ^2, λ^3, in such a way that the sum of the squares of the resulting deviations at these points is a minimum. The values of F, F^*, and of $\delta = F - F^*$ are listed in Table II.

It is of importance to know what degree polynomial is required in order to approximate $F(\lambda)$ over various intervals. In view of the singularity of $F(\lambda)$ at $\lambda = 0$, the degree of the polynomial required to approximate $F(\lambda)$ to the same degree of accuracy increases rapidly as either (a) the interval of approximation nears $\lambda = 0$, (b) the interval of approximation increases in length. In order to obtain a concrete idea of what can be accomplished by way of polynomial approximation, the reader may consult Table II where two additional approximations have been listed, one by a polynomial of degree 4 over the interval $-.40 < \lambda < -.15$ and one by a polynomial of degree 10 over the interval $-1.04 < \lambda < -.13$. The discrepancies alone have been listed. These computations were carried out on SEAC using an orthonormalizing routine which is explained in § 14. See also Davis and Rabinowitz[2].

The following observations may be of some importance if it is desired to approximate F over a large interval in which both small values and large values of F occur. The use of the approximating method described above will in such a case, lead to discrepancies of unusually high percentage at the low values of F. In order to equalize the percent discrepancy over the whole range one may have recourse to the following device. In forming the least square approximation weight each value of $F(\lambda)$ in proportion to $\left(\dfrac{1}{F(\lambda)}\right)^2$. With such a precaution, the percentage discrepancy will be equalized over the whole range. The resulting large spread in the weights means that overflow will be hard to avoid in the least square routines (Cf. § 14) and floating point routines should be employed.

The adoption of (15.12.71) leads to the values $a_0 = .5703$, $a_1 = 5.2679$, $a_2 = 90.618$, $a_3 = 1855.8$ which were used in (15.10.61) and (15.10.62). The following particular solutions were then computed from (15.10.61) and are available at the Computation Laboratory, National Bureau of

[2] P. Davis and P. Rabinowitz: A multiple purpose orthonormalizing code and its uses. U. S. National Bureau of Standards, Rep. No. 3149, March, 1954. — Some SEAC computations of subsonic Fluid Flows by Bergman's method of integral operators., Ibid, Rep. No. 3313,1954. Also: J. Assoc. Comput. Machinery 1, 183—191 (4, Oct. 1954).

Standards, Washington: $Re\,\psi_k^*$ and $Im\,\psi_k^*$, $k = 0, 1, \ldots, 5$. The λ range is $-.25(.01) -.15$ and the θ range is $0(.01).20$. This yielded a total of 11 particular solutions. (The trivial zero solution also occurs.) The total machine time involved exclusive of that used in code checking was approximately $^1/_2$ hour. Over this range, the percentage truncation error is less than 3×10^{-5} (Cf. Table A).

These solutions were computed on SEAC by using an interpretive routine for complex operations. First the coefficients of the polynomials (15.10.61) were computed for a given value of N, starting with $N = 0$, by means of (15.10.62). Then, for fixed θ, starting with $\theta = 0$, Q_N was computed for $Re\,(w^N)$ varying at intervals of .005 between $-.025$ and $+.025$. After Q_N was computed $|Q_N|^2 = Q_N \overline{Q}_N$ was computed and stored and $w^N Q_N = \psi^*_N$ was computed. This process was then continued for θ varying at intervals of .01 until the memory was filled. Then the results were converted to decimals and printed on wire. This continued until θ reached .2. The whole process was repeated for $N = 1, 2, \ldots, 5$.

The code computes the functions ψ_N^* for varying values of $A_N, \ldots, K_N$ and for λ, θ ranging over the interval $-\lambda_1 \leqslant \lambda \leqslant \lambda_1$, $0 \leqslant \theta \leqslant \theta_1$ where λ_1, θ_1 are arbitrary, provided only that two mesh lengths are equal ($\Delta\lambda = \Delta\theta$). This latter quantity which, in the present computation was taken as .005, can also be modified at will.

In computing w^N with $Re\,w^n = -.025(.005).025$ and $Im\,w^N = 0(.005).1$ with a fixed point routine, it turns out that for large N, much significance is lost since these numbers are in the neighborhood of 10^{-2N}. One method of obtaining more significance is by computing $W^N = (10\,w)^N$.

13. Computation of Singular Solutions. Method B

These were computed according to the method outlined in § 11. The function:

$$\chi\,(\lambda, \theta) = -\frac{H\,(\lambda_1)}{l^{-1/2}\,(\lambda_1)}\,\psi^{(L,\,2)}\,(\lambda,\,0\,;\,\lambda_1,\,\theta_1) - \frac{2\,H\,(\lambda_2)}{l^{-1/2}\,(\lambda_2)}\,\psi^{(L,\,2)}\,(\lambda,\,\theta\,;\,\lambda_2,\,\theta_2),$$

is required, and was computed using (15.11.69). The values $\lambda_1 = -.20$, $\theta_1 = 0$, $p_0 = -.63031$; $\lambda_2 = -.15$, $\theta_2 = 0$, $p_0 = -1.27396$ were employed. This computation is simple enough to be carried out by hand. In any future work where it is advisable to utilize more terms, routines for automatic computation similar to that described in § 12 will be necessary. The difficult portion of any extended computation of the singular solutions will be the algebraic work which must precede the actual coding. The utility of the present low-order approximation lies in the fact that with a minimum of work, it provides us with a boundary value problem of the appropriate type to deal with.

14. Method for the Solution of the Boundary Value Problem in the Pseudologarithmic Plane

The pseudograph with which we have dealt numerically has been depicted in Figure 1. We designate its complete contour by b. According to the discussion of § 4, we must obtain that solution of (15.2.14) which on b takes on the values of $\psi^*\,(\lambda_b, \theta_b) - \chi^*\,(\lambda_b, \theta_b)$ where $\psi^*\,(\lambda_b, \theta_b)$ is defined by (15.4.34). We accomplish this numerically in the following way. Designate the total length of the contour b by L and let s be the length parameter which runs along b starting from some fixed point. Let $f_1\,(s)$, $f_2\,(s)$, $\ldots, f_N\,(s)$ designate the values on b of the particular solutions

$\psi_1^*, \psi_2^*, \ldots, \psi_N^*$, and let $g(s) = \psi^*(\lambda_b, \theta_b) - \chi^*(\lambda_b, \theta_b)$ where s is the length parameter at the point (λ_b, θ_b) of b. We now seek that linear combination $\sum_{k=1}^{N} c_k f_k$ such that:

$$\int_0^L [g(s) - \sum_{k=1}^{N} c_k f_k(s)]^2 \, ds = \text{minimum}. \qquad (15.14.72)$$

The approximate solution to the boundary value problem is then given by $\sum_{k=1}^{N} c_k \psi_k^*(\lambda, \theta)$, and, according to (15.4.33), the approximate value of the required stream function is given by:

$$\psi^*(\lambda, \theta) = \chi^*(\lambda, \theta) + \sum_{k=1}^{N} c_k \psi_k^*(\lambda, \theta). \qquad (15.14.73)$$

The numerical solution of the problem (15.14.72) was accomplished as follows. See also Davis and Rabinowitz[2]. At the outset, the integral in (15.14.72) is replaced by an appropriate numerical rule for integration. A fixed set of abscissas $s_1, \ldots, s_q$ are introduced and the values $f(s_1), \ldots, f(s_q)$ are regarded as a q-dimensional vector, f. A code was then written which orthonormalizes a set of vectors $\{f_j\}$ with respect to an inner product of the form:

$$(f, h) = \sum_{i=1}^{q} w_i f(s_i) h(s_i), \qquad (15.14.74)$$

where the weights have been chosen so that:

$$\sum_{i=1}^{q} w_i f(s_i) h(s_i) \approx \int_0^L f(s) h(s) \, ds, \qquad (15.14.75)$$

is sufficiently accurate for the functions in question. This orthonormalization process was carried out by the method of Gram-Schmidt. Having been given a set $f_1, \ldots, f_N$ of linearly independent vectors, a set of linear combinations:

$$\varphi_1 = a_{11} f_1,$$

$$\varphi_2 = a_{12} f_1 + a_{22} f_2, \qquad (15.14.76)$$

$$\varphi_3 = a_{13} f_1 + a_{23} f_2 + a_{33} f_3,$$

is produced which is orthonormal with respect to the inner product (15.14.74), i. e.,

$$(\varphi_i, \varphi_j) = \delta_{ij}. \qquad (15.14.77)$$

The vectors φ_i may be defined by the following set of recursive formulas which are convenient for machine computation:

$$\left. \begin{aligned} D_1 &= (f_1, f_1)^{1/2}, \\ c_{11} &= 1/D_1, \\ \varphi_1 &= c_{11} f; \end{aligned} \right\} \qquad (15.14.78\,a)$$

$$\left. \begin{aligned} D_2 &= ((f_2, f_2) - (f_2, \varphi_1)^2)^{1/2}, \\ c_{12} &= -(f_2, \varphi_1)/D_2, \\ c_{22} &= 1/D_2, \\ \varphi_2 &= c_{12} \varphi_1 + c_{22} f_2, \end{aligned} \right\} \qquad (15.14.78\,b)$$

and, in general:

$$
\left.
\begin{aligned}
D_k &= \{(f_k, f_k) - (f_k, \varphi_1)^2 - (f_k, \varphi_2)^2 - \ldots - (f_k, \varphi_{k-1})^2\}^{1/2}, \\
c_{1k} &= - (f_k, \varphi_1)/D_k, \\
c_{2k} &= - (f_k, \varphi_2)/D_k, \\
&\qquad \cdot \\
&\qquad \cdot \\
&\qquad \cdot \\
c_{k-1,\,k} &= - (f_k, \varphi_{k-1})/D_k, \\
c_{kk} &= 1/D_k, \\
\varphi_k &= c_{1k}\varphi_1 + c_{2k}\varphi_2 + \ldots + c_{k-1}\varphi_{k-1} + c_{kk}f_k.
\end{aligned}
\right\} \tag{15.14.78c}
$$

The code then expands an arbitrary vector f into an orthonormal series of vectors φ_k:

$$
f \sim \sum_{k=1}^{N} (f, \varphi_k)\, \varphi_k, \tag{15.14.79}
$$

which is automatically the best approximation to f in the sense of least squares. If N is the number of functions to be orthonormalized and if q is the number of points used in the quadrature formula (15.14.75), then the present code can handle problems up to the inequality $(q + N)(N + 2) \leqslant 800$.

Table 4 summarizes the data pertaining to the boundary value problem. The contour b in the pseudologarithmic (λ, θ) plane was selected so as to consist of the following segments (Cf. Fig. 3): $(-.25, .00)$ to $(-.15, .00)$, $(-.15, .00)$ to $(-.15, .10)$, $(-.15, .10)$ to $(-.20, .10)$, $(-.20, .10)$ to $(-.20, .20)$, $(-.20, .20)$

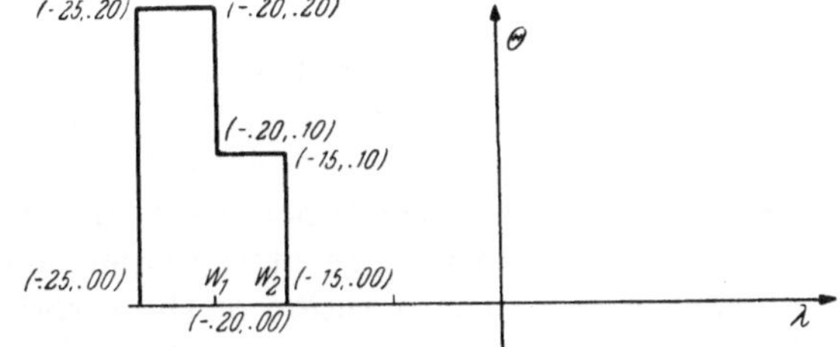

Fig. 3. Contour Selected for Solution of Boundary Value Problem

to $(-.25, .20)$, $(-.25, .20)$ to $(-.25, .00)$. The first column of Table 4 lists the thirty lattice points along this contour which were utilized in the process of numerical integration. The next columns list respectively $-\chi$, ψ, $\psi - \chi$, and $\psi^* - \chi^* = g(s)$, the latter two columns being, by construction, continuous.

Table 5 lists the functions $f_1, \ldots f_9$ which are the values along b of the particular solutions $Re\,\psi_N{}^*$ and $Im\,\psi_N{}^*$. These are followed in turn, by the discrepancies:

$$
\delta = g - \sum_{k=1}^{9} c_k f_k, \tag{15.14.80}
$$

and the coefficients c_k for which (15.14.72) is fulfilled. Only 9 of the 11 particular solutions computed were utilized inasmuch as the last two had been computed to too few significant figures. The weights w_i in (15.14.74) were selected so as to correspond to at least 3rd order integration rules. It is seen from the values of δ listed in Table 5 that:

$$
\max \delta/g = .016, \tag{15.14.81}
$$

and so about 2 percent accuracy has been achieved using the 9 particular solutions. The machine time consumed in this aspect of the boundary value problem was approximately 7 minutes.

15. Sources of Error in Method B

We now discuss the major sources of errors which are incurred in using Method B for the solution of the boundary value problem for flows of type S. There are four principal sources of error as follows: (1) The function $F(\lambda)$ has been approximated by the polynomial $F^*(\lambda)$. (2) The particular solution ψ_N^* have not been obtained exactly, but have been computed from the truncated series (15.3.26). (3) In the computation of $\psi^{*(L, 2)}$, the function $P(\lambda)$ has been approximated, and the series (15.5.37), (15.5.38), (15.5.39) have been truncated. (4) The boundary value problem has, by criterion (15.14.72), been solved only approximately. Each of these errors is independent of the others, and each may be made arbitrarily small by utilizing polynomials of sufficiently high degree and by the retention of sufficiently many terms of the infinite series mentioned above.

The function $F(\lambda)$ embodies (in the pseudologarithmic variables) the equation of state (15.2.1). Conversely, having been given an approximate $F^*(\lambda)$, a corresponding $p(\varrho)$ [or $q(\varrho)$, $M(\varrho)$, $l(\varrho)$] relationship may be determined. The use of an approximation $F^*(\lambda)$ corresponds therefore, to the assumption of a fictitious gas law for the fluid. The $p(\varrho)$ relationship [as well as $q(\varrho)$, $M(\varrho)$, $l(\varrho)$] may be determined from $F^*(\lambda)$ in a fairly straight forward manner using (15.2.9). The reader is referred to von Mises-Schiffer where such relationships have been worked out, for the approximation $F^*(\lambda) = c/\lambda^2$ as well as for the Kármán-Tsien approximation.

The method for estimating truncation errors has been indicated in § 8. A modification of this method can be employed to estimate such errors in the computation of $\psi^{(L, 2)}$. This has not been carried out in the present report. However, since the behavior of the functions F and P is (roughly) the same, one would expect the truncation errors to be of the same magnitude as those for ψ_N^*.

There remains the question of the errors introduced by the deviation of δ from 0 [Eq. (15.14.80)]. In view of the fact that the function $F(\lambda)$ is positive, one does not posses a maximum principle for the differential equation (15.2.14) so that to give an upper bound for this error is a problem of considerable theoretical and numerical difficulty.

16. Concluding Remarks

The method of integral operators for generating particular solutions is most readily adapted to high speed computation when combined with the device of polynomial approximation (Method B). This facilitates the computation tremendously and is especially convenient for any subsequent transformations to the physical plane. It is anticipated also that if computations were carried out over multiply-covered domains (leading to flows around closed bodies) that this method would be best.

17. Tables

Table I. Gas Dynamical Quantities

$-\lambda$	T	q^2	ϱ^2	$\dfrac{1}{\sqrt{\bar{t}}}$	F	$-P$	H
1.04	.96828	.061656	.93985	1.00121	.(2)28260	.(2)28293	1.00061
1.03	.96760	.062943	.93862	1.00126	.(2)29589	.(2)29625	1.00063
1.02	.96690	.064258	.93737	1.00132	.(2)30985	.(2)31025	1.00066
1.01	.96619	.065602	.93610	1.00138	.(2)32452	.(2)32495	1.00069
1.00	.96546	.066974	.93480	1.00144	.(2)33993	.(2)34041	1.00072
.99	.96472	.068377	.93347	1.00150	.(2)35613	.(2)35665	1.00075
.98	.96395	.069810	.93211	1.00157	.(2)37316	.(2)37372	1.00078
.97	.96317	.071274	.93073	1.00163	.(2)39106	.(2)39167	1.00082
.96	.96237	.072770	.92932	1.00171	.(2)40988	.(2)41055	1.00085
.95	.96155	.074299	.92788	1.00178	.(2)42968	.(2)43042	1.00089
.94	.96071	.075861	.92641	1.00186	.(2)45052	.(2)45132	1.00093
.93	.95936	.077458	.92491	1.00194	.(2)47245	.(2)47333	1.00097
.92	.95898	.079089	.92337	1.00203	.(2)49553	.(2)49650	1.00101
.91	.95808	.080757	.92181	1.00212	.(2)51984	.(2)52089	1.00106
.90	.95716	.082461	.92021	1.00221	.(2)54544	.(2)54660	1.00111
.89	.95622	.084205	.91859	1.00231	.(2)57242	.(2)57368	1.00115
.88	.95525	.085984	.91692	1.00241	.(2)60085	.(2)60223	1.00121
.87	.95427	.087804	.91523	1.00252	.(2)63082	.(2)63234	1.00126
.86	.95326	.089665	.91349	1.00264	.(2)66243	.(2)66409	1.00132
.85	.95222	.091567	.91173	1.00275	.(2)69577	.(2)69760	1.00138
.84	.95116	.093511	.90992	1.00288	.(2)73095	.(2)73296	1.00144
.83	.95007	.095500	.90808	1.00301	.(2)76809	.(2)77029	1.00150
.82	.94896	.097532	.90620	1.00314	.(2)80731	.(2)80972	1.00157
.81	.94782	.099611	.90428	1.00329	.(2)84874	.(2)85139	1.00164
.80	.94665	.10174	.90232	1.00344	.(2)89252	.(2)89543	1.00172
.79	.94545	.10391	.90032	1.00359	.(2)93880	.(2)94199	1.00180
.78	.94423	.10613	.89828	1.00376	.(2)98774	.(2)99125	1.00188
.77	.94297	.10841	.89619	1.00393	.010395	.010434	1.00196
.76	.94168	.11073	.89406	1.00411	.010943	.010986	1.00205
.75	.94036	.11311	.89189	1.00430	.011524	.011570	1.00215
.74	.93900	.11554	.88967	1.00450	.012138	.012190	1.00225
.73	.93761	.11803	.88741	1.00471	.012790	.012846	1.00235
.72	.93618	.12058	.88510	1.00493	.013481	.013543	1.00246
.71	.93472	.12319	.88274	1.00516	.014214	.014282	1.00257
.70	.93322	.12585	.88033	1.00540	.014991	.015067	1.00270
.69	.93168	.12858	.87786	1.00565	.015817	.015901	1.00282
.68	.93010	.13137	.87535	1.00592	.016695	.016787	1.00295
.67	.92847	.13423	.87279	1.00620	.017628	.017730	1.00309
.66	.92681	.13716	.87016	1.00649	.018620	.018733	1.00324
.65	.92510	.14015	.86749	1.00680	.019677	.019801	1.00340
.64	.92334	.14322	.86476	1.00713	.020801	.020939	1.00356
.63	.92154	.14636	.86196	1.00747	.022000	.022152	1.00373
.62	.91968	.14957	.85911	1.00783	.023278	.023446	1.00391
.61	.91778	.15286	.85620	1.00821	.024642	.024828	1.00410
.60	.91582	.15623	.85323	1.00861	.026098	.026305	1.00430
.59	.91381	.15969	.85019	1.00903	.027655	.027883	1.00450
.58	.91174	.16322	.84709	1.00947	.029319	.029573	1.00473
.57	.90961	.16685	.84392	1.00994	.031100	.031382	1.00496
.56	.90742	.17056	.84068	1.01043	.033008	.033321	1.00520
.55	.90517	.17437	.83738	1.01095	.035053	.035402	1.00546
.54	.90285	.17827	.83400	1.01150	.037238	.037636	1.00573
.53	.90046	.18227	.83055	1.01208	.039605	.040037	1.00602
.52	.89801	.18637	.82702	1.01269	.042138	.042620	1.00633
.51	.89548	.19057	.82341	1.01334	.044865	.045403	1.00665

$-\lambda$	T	q^2	ϱ^2	$\dfrac{1}{\sqrt{l}}$	F	$-P$	H
.50	.89287	.19488	.81973	1.01402	.047801	.048402	1.00699
.49	.89018	.19930	.81597	1.01475	.050968	.051640	1.00735
.48	.88741	.20384	.81212	1.01551	.054387	.055139	1.00773
.47	.88455	.20849	.80819	1.01632	.058082	.058924	1.00813
.46	.88161	.21327	.80417	1.01718	.062081	.063025	1.00855
.45	.87857	.21817	.80006	1.01809	.066414	.067474	1.00900
.44	.87543	.22320	.79586	1.01906	.071116	.072307	1.00948
.43	.87218	.22837	.79156	1.02008	.076225	.077565	1.00999
.42	.86883	.23368	.78717	1.02117	.081784	.083295	1.01053
.41	.86537	.23913	.78268	1.02233	.087843	.089548	1.01110
.40	.86179	.24473	.77808	1.02356	.094457	.096384	1.01171
.39	.85808	.25049	.77338	1.02487	.10169	.10387	1.01236
.38	.85424	.25641	.76857	1.02627	.10962	.11209	1.01305
.37	.85027	.26250	.76365	1.02776	.11831	.12112	1.01379
.36	.84615	.26876	.75862	1.02935	.12788	.13107	1.01457
.35	.84188	.27521	.75347	1.03105	.13843	.14206	1.01541
.34	.83745	.28184	.74819	1.03287	.15008	.15423	1.01630
.33	.83285	.28867	.74279	1.03482	.16298	.16773	1.01726
.32	.82807	.29571	.73726	1.03691	.17731	.18275	1.01829
.31	.82311	.30296	.73160	1.03916	.19326	.19952	1.01939
.30	.81794	.31043	.72580	1.04157	.21106	.21828	1.02057
.29	.81255	.31814	.71985	1.04417	.23101	.23934	1.02184
.28	.80694	.32610	.71375	1.04697	.25342	.26308	1.02321
.27	.80108	.33431	.70750	1.04999	.27870	.28992	1.02469
.26	.79497	.34279	.70109	1.05327	.30732	.32040	1.02629
.25	.78857	.35156	.69452	1.05682	.33986	.35516	1.02802
.24	.78188	.36063	.68777	1.06068	.37701	.39499	1.02989
.23	.77486	.37002	.68084	1.06488	.41965	.44085	1.03193
.22	.76749	.37974	.67372	1.06946	.46884	.49396	1.03415
.21	.75975	.38982	.66641	1.07448	.52591	.55582	1.03657
.20	.75160	.40027	.65889	1.07999	.59257	.62835	1.03923
.19	.74299	.41113	.65115	1.08606	.67095	.71403	1.04214
.18	.73390	.42242	.64318	1.09278	.76383	.81605	1.04536
.17	.72426	.43416	.63497	1.10022	.87488	.93860	1.04892
.16	.71402	.44641	.62650	1.10853	1.00890	1.08731	1.05287
.15	.70312	.45918	.61776	1.11784	1.17245	1.26981	1.05728
.14	.69147	.47255	.60872	1.12833	1.37450	1.49664	1.06223
.13	.67898	.48654	.59937	1.14023	1.62767	1.78271	1.06782

Table II. Approximation of $F(\lambda)$

$-\lambda$	$F(\lambda)$	$F(\lambda)-F_3^*(\lambda)$	$F(\lambda)-F_4^*(\lambda)$	$F(\lambda)-F_{10}^*(\lambda)$
1.03	.00283			.010
1.01	.00325			.006
.99	.00356			.000
.97	.00391			— .005
.95	.00430			— .009
.93	.00472			— .011
.91	.00520			— .011
.89	.00572			— .010
.87	.00631			— .007
.85	.00696			— .003
.83	.00768			.002
.81	.00849			.006
.79	.00939			.010
.77	.01039			.013
.75	.01152			.015

$-\lambda$	$F(\lambda)$	$F(\lambda)-F_3^*(\lambda)$	$F(\lambda)-F_4^*(\lambda)$	$F(\lambda)-F_{10}^*(\lambda)$
.73	.01279			.015
.71	.01421			.014
.69	.01582			.012
.67	.01763			.009
.65	.01968			.004
.63	.02200			— .001
.61	.02464			— .006
.59	.02766			— .011
.57	.03110			— .016
.55	.03505			— .019
.53	.03960			— .022
.51	.04487			— .022
.49	.05097			— .021
.47	.05808			— .017
.45	.06641			— .012
.43	.07622			— .004
.41	.08784			.004
.40	.09446		.0018	
.39	.10169		.0001	.013
.38	.10962		— .0005	
.37	.11831		— .0006	.002
.36	.12788		— .0005	
.35	.13843		— .0002	.030
.34	.15008		.0001	
.33	.16298		.0003	.035
.32	.17731		.0004	
.31	.19316		.0004	.036
.30	.21106		.0003	
.29	.23101		.0002	.032
.28	.25342		.0001	
.27	.27870		— .0001	.023
.26	.30732		— .0002	
.25	.33986	.038	— .0003	.006
.24	.37701	— .009	— .0003	
.23	.41965	— .024	— .0003	— .016
.22	.46884	— .018	— .0002	
.21	.52591	.001	.0001	— .042
.20	.59257	.022	.0001	
.19	.67095	.037	.0002	— .064
.18	.76383	.037	.0002	
.17	.87488	.015	.0002	— .070
.16	1.00890	— .036	.0000	
.15	1.17245	— .120	— .0005	— .030
.13	1.62767			.110

Table III. $-\chi(\lambda, \theta)$

$\theta \backslash \lambda$	—.25	—.24	—.23	—.22	—.21	—.20
.20	— .120405	— .137008	— .156291	— .178359	— .203283	— .231090
.19	— .010085	— .025775	— .044407	— .066121	— .091027	— .119169
.18	+ .103006	+ .088404	+ .070569	+ .049312	+ .024481	.004010
.17	.219105	.205816	.188979	.168329	.143660	+ .114831
.16	.338463	.326784	.311206	.291380	.267013	.237892
.15	.461344	.451651	.437690	.418989	.395138	.365821
.14	.588009	.580793	.568921	.551767	.528756	.499419
.13	.718702	.714588	.705438	.690421	.668728	.639671
.12	.853621	.853411	.847829	.835767	.816102	.787816
.11	.992884	.997590	.996692	.988734	.972140	.945411

θ \ λ	—.25	—.24	—.23	—.22	—.21	—.20
.10	1.136445	1.147332	1.152602	1.150341	1.138362	1.114440
.09	1.284014	1.302629	1.315996	1.321655	1.316576	1.297453
.08	1.434909	1.463063	1.486998	1.503651	1.508873	1.497732
.07	1.587863	1.627541	1.665088	1.696932	1.717523	1.719505
.06	1.740802	1.793884	1.848534	1.901137	1.944640	1.968146
.05	1.890634	1.958282	2.033436	2.113743	2.191290	2.250244
.04	2.033214	2.114756	2.212255	2.327678	2.455261	2.573264
.03	2.163816	2.255063	2.371989	2.526619	2.725362	2.944127
.02	2.278591	2.370323	2.493653	2.676656	2.964807	3.365871
.01	2.377138	2.456184	2.559503	2.720956	3.052118	3.832029
0	2.464862	2.520543	2.576672	2.633246	2.690270	2.747739/ 4.318536

θ \ λ	—.19	—.18	—.17	—.16	—.15
.20	— .261753	— .295189	— .331253	— .369749	— .410418
.19	— .150539	— .185050	— .222543	— .262789	— .305486
.18	— .036168	— .071914	— .111070	— .153368	— .198450
.17	+ .081804	+ .044649	+ .003559	— .041150	— .089048
.16	.203925	.165166	.121823	.074273	+ .023037
.15	.330868	.290293	.244322	.193404	.138178
.14	.463467	.420851	.371805	.316866	.256836
.13	.602779	.557889	.505231	.445452	.379583
.12	.750150	.702763	.645844	.580186	.507121
.11	.907339	.857266	.795298	.722393	.640339
.10	1.076656	1.023818	.955822	.873861	.780369
.09	1.261217	1.205749	1.030522	1.037019	.928657
.08	1.465289	1.407781	1.323840	1.215279	1.087074
.07	1.694834	1.636804	1.542333	1.413572	1.258044
.06	1.958363	1.903183	1.796064	1.639294	1.444680
.05	2.268246	2.223056	2.101131	1.904057	1.650906
.04	2.642853	2.622274	2.484469	2.227305	1.881400
.03	3.110451	3.142806	2.992839	2.645156	2.141146
.02	3.719827	3.849773	3.707277	3.237892	2.434351
.01	4.586246	4.820702	4.744194	4.238447	2.763412
0	5.967084	6.045344	6.124114	6.203394	6.283185/3.141593

Table IV

λ	θ	$-\chi(\lambda, \theta)$	$\psi(\lambda, \theta)$	$\psi - \chi$	$\psi^* - \chi^*$
— .25	.00	2.464862	π	5.606455	5.453644
— .23	.00	2.576672	π	5.718265	5.541330
— .21	.00	2.690270	π	5.831863	5.626116
— .19	.00	5.967084	0.000000	5.967084	5.725799
— .17	.00	6.124114	0.000000	6.124114	5.838495
— .15	.00	6.283185	0.000000	6.283185	5.942782
— .15	.02	2.434351	π	5.575944	5.273857
— .15	.04	1.881400	↑	5.022993	4.750864
— .15	.06	1.444680		4.586273	4.337804
— .15	.08	1.087074		4.228667	3.999572
— .15	.10	.780369		3.921962	3.709483
— .17	.10	.955822		4.097415	3.906318
— .19	.10	1.076656		4.218249	4.047630
— .20	.11	.945411		4.087004	3.932723
— .20	.13	.639671		3.781264	3.638525
— .20	.15	.365821		3.507414	3.375012
— .20	.17	+ .004831	↓	3.256424	3.133497
— .20	.19	— .119169	π	3.022424	2.908330

λ	θ	$-\chi(\lambda, \theta)$	$\psi(\lambda, \theta)$	$\psi - \chi$	$\psi^* - \chi^*$
— .21	.20	.203283	π	2.938310	2.834647
— .23	.20	— .156291		2.985302	2.892931
— .25	.20	— .120405		3.021188	2.938842
— .25	.18	+ .103006		3.244599	3.156163
— .25	.16	.228463		3.480056	3.385203
— .25	.14	.588009		3.729602	3.627947
— .25	.12	.853621		3.995214	3.886319
— .25	.10	1.136445		4.278038	4.161435
— .25	.08	1.434909		4.576502	4.451764
— .25	.06	1.740802		4.882395	4.749319
— .25	.04	2.033214		5.174807	5.033761
— .25	.02	2.278591	π	5.420184	5.272450

Table V. Solution of Boundary Value Problem

λ	θ	f_1	f_2	f_3	f_4	f_5
— .25	0	0.99884796	— 0.02498617	0.00000000	0.00062478	0.00000000
— .23	0	0.99954605	— 0.01499663	0.00000000	0.00022497	0.00000000
— .21	0	0.99994659	— 0.00499987	0.00000000	0.00002500	0.00000000
— .19	0	0.99993896	0.00499985	0.00000000	0.00002500	0.00000000
— .17	0	0.99938583	0.01499525	0.00000000	0.00022495	0.00000000
— .15	0	0.99797058	0.02497282	0.00000000	0.00062453	0.00000000
— .15	.02	0.99765778	0.02496897	0.00998659	0.00052456	0.00049955
— .15	.04	0.99673462	0.02495815	0.01996277	0.00022482	0.00099877
— .15	.06	0.99525833	0.02494232	0.02991892	— 0.00027421	0.00149744
— .15	.08	0.99332428	0.02492510	0.03984756	— 0.00097183	0.00199547
— .15	.10	0.99105453	0.02491016	0.04974451	— 0.00186722	0.00249300
— .17	.10	0.99335098	0.01496536	0.04982286	— 0.00226922	0.00149735
— .19	.10	0.99444199	0.00499535	0.04985907	— 0.00247031	0.00049954
— .20	.11	0.99365234	0.00000923	0.05482544	— 0.00301865	0.00000051
— .20	.13	0.99144363	0.00001265	0.06472179	— 0.00421311	0.00000082
— .20	.15	0.98908233	0.00001144	0.07459059	— 0.00560489	0.00000086
— .20	.17	0.98667526	0.00000162	0.08443354	— 0.00719366	— 0.00000014
— .20	.19	0.98434067	0.00003878	0.09425600	— 0.00897935	— 0.00000368
— .21	.20	0.98229980	— 0.00505308	0.09911127	— 0.00991667	— 0.00100411
— .23	.20	0.97980881	— 0.01496931	0.09897369	— 0.00970586	— 0.00299171
— .25	.20	0.97696304	— 0.02481947	0.09883041	— 0.00929782	— 0.00497082
— .25	.18	0.98162079	— 0.02481005	0.08917372	— 0.00742793	— 0.00447539
— .25	.16	0.98558807	— 0.02482777	0.07943277	— 0.00574766	— 0.00398129
— .25	.14	0.98893356	— 0.02485775	0.06962446	— 0.00426013	— 0.00348692
— .25	.12	0.99171448	— 0.02489046	0.05976225	— 0.00296758	— 0.00299154
— .25	.10	0.99398422	— 0.02492056	0.04985781	— 0.00187177	— 0.00249494
— .25	.08	0.99578095	— 0.02494535	0.03992105	— 0.00097391	— 0.00199724
— .25	.06	0.99714279	— 0.02496395	0.02996023	— 0.00027485	— 0.00149864
— .25	.04	0.99809647	— 0.02497667	0.01998243	0.00022484	— 0.00099941
— .25	.02	0.99866104	— 0.02498386	0.00999388	0.00052477	— 0.00049980

λ	θ	f_6	f_7	f_8	f_9	$\delta = g - \sum\limits_{k=1}^{9} c_k f_k$
— .25	0	— 0.00001562	0.00000000	0.00000039	0.00000000	— 0.04279779
— .23	0	— 0.00000337	0.00000000	0.00000005	0.00000000	0.01926395
— .21	0	— 0.00000013	0.00000000	0.00000000	0.00000000	0.01833401
— .19	0	0.00000013	0.00000000	0.00000000	0.00000000	0.00036980
— .17	0	0.00000337	0.00000000	0.00000005	0.00000000	— 0.00305759
— .15	0	0.00001562	0.00000000	0.00000039	0.00000000	0.02706501
— .15	.02	0.00000812	0.00001774	0.00000003	0.00000052	— 0.02898842

Continued on page 172

λ	θ	f_6	f_7	f_8	f_9	$\delta = g - \sum\limits_{k=1}^{9} c_k f_k$
— .15	.04	— 0.00001436	0.00002947	— 0.00000095	0.00000045	— 0.02485462
— .15	.06	— 0.00005180	0.00002922	— 0.00000217	— 0.00000082	0.00021309
— .15	.08	— 0.00010418	0.00001102	— 0.00000305	— 0.00000389	0.01211302
— .15	.10	— 0.00017147	— 0.00003107	— 0.00000273	— 0.00000935	— 0.00839962
— .17	.10	— 0.00010897	— 0.00009106	0.00000292	— 0.00000682	0.02221595
— .19	.10	— 0.00003734	— 0.00012108	0.00000587	— 0.00000247	0.05089721
— .20	.11	— 0.00000002	— 0.00016612	0.00000914	0.00000000	0.02674063
— .20	.13	— 0.00000005	— 0.00027405	0.00001782	0.00000000	— 0.01888320
— .20	.15	— 0.00000006	— 0.00042076	0.00003158	0.00000000	— 0.04697001
— .20	.17	0.00000002	— 0.00061218	0.00005207	0.00000000	— 0.05169445
— .20	.19	0.00000033	— 0.00085423	0.00008122	0.00000003	— 0.01894728
— .21	.20	0.00015016	— 0.00098819	0.00009816	0.00001997	0.04186231
— .23	.20	0.00044542	— 0.00092732	0.00008614	0.00005850	0.03835067
— .25	.20	0.00073079	— 0.00080709	0.00006252	0.00009335	— 0.01378577
— .25	.18	0.00058933	— 0.00055748	0.00003547	0.00006703	— 0.01712891
— .25	.16	0.00046273	— 0.00036071	0.00001730	0.00004607	— 0.00905787
— .25	.14	0.00035090	— 0.00021122	0.00000602	0.00002986	0.00037098
— .25	.12	0.00025385	— 0.00010333	— 0.00000015	0.00001782	0.00576043
— .25	.10	0.00017163	— 0.00003123	— 0.00000273	0.00000936	0.00743948
— .25	.08	0.00010427	0.00001098	— 0.00000305	0.00000390	0.00730900
— .25	.06	0.00005184	0.00002923	— 0.00000217	0.00000082	0.00795910
— .25	.04	0.00001437	0.00002949	— 0.00000095	— 0.00000045	0.00315581
— .25	.02	— 0.00000812	0.00001774	0.00000003	— 0.00000052	— 0.01849483

k	$- c_k$
1	— 5.68
2	— 11.91
3	33.93
4	— 95.29
5	508.89
6	5467.98
7	3812.95
8	22826.48
9	— 18945.97

Final Remarks*

The first manuscript of the present book was completed in 1951. Since then various developments have taken place and some new methods have appeared. The book on transonic flow by Guderley[1], the book on Southwell's method by Shaw[2], the aerodynamic encyclopedia published by Princeton[3], and its counterpart in Great Britain[4], etc., should be mentioned in any bibliography. There are many other methods available at the present time, both exact and approximate, as well as reviews and résumés of the available methods. But strange as it may seem, these books and encyclopedias neither accurately describe nor properly use Bergman's method. This may be partly due to the fact that an acquaintance with this method requires slightly more mathematical knowledge then is usually assumed. Thus the present book may well serve the purpose of making the theory of compressible flow more comprehensible from a physical point of view by the use of the appropriate mathematical techniques.

This book would be incomplete without some general comments on the value of mathematical methods (like the one developed by Bergman) in both theoretical and applied mechanics. There are voices here and there that such methods are highly theoretical and too mathematical. Thus, obviously, some remarks are in order. Before the 19th Century a mathematician, a physicist, and even an engineer were not necessarily distinct individuals. A scientist was engaged in all these activities. The 19th Century produced the specialist, the twentieth century more so. Many fluid-dynamists maintain today that fluid dynamics belong to mathematical physics. But they are fundamentally wrong. We distinguish today a few fields which are becoming increasingly independent—i. e., theoretical physics, mathematical physics, physico-mathematics, applied mathematics, experimental physics, engineering sciences, etc. The methods used in mathematical physics are different from those of physico-mathematics or applied mathematics. The mathematical physicist is satisfied with the derivation of fundamental equations and the proposal of some formal methods of solution. Such formal procedures may be meaningless to mathematicians especially when the existence of a nonexisting solution is asserted or when a divergent series is used. An applied mathematician attempts to go further towards constructing the mathematical proofs for the existence of physical flow itself, and of the solutions of the equations governing the flow. A branch of theoretical and/or applied mechanics,

---

* The author wishes to express his deepest thanks to Dr. E. F. Trombley, Cincinnati, Ohio, U. S. A., for many valuable comments and remarks referring to this section.

[1] K. Guderley: Theorie schallnaher Strömungen. Berlin: Springer-Verlag. 1957.

[2] F. S. Shaw: An Introduction to Relaxation Methods. New York: Dover Publications.

[3] High Speed Aerodynamics and Jet Propulsion, Volume I to XII. Princeton, New Jersey: Princeton University Press. 1951.

[4] Leslie Howarth, Editor: Modern Developments in Fluid Dynamics. High Speed Flow. Oxford: Clarendon Press. 1953.

like the theory of elasticity or fluid dynamics actually should be treated in all those fields mentioned above, where the choice depends on the kind of analytical tool required. There is absolutely no justification in maintaining that one field is superior to another; rather, all fields complete one each other to give a full picture of the field of mechanics in question (elasticity, fluid dynamics, etc.). A classical example of misunderstanding which results from a one-directional point of view, was the theory of the boundary layer. Constructed by the mathematical physicist with the use of the most sloppy mathematical methods (from the rigorous mathematical standpoint), it left the impression for a half century that the results obtained were mathematically correct. It took a relatively short period of time for applied mathematicians to show that there is no rigorous mathematical justification for these results. After this was published there was no real justification in maintaining that a rigorous mathematical theory of the boundary layer existed. There was a physical phenomenon with some sort of an analytical description of a more physical than mathematical nature*. Obviously the basic differences in the interpretation of these concepts suggest that their content is purely subjective. It depends upon intelligence, intellect, broadness of the mind, etc., of the persons in question.

Returning to Bergman's method, it is worth noticing that some of the past reviews should be critically inspected. For example, Lighthill[5] in his review of Bergman's paper stated: "It is difficult to assess the work as a piece of mathematical physics until some applications of it have been described." It is difficult to assess Lighthill's assertion as a piece of logic, or even as an intelligent understanding of the current state of affairs. It is unimportant to show that the order in which a problem is attacked is the sequence: engineering, mathematical physics, applied mathematics, etc., or the opposite sequence, or any other sequence. The important thing is that the problem attacked should be solved correctly, and it is sufficient if it is contained in only one of the above mentioned fields: theoretical physics, mathematical physics, applied mathematics, etc. A different attack may transfer it from one field to another. Bergman ist not a physicist. He used some very advanced techniques of mathematics to obtain a correct solution of the problem in question. That cannot be said about the works of Lighthill as evidenced by his admission that some of his papers in the past were failures. In [6] he stated: "The theory in Lighthill's 1945 paper is altogether wrong." The reason is that he overemphasized intuition in the development of his methods and neglected the mathematical techniques needed for establishing the validity of the methods. Such a neglect of mathematics is most inappropriate in a time when the technological sciences are demanding more advanced mathematical techniques than are now available. Thus, the reviews by Lighthill of works (or rather attacks on) by major contributors in the field of mechanics (Truesdell, etc.) miss the point by failing to mention the essential ideas involved due to the mathematical content of these ideas. This requires some thinking and this may be a difficult item. There are some schools of thought in various parts of the world, expressing the conviction that the less mathematics and the more speculation is involved, the better the method from the "physical" point of view. This lack of maturity in thinking has usually its origin in the lack of knowledge of mathematics. That Lighthill is

---

* This situation was recently improved.

[5] M. J. Lighthill (Manchester): Review of paper: Stefan Bergman: Two-dimensional transonic flow patterns. Amer. J. Math. **70**, 856—891 (1948). In: Mathem. Reviews **10**, No. 10, 752—753 (November 1949).

[6] W. R. Sears: General Theory of High Speed Aerodynamics, Vol. VI of: High Speed Aerodynamics and Jet Propulsion. Princeton, New Jersey: Princeton University Press. 1954. In this Vol. VI: Section E, Higher Approximations, by M. J. Lighthill, pp. **345—489**. See p. **457**.

not alone is seen by noting a "review" by Cohen Hirsh[7] of a paper, pertinent to the contents of this book (the potential function of a source in compressible fluid). The reviewer did not actually review the paper; he only stated that the work is too mathematical: "Their discussion is purely formal and would seem to be more an exercise in algebra than a contribution to the mechanics or the mathematics of compressible flow." Here, the reviewer had shown a complete lack of an intelligent understanding of the problem. A source is a mathematical concept, not a physical one; clearly, when treated in the field of applied mathematics, there is no limit to the number of formal mathematical techniques which can be used. Every field of mathematics can be used since applied mathematics is mathematics. An assertion that there is too much formal mathematics is not only absurd but also shows ignorance, because the truth is that not enough formal mathematics is known for the solution of the problems accumulated today in both theoretical and mathematical physics. Every mathematical approach must begin with formal mathematics, in particular, in problems concerning sources where there are no physical aspects (except of the conservation of mass). But evidently there is no remedy for those reviewers whose minds are in the 18th Century and who do not understand the state of today's affairs in engineering and the fact that many fields of mathematics which were recently considered "pure mathematics" are used today in engineering and mechanics. The use and the enjoyment of sharper mathematical methods requires mathematics beyond advanced calculus, which seems to be the upper limit of knowledge of many reviewers. Another example is a review by Germain[8], whose obvious difficulty lies in understanding papers written in English. There seems to be no real evidence that he appreciates the interesting results obtained in diabatic flow. In Chapter XIII it was assumed that the function $Q(x, y)$ denotes the total amount of energy (heat) added to a particle moving along a streamline. A streamline is given by the equation $\psi = \psi(x_{st}, y_{st}) = $ const., or $y_{st} = y_{st}(x_{st})$, which implies that the function Q is defined in terms of one coordinate only, x_{st} or y_{st}. A similar situation exists in the hodograph $\{q, \theta\}$-plane, i. e., $Q = Q(q)$ only. All the dependance variables, (p, ϱ, T, Q), taken along a streamline, depend only upon one coordinate in any of the two planes: $\{x, y\}$, $\{q, \theta\}$. This is valid in all kinds of flow: isentropic irrotational (homentropic, the same amount of entropy on all the streamlines), isentropic rotational (entropy constant along each streamline but its value varies from streamline to streamline), diabatic, etc. Thus, one can choose for $Q(x, y)$ [or $Q(q, \theta)$] any arbitrary function of both $\{x, y\}$ or $\{q, \theta\}$. But, since Q defines the energy (heat) input function along a streamline, one of the independent variables, y or θ, must be represented in terms of $x = x_{st}$ or $q = q_{st}$ from the streamline equation $\psi(x_{st}, y_{st}) = $ const., or $\psi(q_{st}, \theta_{st}) = $ const., and so finally Q is representable in terms of $x = x_{st}$ or $q = q_{st}$ only. This is a property of the flow in question. It is not an easy problem to find a solution of the stream-function equation (13.4.14) in its general form. For this reason and for illustrative examples a simplified example was calculated. Germain wrote: "Il semble, en fait, que les auteurs — qui n'insistent guère sur les hypothèses faites, ni sur leur signification physique — se restreignent au cas où la chaleur totale introduite en un point est uniquement fonction de la vitesse." It is a real pity that such a confusion of objectives exists in mind of the reviewer.

[7] Cohen Hirsh (Rennselaer): Review of the paper: M. Z. v. Krzywoblocki, L. Pottsepp and M. Saarlas: On plane and spatial source in compressible flow. Bull. Techn. Univ. Istanbul **10**, No. 2, 19—27 (1957). In: Mathem. Reviews **20**, No. 1, 94, Review 578 (Jan. 1959).

[8] Germain (Paris): Review of the paper: M. Z. v. Krzywoblocki, and H. A. Hassan: On the limiting lines in diabatic flow. Comment. Math. Univ. St. Paul. **6**, 115—139 (1958). In: Mathem. Reviews **19**, No. 11, 1222 (Dec. 1958).

Kant said once that there is as much truth in a field of science as there is mathematics in it. It seems possible that the 19th Century developers of some fields of mechanics subconsciously followed this statement. A beautiful application of the theory of functions of a complex variable to incompressible fluid dynamics was one of main factors—if not the main factor—of the rapid development of aerodynamics of an incompressible medium at the beginning of the 20th Century. Rigorous mathematical methods were, and still are, abandoned (or used inadequately) in compressible fluid flow in favor of a most barbaric approach which consists of speculations combined with unjustified or even inconsistent assumptions. The advocates of such an approach are paying for it in terms of failures in various parts of the world. In the report of the Science Advisory Committee of the U. S. A. President under the Chairmanship of James R. Killian, there are the following sentences[9]: "Also undernourished, along with basic research, is 'analytical engineering' . . . while difficult to prove, there is little doubt that a more vigorous support of basic research—and 'analytical engineering' over the past decade would now be paying rich dividends in making 'things' easier and less costly to develop." Bergman's method must be considered to be one of the landmarks among the approaches to compressible fluid flow problems. In a period of 20 years every single step of this method, including analytical continuation, was rigorously proved. From this point of view this method should be justly considered to be an honest attempt to transfer the beauty of the 19th Century mathematical rigorous approach to the incompressible fluid flow into the barbarism of the 20th Century jungle prevailing in the compressible fluid flow field. Clearly such methods must be constructed before they can be reduced to a form suitable for practical applications. This path of development, if not followed in some parts of the world, will be followed in other parts, regardless of the voices of dissatisfied people.

---

[9] Strengthening American Science. A report of the President's Science Advisory Committee, James R. Killian, Jr., Chairman. Amer. Scientist **47**, No. 2, 264—287 (June 1959).

Additional Bibliography Pertinent to the Subject

Abramovich, G. M.: The theory of a free jet of a compressible gas. N. A. C. A., T. M. No. 1058 (1939) (translated from CAHI Report No. 377).

Akita, Y.: On the nonlinearity of gas dynamics. J. Soc. Appl. Mech., Japan 1, 48—56 (1948) (in Japanese).

— Non-linear character of the compressible aerodynamics. Jap. Sci. Rev. Ser. 1, 1, No. 2, 5—10 (1950).

Allen, D. N., de G. and S. C. R. Dennis: The application of relaxation methods to the solution of differential equations in three dimensions. I. Boundary value potential problems. Quart. J. Mech. Appl. Math. 4, part 2, 199—208 (1951).

Bailey, W. N.: A note on the paper by Tempest and Rosenhead. Proc. London Math. Soc. 51, 213—214 (1949).

Behrbohm, H.: An approximate theory of the unsymmetrical transonic flow through a Laval nozzle. Z. angew. Math. Mechan. 30, 4, 101—112 (1950).

Behrbohm, H. and M. Pinl: Zur Theorie der kompressiblen Potentialströmungen II. — Intermediäre und singuläre Integrale der Grundgleichung der ebenen adiabatisch kompressiblen Potentialströmung. Z. angew. Math. Mechan. 21, 341—350 (1941), Also R. T. P. Translation No. 1625.

Bergman, S.: Boundary values of functions satisfying a linear partial differential equation of elliptic type. Proc. Nat. Acad. Sci. 26, 668—671 (1940).

— Linear operators in the theory of partial differential equations. Trans. Amer. Math. Soc. 53, 130—155 (1943).

— Solutions of linear partial differential equations of fourth order. Duke Math. J. 11, 617—649 (1944).

— Construction of a complete set of solutions of a linear partial differential equation in two variables, by use of punch card machines. Quart. Appl. Mathem. 4, 233—245 (1946).

— On functions satisfying certain partial differential equations of elliptic type and their representation. Duke Math. J. 14, 349—366 (1947).

— Operator methods in the theory of compressible fluids. Proceedings of the First Symposium of Applied Mathematics, Amer. Math. Soc., pp. 19—40 (1949).

— On solutions with algebraic character of linear partial differential equations. Trans. Amer. Math. Soc. 68, 461—507 (1950).

— Operatorenmethoden in der Gasdynamik. Z. angew. Math. Mechan. 32 (2/3), 33—45 (1952).

— Methods for the determination of subsonic flows around profiles. Proc. First U. S. Nat. Congr. Appl. Mech., pp. 705—713 (1952).

— The coefficient problem in the theory of linear partial differential equations. Trans. Amer. Math. Soc. 73, 1—34 (1952).

— The kernel function and conformal mapping. Mathem. Surveys, No. 5, American Math. Society, New York, 1950, VII + 161.

— Über die Entwicklung der harmonischen Funktionen der Ebene und des Raumes nach Orthogonalfunktionen. Math. Ann. 86, 238—271 (1922).

— Zur Theorie der algebraischen Potentialfunktionen des dreidimensionalen Raumes. Math. Ann. 99, 629—659 (1928); also: Ibid. 101, 534—558 (1929).

— Über die Kurvenintegrale zweier komplexer Veränderlichen, die Differentialgleichung $\Delta V + V = 0$ befriedigen. Math. Zeit. 32, 386—406 (1930).

— A class of harmonic functions in three variables and their properties. Trans. Amer. Math. Soc. 59, 216—247, 419—458 (1946).

— Classes of solutions of linear partial differential equations in three variables. Duke Math. J. 13, 216—249, 419—458 (1946).

— The kernel function and conformal mapping. Math. Surveys 5 (1950).

Bergmän, S.: On solutions with algebraic character of linear partial differential equations. Trans. Amer. Math. Soc. **68**, 461—507 (1950).

— Methods for the determination of subsonic flows around profiles. First U. S. National Congress of Applied Mechanics, pp. 705—713 (1951/52).

— Operatorenmethoden in der Gasdynamik. Z. angew. Math. Mechan. **32**, 33—45 (1952).

— On solutions of linear partial differential equations of mixed type. Amer. J. Math. **74**, 444—474 (1952).

— The coefficient problem in the theory of linear partial differential equations. Trans. Amer. Math. Soc. **73**, 1—34 (1952).

— Some methods for solutions of boundary value problems of linear partial differential equations. Proceed. Sixth Symp. Appl. Mathem., Amer. Math. Soc., Numerical Analysis, No. 6, Santa Monica City College, pp. 11—29 (August 1953).

— Multivalued solutions of linear partial differential equations; Contributions to the theory of Rieman surfaces. Ann. of Math. Studies, No. 30, 229--245 (1953).

— A method for solving boundary value problems of mathematical physics on punch card machines. Jour. Association for Computing Machinery 1, No. 3, 101—104 (July 1954).

— Essential singularities of solutions of a class of linear partial differential equations in three variables. J. Rational Mechanics and Analysis **3**, No. 5, 539—560 (Sept. 1954).

— On representation of stream functions of flows of compressible fluids in sub and supersonic cases. J. of Rational Mechanics and Analysis **4**, 883—906 (1955).

— New methods for solving boundary value problems. Z. angew. Math. Mechan. **36**, No. 5/6, 182--191 (May/June 1956).

— Multivalued harmonic functions in three variables. Communic. Pure and Appl. Mathem. **9**, No. 3, 327—338 (Aug. 1956).

— Bounds for solutions of a system of partial differential equations. J. of Rational Mechanics and Analysis **5**, No. 6, 993—1002 (Nov. 1956).

— Some methods for solutions of boundary-value problems of linear partial differential equations. Proc. Symposia in Appl. Math., Vol. VI. Numerical Analysis, 11—29. Amer. Math. Soc. Providence 1956. Publ. by McGraw-Hill Book Company.

— On singularities of solutions of certain differential equations in three variables. To appear.

Bergman, S. and M. Schiffer: Kernel functions and conformal mapping. Compositio Math. **8**, 205—249 (1951).

— — Kernel functions and elliptic differential equations in Mathematical Physics. New York: Academic Press, Inc. 1953.

— — Properties of solutions of a system of partial differential equations. Studies in Mathematics and Mechanics presented to R. von Mises, pp. 79—87. New York: Acad. Press. (1954).

Bers, L.: Velocity distribution on wing sections of arbitrary shape in compressible potential flow. II. Subsonic symmetric adiabatic flows. N. A. C. A., T. N. No. 1012 (1946).

— An existence theorem in two-dimensional gas dynamics. Proc. Symposia Appl. Mathem. 1, 41—46, Amer. Math. Soc. (1949).

Bers, L. and A. Gelbart: On generalized Laplace transformations. Ann. of Math. **48**, 342—357 (1947).

Betz, A. and E. Krahn: Berechnung von Unterschallströmungen kompressibler Flüssigkeiten um Profile. Ing.-Arch. **17**, 403—417 (1949).

Bilharz, H. and E. Hölder: Zur Berechnung der Druckverteilung an Rotationskörpern in der Unterschallströmung eines Gases. Teil I. Achsensymmetrische Strömung. Forschungsbericht No. 1/159/1, Deutsche Luftfahrtforschung (Braunschweig, 1940).

Blank, H.: Application of Ritz's method to the calculation of compressible flow around a circular cylinder (in Italian). Monogr. Sci. Aero. No. 11, 11 pp. (Nov. 1951).

Buseman, A.: Gasdynamik. Beitrag im Handbuch der Experimentalphysik (Wien-Harms). Vol. 4, Part 1, p. 421 (Leipzig, 1931).

— Die Expansionsberichtigung der Kontraktionsziffer von Blenken. Forschung **4**, 186—187. (1933).

— Atti dei convegni fondazione Allessandro Volta (Roma) **5**, 328--360 (1935).

— Infinitesimale kegelige Überschallströmung. Sonderdruck, Jahrbuch 1942/43 der Deutschen Akademie der Luftfahrtforschung 7 B, No. 3, 105—122 (1943). Also: Luftfahrtforsch. **20**, 105 (1943).

Cabannes, H.: On compressible flow in the neighborhood of sonic speed. C. R. Acad. Sci. Paris **229**, 102—104 (1949).

Chang, C. C.: General consideration of problems in compressible flow using the hodograph method. N. A. C. A., T. N. No. 2582 (1952).

Chang, C. C. and V. O'Brien: Some exact solutions of two dimensional flows of compressible fluid with hodograph method. N. A. C. A., T. N. No. 2885 (February 1953).

Cherry, T. M.: Tables and approximate formulae for hypergeometric functions, of high order, occurring in gas flow theory. Proc. Roy. Soc. London, Ser. A 217, 222—234 (Apr. 1953).

— A transformation of the hodograph equation and the determination of certain fluid motion. Philos. Transact. Roy. Soc. London, Ser. A 245, 904, 583—626 (Apr. 1953).

Christianovich, S. A.: The flow of gases around bodies at high speeds, approaching sonic speeds. Trudy Central. Aero-Gidrodinam. Inst. (CAHI) No. 481 (1946, in Russian).

— Approximate integration of the equations of supersonic gas flow. Prikl. Mat. i Mekh. 11, 215—222 (1947) (in Russian). Also: Tech. Rep. No. F-TS-1220-IA (GDAM A 9-T-39) (1949), Headquarters Air Material Command Wright-Patterson Air Force Base, Dayton, Ohio.

Christianovich, S. A. and I. M. Yuriew: Subsonic gas flow past a wing profile. N. A. C. A., T. M. No. 1250, 29 pp. (1950).

Craggs, J. W.: The application of the hodograph method to problems of subsonic compressible flow in two-dimensions. A. C. A. Report No. 9801 (1946).

— The breakdown of the hodograph transformation for irrotational compressible flow in two dimensions. Proc. Camb. Phil. Soc. 44, 360—376 (1948).

— The compressible flow corresponding to a line doublet. Quart. Appl. Math. 10, 1, 88—93 (Apr. 1952).

Demtchenko, B.: Variation de la résistance aux faibles vitesses sous l'influence de la compressibilité. C. R. Acad. Sci. Paris 194, 1720—1723 (1932).

— Quelques problems d'hydrodynamique bidimensionelle des fluides compressibles. Pub. Scientifiques et Techniques du Ministere de l'Air, No. 144 (1938).

— Sur la relation entre la dynamique des fluides compressibles et celle des fluides incompressibles. Publications mathematiques de l'Universite de Belgrad 2, 85—106 (1933).

Des Clers, B. and C. C. Chang: On some special problems in linearized axially symmetric flow. J. Aeron. Sci. 18 (2), 127—138 (1951).

Diaz, J. B. and G. S. S. Ludford: On the integration methods of Bergman and Le Roux. Quart. Appl. Math. 14, 4, 428—432 (Jan. 1957).

Eggers, A. J. and R. C. Savin: Approximate methods for calculating the flow about nonlifting bodies of revolution at high supersonic airspeeds. N. A. C. A., T. N. No. 2579, 40 pp. (Dec. 1951).

Eichler, M. M. E.: On the differential equation $U_{xx} + U_{yy} + N(x) U = 0$. Trans. Amer. Math. Soc. 65, 259—278 (1949).

— Analytic functions in three-dimensional Riemannian spaces. Duke Math. J. 16, 339—349 (1949).

Eser, F.: Zur Strömung kompressibler Flüssigkeiten um feste Körper mit Unterschallgeschwindigkeit. Luftfahrtforsch. 20, 220—230 (1943).

Falkovich, S. V.: Two-dimensional motion of a gas at large supersonic velocities. N. A. C. A., T. M. 1239 (1949).

Ferrari, C.: Aerotecnica 18, 400—411 (1938).

— On the determination of some fields of supersonic flow (in Italian). Atti Accad. Naz. Lincei Rend., Cl. Sci. Fis. Mat. Nat. 7, 6, 277—283 (1949).

Frankl, F. I.: On the theory of the Laval nozzle. Bull. Acad. Sci. URSS. 9, 387—422 (1945, in Russian).

Frankl, F. and R. Aleksyiva: Zwei Randwertaufgaben aus der Theorie der hyperbolischen partiellen Differentialgleichungen zweiter Ordnung mit Anwendungen auf Gasströmungen mit Überschallgeschwindigkeit. Rec. Math. Moscow 41, 483—499 (Russian, German summary).

Frankl, F. and M. Keldysch: Die äußere Neumannsche Aufgabe rür nichtlineare elliptische Differentialgleichungen mit Anwendung auf die Theorie des Flügels im kompressiblen Gas. Bull. Acad. Sci. URSS. 7, 561—599 (Russian, German summary).

Gelbart, A. and D. Resh: A method of computing subsonic flows around given airfoils. N. A. C. A., T. N. No. 2057 (1950).

Germain, P.: Application of homographic approximation to the study of transonic flow (in French). C. R. Acad. Sci. Paris 232 (20), 1811—1813 (1951).

— Researches on an equation of the mixed type. Introduction to the mathematical study of transonic flow (in French). Rech. Aéro. No. 22, 7—20 (July-Aug. 1951).

Hypotheses and general methods of linearised supersonic aerodynamics (in French). Publ.

— sci. tech. Min. Air, France, No. 250, 217—250 (1951).

Germain, P.: Homographic approximation in the study of compressible fluids (in French). Rech. Aéro. No. 25, 9—17 (Jan.-Febr. 1952).
— Elementary solutions of equations governing the flow of compressible fluids (in French). C. R. Acad. Sci. Paris 234, 12, 1248—1250 (March, 1952).
Germain, P. and R. Bader: Problème de Dirichlet pour une équation du type mixte. C. R. Acad. Sci. Paris 230, 1824—1825 (1950).
— — Sur quelques problèmes aux limites, singuliers, pour une équation hyperbolique. C. R. Acad. Sci. Paris 231, 268 (1950).
Germain, P. and M. Liger: A new approximation for the study of subsonic and transonic flow (in French). C. E. Acad. Sci. Paris 234, 19, 1846—1848 (May 1952).
Ghaffari, A. G.: Velocity-correlation factors and the hodograph method in gas dynamics. Ph. D. Thesis, London University (1948).
— The hodograph method in gas dynamics. University of Tehran, Faculty of Science Publication No. 85, Taban Press, Tehran (1950).
Giese, J. H.: Stream-function for three-dimensional flows. J. Math. Phys. 30, 1, 31—35 (Apr. 1951).
Görtler, H.: Zum Übergang von Unterschall- zu Überschallgeschwindigkeiten in Düsen. Z. angew. Math. Mechan. 19, 325—337 (1939).
Göthert, B.: Wind tunnel corrections at high subsonic speeds. Jahrb. der deutschen Luftfahrt-forschung, p. 684 (1941, in German).
Göthert, B. and K. H. Kawalki: The calculation of compressible flows with local regions of supersonic velocity. N. A. C. A., T. M. No. 1114.

Hansen, A. G. and M. H. Martin: Some geometrical properties of plane flows. Proc. Cambr. Philos. Soc. 47, 763—776 (1951).
Harder, K. C. and E. B. Klunker: On a source-sink method for the solution of the Prandtl-Busemann iteration equations in two-dimensional compressible flow. N. A. C. A., T. N. No. 2253 (1950).
— — On folds occurring in the mapping from the physical to the hodograph plane. Readers' Forum, J. Aer. Sci. 19, 10, 719 (Oct. 1952).
Hasimoto, H.: On the subsonic flow of a compressible fluid past a Rankine ovoid. J. Phys. Soc. Japan 6, 175—178 (1951).
— Application of the thin-wing-expansion method to the compressible flow past an elliptic cylinder. J. Phys. Soc. Japan 7, 322—328 (May/June 1952).
— On the solution of some boundary value problems of compressible fluid flow (in Japanese). Stud. Math. Phys., Tokyo 2, 141—187 (June 1952).
Hayes, W. D.: Linearized supersonic flow. Report No. A. L. 222, North American Aviation (1947).
— Reversed flow theorems in supersonic aerodynamics. VIIth International Congress of Applied Mechanics, London (1948).
Heaslet, M. A. and H. Lomax: The use of source-sink and doublet-distributions extended to the solution of boundary value problems in supersonic flow. N. A. C. A., Report No. 900 (1948).
Henrici, P.: Zur Funktionentheorie der Wellengleichung. Comm. Math. Helv. 27, 235—293 (1953).
Hida, Kinzo: On the subsonic flow of a compressible fluid past a prolate spheroid. J. Phys. Soc. Japan 8, 2, 257—264 (Mar/Apr. 1953).

Imai, I.: On the theory of arbitrary wing sections (in Japanese). J. Soc. Aero. Sci. Nippon 9, 865 (1942).
— Two-dimensional aerofoil theory for compressible fluids, II (in Japanese). Kagaku (Science) 14, 44 (1944).
— Application of the hodograph method to the flow of a compressible fluid past a profile, with special reference to its lift and moment (in Japanese). Rep Aer. Res. Inst. Tokyo Imp. Univ., No. 323 (1945).
— Tables useful for the numerical calculation of the air stream at high speeds. J. Phys. Soc. Japan 3, 342 (1948).
— On a new method of approximation for treating compressible fluid flow. J. Phys. Soc. Japan 3, 352 (1948).
— On a new method of approximation for treating compressible fluid flow. J. Phys. Soc. Japan 3, 352—356 (1948).
— An approximate method of calculating compressible fluid flow past a thin aerofoil. J. Phys. Soc. Japan 3, 346—351 (1948).

Imai, I.: On the asymptotic behaviour of compressible fluid flow at a great distance from a cylinder in the absence of circulation. J. Phys. Soc. Japan 8, 4, 537—544 (July/Aug. 1953).
— Some particular solutions of compressible flow equations for arbitrary pressure-density relation. J. Phys. Soc. Japan 8, 799—801 (1953).
Imai, I. and D. Kawada: On the velocity distribution round an elliptic cylinder at high speeds (in Japanese). Rep. Aeron. Res. Inst., Tokyo Imp. Univ., No. 325 (1945).
Imai, I. and K. Sato: On a numerical method of finding the analytic function which maps conformally the region outside a given profile on to the region outside a circle (in Japanese). J. Aeron. Res. Inst., Tokyo Imp. Univ., No. 247 (1945).

Jacob, C.: Etude d'un jet gazeau. Bulletin Scientifique de l'École Polytechnique de Temisoara 7, 47—59, 224—244. (1931).
— Sur un problème concernant les jets gazeaux. Mathematica 8, 205—211 (1934).
— Sur quelques problèmes concernant l'écoulement des fluides parfaits compressibles. C. R. Acad. Sci. Paris 197, 125—126 (1939).
— La methode hodographique en mécanique des fluides compressibles. Sixth International Congress for Applied Mechanics (1946).

Kaplan, C.: On a new method for calculating the potential flow past a body of revolution. N. A. C. A., Rep. No. 752 (1943).
— On the particular integrals of the Prandtl-Buseman iteration equations for the flow of a compressible fluid. N. A. C. A., T. N., No. 2159 (1950), or N. A. C. A. Rep. 1039 (1951).
— On a solution of the nonlinear differential equation for transonic flow past a wave-shaped wall. N. A. C. A., T. N. No. 2383 (1951).
Kawaguti, M.: Note on the velocity distribution over an elliptic cylinder submerged in a uniform flow of compressible fluid. J. Phys. Soc. Japan 7, 3, 313—315 (May/June 1952).
Kiebel, J. A.: Sur les équations différentielles qui servent pour déterminer la densité d'un fluide compressible en mouvement. Rec. Math. Moscow 39, 141—151 (Russian, French summary).
— Exact solutions of equations of gasdynamics. Prikl. Mat. i Mekh. 9, 193—198 (1947). Also: N. A. C. A., T. M. No. 1260.
— Sur quelques mouvements plans d'un fluid pésant compressible. Prikl. Mat. i. Mekh. 1, 51—55 (Russian, French summary).
— Some studies on the flow of a gas in the region of transition through the velocity of sound. Izvestia Akademii Nauk SSSR, No. 3, 253—259 (1947). Also: N. A. C. A., T. M. No. 1252 (1950).
Kiselev, B. M.: Calculation of one-dimensional gas flow. Headqtrs. Air Mat. Comm. Dayton, Transl. Tech. Rep. No. F-TS-1209-1 A (1949).
Klein, M. M. and W. Perl: Theoretical Investigation of transonic similarity for bodies of revolution. N. A. C. A., T. N. No. 2239 (1950).
— Calculation of compressible potential flow past slender bodies of revolution by an integral method. N. A. C. A., T. N. No. 2245 (1950).
Kolodner, I.: On the linearized theory of supersonic flows through axially symmetrical ducts. Comm. pure appl. Math. 3, 2 (1950), 133—152.
v. Koppenfels, W.: Two-dimensional potential flow past a smooth wall with partly constant curvature. N. A. C. A., T. M. No. 996 (1941).
Kraft, H.: The calculation and use of logarithmic singularities in the compressible hodograph plane. Sixth International Congress for Applied Mechanics (1946).
Krahn, E.: An approximate method for the calculation of compressible subsonic flow (in German). Z. angew. Math. Mechan. 29, 2/3 (1949).
Kreyszig, E.: On a class of partial differential equations. J. Rat. Mech. Analysis 4, 907—923 (1955).
— On certain partial differential equations and their singularities. J. Rat. Mech. Analysis 5, 805—820 (1956).
— Zum Koeffizientenproblem der Lösungen partieller Differentialgleichungen. Tl appear.
Krzywoblocki, M. Z. v.: Bergman's Linear Integral Operator Method in the Theory of Compressible Fluid Flow. Bureau of Ordnance, Operator Method in the Theory of Compressible Fluids, (Report 18), Contract Nord 10-449, Task 3, Harvard University, Cambridge, Massachusetts (March, 1951).

Legendre, R.: Ecoulement isentropique plan d'un fluid compressible. C. R. Acad. Sci. Paris 237, 595—597 (1953).
Leibenson, L. S.: On the theory of motion of gases. Centr. Aero-Gidrod. Inst., Moscow, or C. R. Acad. Sci. URSS. N. S. 3, 397—398 (1935).

Levey, H. C.: High speed flow of a gas past an approximately elliptic cylinder. Cambridge Philosophical Soc. Proceed. **46**, 3, 479—490 (July 1950).

Lighthill, M. J.: Supersonic flow past slender bodies of revolution the slope of whose meridian section is discontinuous. Quart. J. Mech. Appl. Mathem. 1, 90—102 (1948).

— A new approach to thin aerofoil theory. The Aero. Quart. **3**, 3, 193—210 (Nov. 1951).

Lock, C. N. H.: The interference of a windtunnel on a symmetrical body. R. and M. No. 1275, British A. R. C. (1929).

Lotkin, M.: Supersonic flow over bodies of revolution. Quart. Appl. Mathem. 7, 65—74 (1949).

Ludford, G. S. S.: On an extension of Riemann's method of integration, with applications to one-dimensional gas dynamics. Camb. Phil. Soc. Proceed. 48, 3, 499—510 (1952).

— The extension to a two-dimensional subsonic flow of the Joukowski force of ideal fluid theory. Proc. U. S. First Nat. Congr. Appl. Mech. (1952), pp. 715—721.

Manwell, A. R.: A Note on The Hodograph Transformation. Quart. Appl. Math. 10, 2, 177—184 (July 1952).

— The variation of compressible flows. Quart. J. Mech. Appl. Math. 7, 40—50 (1954).

Martin, J. C.: Retarded potentials of supersonic flow. Quart. Appl. Math. 8, 358—364 (1951).

Martin, M. H.: A new approach to problems in two dimensional flow. Quart. Appl. Math. 8, 137—150 (1950).

— Plane rotational Prandtl-Meyer flows. J. Math. Phys. **29**, 76—89 (1950).

— Steady, rotational, plane flow of a gas. Amer. J. Math. **72**, 465—484 (1950).

Mitchell, A. R. and D. E. Rutheford: Application of relaxation methods to compressible flow past a double wedge. Proc. Roy. Soc. Edinburgh, Ser. A **63**, part II (11), 139—154 (1951).

Mitchell, J.: Some properties of solutions of partial differential equations given by their series development. Duke Math. J. 13, 87—104 (1946).

Moore, F.: Linearized supersonic, axially symmetric flow about open-nosed bodies obtained by use of stream function. N. A. C. A., T. N. No. 2116 (1950).

Mughin, S. G.: Theory of flow of gas past a body at high subsonic velocities. Prikl. Mat. i Mekh. 10, No. 5/6 (1946).

Nielsen, K. L.: On the Bergman Operators for Linear Partial Differential Equations. Bull. Amer. Math. Soc. 50, 195 (1944).

— Some properties of functions satisfying partial differential equations of elliptic type. Duke Math. J. 11, 121—137 (1944).

Nielsen, K. L. and B. Ramsey: On particular solutions of linear partial differential equations. Bull. Amer. Math. Soc. 49, 156—162 (1943).

Nikolskij, A. A. and G. I. Taganov: Gas motion in a local supersonic region and conditions of potential-flow breakdown. Prikl. Mat. i Mekh. 10, 481—502 (1946) (in Russian). Also: N. A. C. A., T. M. No. 1213 (1949).

Nitsche, J. C. C.: On solutions of differential equations of mixed type generated by a linear operator and their properties. To appear.

Opatowski, I.: Two-dimensional compressible flows. Proc. Symposia Appl. Math. 1, 87—93. Amer. Math. Soc. (1949).

Oswatitsch, K.: Laws in transonic flow (in German). Z. angew. Math. Mechan. 29, 4—5 (1949).

— The effect of compressibility on the flow around slender bodies of revolution. Kungl. Tekniska Högskolan, Institutionen för Flygteknik. Kth-Aero TN 12, Sweden (1949).

Oswatitsch, K. and S. B. Berndt: Aerodynamic similarity at axisymmetric transonic flow around slender bodies. Kungl. Tekniska Högskolan Institutionen för Flygteknik, KTH-Aero TN 15.

Ovsyannikov, L. V.: On a gas flow with a straight transition line (in Russian). Prikl. Mat. Mekl. 13, 5, 537—542 (1949).

Pack, D. C.: Hodograph methods in gas dynamics. (Lecture series no. 17, edited by Roth, H.) Univ. Maryland, Inst. Fluid Dynam. Appl. Math., 59 pp. 1951/52.

Perl, W.: Calculation of compressible flow past aerodynamic shapes by use of streamline curvature. N. A. C. A., T. M. No. 1328 (1947).

Pinl, M.: On the theory of compressible flows III. Characteristics, bicharacteristics, and eiconal (geodetic distance) of the linearized fundamental equation. R. T. P. Translation No. 1094 from Z. angew. Math. Mechan. 22, 305 (1942).

Prandtl, L.: Luftfahrtforsch. **13**, 313—319 (1936).

— Allgemeine Betrachtungen über die Strömung zusammendrückbarer Flüssigkeiten. Z. angew. Math. Mechan. **16**, 129—142 (1936).

Prandtl, L.: General considerations on the flow of compressible fluids. N. A. C. A., T. M. No. 805 (1936). Also British R. T. P. Translation No. 1872 (translated from Proceedings of 5th Volta Congress, 1935).
— Theory of the airplane wing in compressible media. Luftfahrtforsch. 12, No. 10 (1936, in German).
Protter, M. H.: A boundary value problem for an equation of mixed type. Trans. Amer. Math. Soc. 71, 416—429 (1951).

Rabineau, B. A.: Compressibility corrections for bodies of revolution. J. Aer. Soc. 19, 3, 197 to 200, 206 (1952).
Rao, G. V. R.: Subsonic compressible flow past elliptic cylinders. N. A. C. A., T. N. (to be published).
Rey, P. J.: On the linearized equation of supersonic flow. (In Spanish.) Ann. Mat. Pura Appl. (4), 30, 91—96 (1949).
Riabouchinsky, D.: Mouvement d'un fluide compressible autour d'un obstacle. C. R. Acad. Sci. Paris 194, 1215 (1932).
— Quelques nouvelles remarques sur l'analogie hydraulique des mouvements d'un fluide compressible. C. R. Acad. Sci. Paris 199, 632—634 (1934).
Ringleb, F.: Exakte Lösungen der Differentialgleichungen einer adiabatischen Gasströmung. Z. angew. Math. Mechan. 20, 185—198 (1940). Also: Roy. Aero. S. J. (Nov. 1942).
Robinson, A. and A. D. Young: Note on the application of the linearized theory for compressible flow to transonic speeds. Minist. of Sup., Aero. Res. Coun., Rep. and Mem., No. 2399 (10,474) (1951), pp. 6.

Saito, O.: On some exact solutions of the two-dimensional steady flow of compressible fluid. J. Phys. Soc. Japan 6, 243—254 (1951).
Saito, O. and A. Amemiya: On the solution of differential equations of the two-dimensional steady flow of compressible fluid. J. Phys. Soc. Japan 5, 201—202 (1950).
Sauer, R.: Geometrical relationship between two-dimensional fields of compressible flow. Z. angew. Math. Mechan. 312—315 (1941), or. R. T. P. Translation No. 1776.
— Introduction to theoretical gas dynamics. J. W. Edwards, Ann Arbor. 1947.
— Unterschallströmungen um Profile bei quadratisch approximierter Adiabate. S.-B. Math. Nat. Kl. Bayr. Akad. Wiss. 1951, 65—71, 1952.
Schmieden, C.: Beiträge zum Umströmungsproblem bei hohen Geschwindigkeiten. Bericht s 1311. Lilienthal-Gesellschaft für Luftfahrtforschung (1942), pp. 40—68.
Schmieden, C. and K. H. Kawalki: Contribution to the problem of flow at high speed. N. A. C. A., T. M. No. 1233 (1949).
Sears, W. R.: The linear-perturbation theory for rotational flow. J. Mathem. Physics 28, 268 to 271 (1950).
Shiffman, M.: On the existence of subsonic flows of a compressible fluid. Proc. Nat. Acad. Sci. Washington 38, 5, 434—438 (May 1952).
— On the existence of subsonic flows of a compressible fluid. J. Rational Mech. and Anal. 1, 4, 605—652 (Oct. 1952).
Slioskin, N. A.: On the problem of motion of a gas in two-dimensions. C. R. Acad. Sci. URSS., N. S. 3, 419—421 (1936).
Steichen, A.: Beiträge zur Theorie der zweidimensionalen Bewegungsvorgänge in einem Gase, das mit Überschallgeschwindigkeit strömt. Diss. Göttingen (1909).

Tamada, K.: Application of the hodograph method to the flow of a compressible fluid past a circular cylinder. Proc. Physico-Mathem. Soc. of Japan 22 (3rd Ser.), 208—219 (1940).
— Studies of the two-dimensional flow of gas, with special reference to the singularities of the solutions of a certain partial differential equation of mixed type (in Japanese). Studies in Mathematical Physics, Iwanami Shoten Publ. Tokyo 1, 82—92 (1950).
— Studies of two-dimensional flow of gas with special reference to the flow through various nozzles (in Japanese). Studies in Mathematical Physics, Iwanami Shoten Publ. Tokyo 1, 56—80 (1950).
— On the hodograph method and analytic continuation of solution in the theory of compressible fluid flow. Mem. Coll. Sci. Univ. Kyoto, Ser. A 26, 21—30 (1950).
Taylor, G. I.: Recent work on the flow of compressible fluids. J. London Math. Soc. 5, 224—240 (1930).
— Strömung um einen Körper in kompressibler Flüssigkeit. Z. angew. Math. Mechan. 10, 334 (1930).
— Some cases of flow of compressible fluids. R. and M. No. 1382 (1930).

Taylor, G. I.: Application to aeronautics of Ackeret's theory of aerofoils moving at speeds greater than that of sound. R. and M. No. 1467, British A. R. C. (1932).

Tempest, R. K.: The supersonic flow of compressible fluid through axially symmetric tubes of uniform and varying section. Roy. Soc. London Proc., Ser. A 200, 511—523 (1950).

Tempest, R. K. and L. Rosenhead: Notes on the linearized equation for the velocity potential of the steady supersonic flow of a compressible fluid. Proc. London Mathem. Soc. 51, 197—212 (1949).

Theodorsen, Th.: Extension of the Chaplygin Proofs on the Existence of Compressible-Flow Solutions to the Supersonic Region. N. A. C. A., T. N. No. 1028 (1946).

Tollmien, W.: Steady two-dimensional and rotationally symmetric supersonic flows. Translation, Brown Univ. (1948).

Tomotika, S. and H. Hasimoto: On the transonic flow of a compressible fluid through an axially symmetrical nozzle. J. Math. Phys. 29, 105—117 (1950).

Tomotika, S. and K. Tamada: Studies on two-dimensional transonic flows of compressible fluid. — Part I. Quart. Appl. Math. 7, 381—396 (1949).

— — Studies on two-dimensional transonic flows of compressible fluid. — Part II. Quart. Appl. Math. 8, 127—136 (1950).

— — Studies on two-dimensional transonic flows of compressible fluid. III. Quart. Appl. Math. 9, No. 2, 129—147 (July 1951).

Ursell, F. and G. N. Ward: On some general theorems in the linearized theory of compressible flow. Quart. J. Mech. Appl. Math. 3, part 3, 326—348 (1950).

Van Driest, E. R.: Die linearisierte Theorie der dreidimensionalen kompressiblen Unterschall-strömung und die experimentelle Untersuchung von Rotationskörpern in einem geschlossenen Windkanal. Zürich, Eidgenössische Technische Hochschule, Institut für Aerodynamik, Mitteilung Nr. 16 (1949), 31 p.

Van Dyke, M. D.: A study of second-order supersonic flow theory. N. A. C. A., T. N. No. 2200 (1951).

— First- and second-order theory of supersonic flow past bodies of revolution. J. Aero. Sci. 18, 161—178 (1951).

Walther, P. A.: C. R. Acad. Sci. URSS, N. S. 4, 117—120 (1936).

— Trans. Centr. Aero. Hydrodyn. Inst. (Moscow), No. 301, 7—13 (1937).

Wang, C. T.: Two-dimensional subsonic compressible flows past arbitrary bodies by the variational method. N. A. C. A., T. N. No. 2326 (1951).

— A variational method for transonic flows with shock waves. N. A. C. A., T. N. (to be published).

— A note on Bateman's variational principle for compressible fluid flow. Quart. Appl. Math. 9, 1, 99—102 (Apr. 1951).

Wang, C. T. and R. F. Brodsky: Application of Galerkin's method to compressible fluid flow problems. J. Appl. Physics 20, 1255—1256 (1949).

— — Approximate solutions of compressible fluid flow problems by Galerkin's method. J. Aero. Sci. 17, 660—666 (1950).

Ward, G. N.: The supersonic flow past a slender body of revolution with an annular intake. Aeron. Quart. 1, 305—318 (1950).

Weinstein, A.: The method of singularilies in the physical and in the hodograph plane. Proc. Symposia Appl. Math. 4, Fluid Dynam. 167—178. McGraw-Hill Co. 1953.

Recommended:

Bergman, Stefan: Integral Operators in the Theory of Linear Partial Differential Equations. Ergebnisse der Mathematik, Neue Folge, No. 23. Berlin: Springer-Verlag. 1960.

Subject Index

The numbers refer to the pages

Manzsche Buchdruckerei, Wien IX